INFRARED SPECTRA OF LABELLED COMPOUNDS

INFRARED SPECTRA OF LABELLED COMPOUNDS

S. PINCHAS

The Weizmann Institute of Science
Rehovoth, Israel

with

I. LAULICHT

Bar-Ilan University
Ramat-Gan, Israel

1971

ACADEMIC PRESS · London and New York

ACADEMIC PRESS INC. (LONDON) LTD.
Berkeley Square House
Berkeley Square,
London, W1X 6BA

U.S. Edition published by

ACADEMIC PRESS INC.
111 Fifth Avenue,
New York, New York 10003

SBN: 12-556650-6
Library of Congress Catalog Card Number: 76-117142

Printed in Great Britain by
William Clowes & Sons, Limited
London, Beccles and Colchester

Acknowledgements

We gratefully acknowledge permission from the following for reproduction of original and redrawn versions of figures:

Pergamon Press Limited for 10 figures from *Spectrochimica Acta* **19–23**, The Society for Analytical Chemistry for Fig. 2 (1967) from *The Analyst* **92**, 717 and Academic Press for Fig. 1 from (1957) *J. Mol. Spectroscopy* **1**, 73.

Contents

CHAPTER 6. NITROGEN LABELLED COMPOUNDS

CHAPTER 7. OXYGEN-18 COMPOUNDS

Organic Compounds

"Blessed be He who varies the creations"
(*Babylonian Talmud, Tractate B'raḣoth 58:2*)

CHAPTER 1

Main Aspects of Infrared Spectroscopy of Isotopic Molecules

Since the characteristic light absorption of molecules in the infrared (30–12,500 cm^{-1}) region depends strongly upon the masses of their constituent atoms, labelled molecules usually show an infrared absorption spectrum which is distinctly different from that of the respective normal molecules. Evidently, the spectral differences observed for the isotopic compounds are larger the higher the respective m^i/m ratio, where m^i is the (higher) mass of the labelling atom and m is the mass of the normal atom exchanged for it. The strongest isotope effect on the infrared spectrum of labelled compounds can thus be expected for tritiated and deuterated molecules which indeed show absorption band shifts *up* to *ca.* 1300 and 1000 cm^{-1}, respectively. The isotope shifts in the case of lower m^i/m ratios, although appreciably smaller than for these compounds, are still considerable and for ^{18}O-labelled molecules, e.g. can be as high as 40 cm^{-1}. Figures 1.1 and 1.2 show the effect of such a partial labelling on the nitrobenzene characteristic absorption band due to a symmetrical stretching vibration of its $-NO_2$ groups.[1]

It can be seen from Figs. 1.1 and 1.2 that the 1349 cm^{-1} band of normal nitrobenzene shifts, in part, to 1335 cm^{-1} (in the $-N^{16}O^{18}O$ modification) and 1322 cm^{-1} (in the $-N^{18}O_2$ isotopomer) as a result of its ^{18}O-labelling.

Even in the normal abundance HBr case ($H^{81}Br$ and $H^{79}Br$), where the m^i/m ratio is only $81/79 = 1{\cdot}02$, the isotopic shift of the (vibration–rotation) absorption bands, amounting to 0·33–0·37 cm^{-1}, can still be resolved with a good commercial instrument.[2]

It is therefore usually possible to observe directly the infrared absorption spectra of the individual isotopic modifications constituting a given labelled sample. This possibility is very important in a number of ways, since it can be used to:

(*a*) analyse qualitatively for, and in many cases also determine quantitatively, the various individual isotopomers liable to be present in an isotopic mixture, by measuring its infrared spectrum and comparing the results with comparable spectra of samples of known isotopic composition,

(*b*) obtain spectral data for various isotopic modifications which are not readily available in a pure state, and

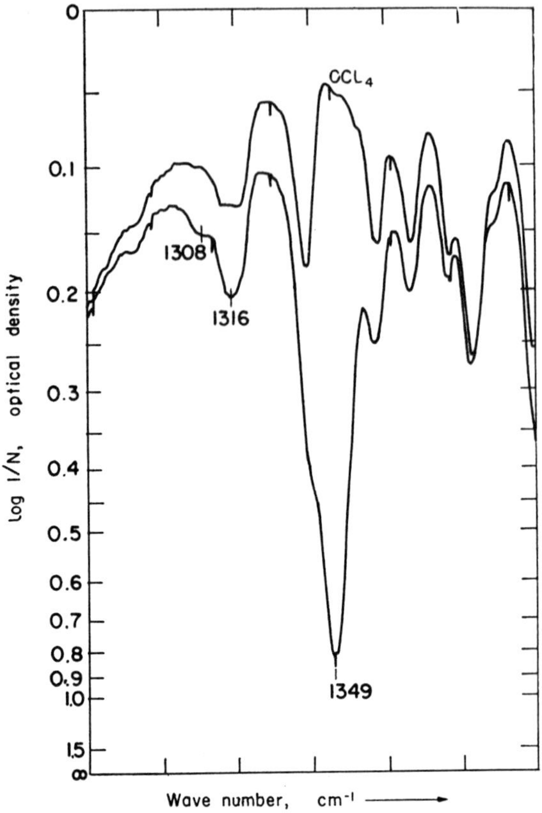

Fig. 1.1 The symmetrical NO_2 stretching band of normal nitrobenzene (0·01 g/ml CCl_4; 0·2 mm cell).

(*c*) find new isotopes, judge upon the isotopic purity of investigated samples, etc.

The spectral data obtained for the various individual isotopic molecules, mainly when derived from isotopically pure samples, are very valuable from several aspects. Their highest significance seems to be in respect to the problem of assigning the individual absorption bands, observed for a certain molecule, to the respective specific modes of its vibration which are excited by the absorbed infrared light (see Section 2.6). The frequency shift of each of these bands as a result of labelling this molecule by one (or more) isotopic atom(s) (the *isotopic shift*) depends directly upon the extent to which this (these) atom(s) participates (participate) in the specific vibration which brings about this band. The more this part of the molecule is involved in this vibration, the more is the respective isotope shift affected by the change in its mass as a result of the isotopic exchange. This dependence makes it usually

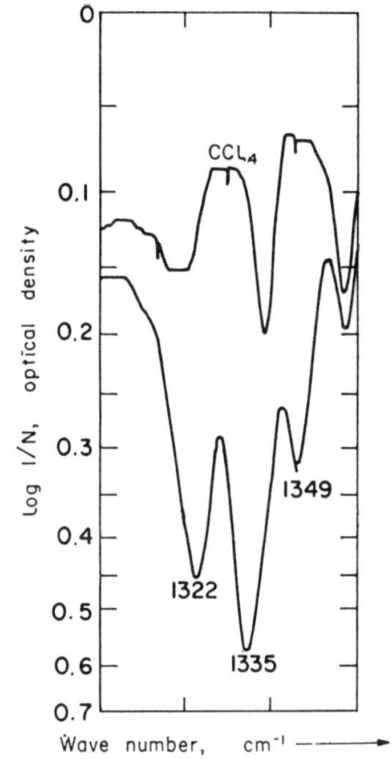

Fig. 1.2 The symmetrical NO_2 stretching band of ^{18}O-labelled nitrobenzene (0·01 g/ml CCl_4; 0·2 mm cell).

possible to identify unambiguously many observed absorption bands on basis of their shifts in the spectra of the appropriately labelled molecules.

The exact frequencies of the various characteristic modes of vibration of labelled molecules (equal to their absorption band frequencies) can also yield information as to the respective molecular structure. If, e.g. an Y–X–Y molecule shows a significant isotope shift for *all* of its three main characteristic vibrations, when X is exchanged for $X^{i(\text{isotopic})}$, then it must be bent since if this molecule were linear its in-phase stretching vibration would leave the X atom unaffected (see Fig. 1.3). The frequency of this mode would then be the same both in the Y–X–Y and in the Y–X^i–Y molecules resulting in no isotopic shift at all.

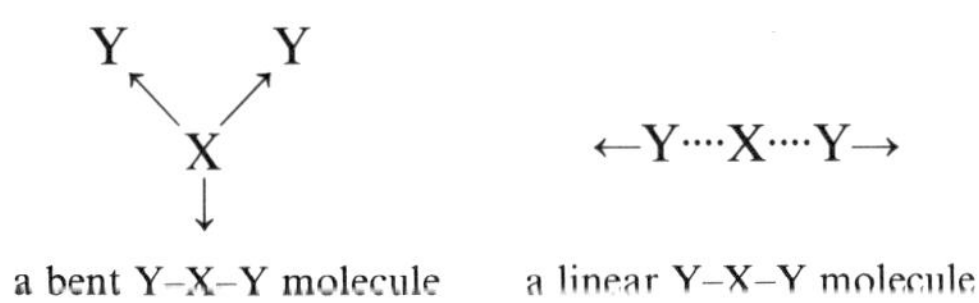

Fig. 1.3 The in-phase stretching mode of the bent and linear Y–X–Y molecules.

The analysis of the highly resolved (vibration–rotation or pure rotation) spectra of gaseous isotopic molecules yields the values of their moments of inertia about their principal axes running through their centres of mass (see Sections 2.2–2.4). These values, together with the corresponding normal moments of inertia, can often be used to obtain the interatomic distances of these molecules[3] or the exact mass of their isotopic atoms.[4]

The precise frequency of the fundamental modes of vibration of labelled molecules, obtained from their infrared absorption and Raman spectra, is also very important for the evaluation of the molecular force constants of the respective normal molecules (which are practically the same as in the labelled ones).

A knowledge of the fundamental vibration frequencies of isotopic molecules is further necessary in order to be able to calculate the partition function for these molecules which is needed for the estimation of the *kinetic* isotope effect on the reactions involving them.[5–7]

The fact that deuterated molecules show infrared absorption spectra which are very different from the respective normal spectra can be made use of when the spectrum of a solvent, such as $CHCl_3$ or H_2O, absorbs too strongly at the region of interest and cannot be used. In this case, the respective deuterated solvents, which are otherwise practically (but for their price, of course . . .) equivalent to the normal ones, can usually be substituted successfully instead. The spectra of sodium permanganate,[8] periodates[9] and various nucleosides and nucleotides[10–14] were thus, e.g. examined in D_2O solutions and the absorption of a number of steroids[15,16] in the C—H stretching region was investigated in $CDCl_3$ solutions.

The spectrum of a labelled compound can be used to clarify some more problems connected with the interpretation of the respective normal spectrum. Thus, e.g. if a doubt arises whether a certain pair of its absorption bands are not due to two interacting modes of vibration (in "Fermi resonance", see Section 2.10) their location in the labelled spectrum can assist to solve this problem.[16]

References

1. Pinchas, S., Samuel, D. and Silver, B. L. (1964). *Spectrochim. Acta* **20**, 179.
2. The Perkin-Elmer Corporation, Norwalk, Connecticut (1965). *Instrument News* **16**, 2, p. 6.
3. Herzberg, G., Patat, F. and Spinks, J. W. T. (1934). *Z. Physik* **92**, 87.
4. Townes, C. H., Merrit, F. R. and Wright, B. D. (1948). *Phys. Rev.* **73**, 1334.
5. Bigeleisen, J. and Wolfsberg, M. (1958). *In* Prigogine, I., (ed.) "Advances in Chemical Physics", p. 17. Interscience Publishers, New York.
6. Wiberg, K. B. and Motell, E. L. (1963). *Tetrahedron* **19**, 2009.
7. Staschevski, D. (1965). *Ber. Bunsenges. phys. Chem.* **69**, 426.
8. Pinchas, S. (1968). *Israel J. Chem.* **6**, 52.
9. Kyrki, J. R. (1965). *Suomen Kem.* **38B**, 192.
10. Miles, H. T. (1958). *Biochim. Biophys. Acta* **27**, 46.

11. Miles, H. T. *Chem. and Ind.* (1958), 591.
12. Miles, H. T. and Frazier, J. (1964). *Biochim. Biophys. Acta* **79**, 216.
13. Miles, H. T. and Frazier, J. (1964). *Biochim. Biophys. Res. Comm.* **14**, 21.
14. Fritzsche, H. (1966). *Biochim. Biophys. Acta* **119**, 645.
15. Hampel, B. (1963). *Ber.* **96**, 2281.
16. Fritzsche, H. (1964). *Abh. dt. Akad. Wiss. Berl. Kl. Chem. Geol. Biol.* **7**, 399.

CHAPTER 2

A Résumé of the Theory of Infrared Absorption Spectroscopy

I. LAULICHT

2.1 Introduction

To a first approximation, the total energy of a molecule can be resolved into the sum of three, more important, additive components: (*a*) that associated with the rotation of the molecule as a whole; (*b*) its vibrational energy; and (*c*) its electronic energy. The basis for this separation lies in the fact that generally the electronic mobility is much greater than that of the atomic nuclei, the vibration frequency of which is, in its turn, considerably higher than that of the molecular rotations.

If a molecule is placed in an electromagnetic field an energy transfer from the field to this molecule will occur only when Bohr's frequency condition (2.1) is satisfied:

$$E_1 - E_2 = h\nu \tag{2.1}$$

where E_1 and E_2 are two quantized energy states of this molecule, h is Planck's constant and ν is the frequency of the electromagnetic field, in sec^{-1}.

Since the quantized rotational energy states are relatively closely spaced, transitions between these levels (which give rise to "pure rotation" spectra) occur usually at low electromagnetic frequencies, i.e. under the impact of far infrared light or in the microwave region ($\nu = 1\text{–}100\ \text{cm}^{-1}$).

The separation between vibrational levels is generally appreciably greater, hence vibration spectra of molecules are observed, as a rule, in the infrared region ($\nu = 30\text{–}12{,}500\ \text{cm}^{-1}$) while the electronic spectra appear almost always above these frequencies.

In order that such a transition from level E_2 to level E_1 should in fact take place, certain "selection rules" must also be fulfilled in each case.

2.2 Rotational Spectra of Linear Molecules

A linear molecule in the gas phase has only two rotational degrees of freedom, of rotation around the two rotation axes perpendicular to the molecular axis (passing through its centre of mass), since rotation about the latter does not change the position of its atoms.

If it possesses a permanent dipole moment, its rotation will produce an oscillating dipole which, as such, will absorb or emit electromagnetic waves with frequencies equal to the frequencies of these oscillations, provided this is allowed by the selection rules. Such molecules will thus show a pure rotation spectrum.

Solving Schroedinger's equation for this case gives its rotational energy levels as

$$E_J = (h^2/8\pi^2 I) \cdot J \cdot (J + 1) \text{ (erg)}, \qquad (2.2)$$

where I is its moment of inertia about its rotation axis and J is the rotational quantum number of the level ($J = 0, 1, 2, \ldots$). Going over from ergs to cm^{-1} units, according to the transformation:

$$1 \text{ erg} \equiv 1/hc \text{ cm}^{-1}, \qquad (2.3)$$

(where c is light velocity), and defining

$$B = h/8\pi^2 cI(\text{cm}^{-1}), \qquad (2.4)$$

gives

$$F_J = BJ(J + 1) \text{ (cm}^{-1}), \qquad (2.5)$$

where $F_J = E_J/hc$.

The selection rule for the absorption of linear molecules is

$$\Delta J = 1. \qquad (2.6)$$

Usually, even at room temperature, many excited rotational levels ($J > 0$) are already populated to an appreciable extent. This results in the appearance of "hot" bands, i.e. bands brought about by the absorption of excited molecules, in room temperature molecular absorption spectra.

The pure rotation spectrum of linear molecules will therefore generally consist of a series of lines ν_i, due to the transitions $E_0 \rightarrow E_1$, $E_1 \rightarrow E_2, \ldots,$ $E_i \rightarrow E_{i+1}$, which according to eqns. (2.1) and (2.5) must have the frequencies of $2B, 4B, \ldots, 2B \cdot (i + 1)$ cm^{-1}, respectively.

Since real molecules are not rigid but are flexible and therefore stretch slightly under the influence of the centrifugal forces that rise with the rotation, their moment of inertia I also increases somewhat with the extent of this rotation, i.e. with their J value. This means that for each J value there also exists a different B value, B_0 being somewhat higher than B_1 and so on. The highest B value is that of B_e which is the value calculated for the molecule at the equilibrium position (while B_0 is the mean B value for the molecule in the ground state).

In order to correct for this effect another term must be added in eqn. 2.5 which now becomes:

$$F_J = BJ(J + 1) - DJ^2(J + 1)^2 \text{ (cm}^{-1}), \tag{2.7}$$

where B and D are now constant within each vibrational level. The value of D is always much less than that of the corresponding B. The expression for the energy transitions responsible for the pure rotation absorption spectrum of a non-rigid rotor is accordingly:

$$\Delta F_{(J \to J+1)} = 2B(J + 1) - 4D(J + 1)^3 \text{ (cm}^{-1}). \tag{2.8}$$

2.3 Symmetric Top Molecules

A non-linear molecule which two of its principal moments of inertia (about its three principal rotation axes) are equal is called a *symmetric top*.

The energy of a rotational level of a rigid symmetric top is given by the equation:

$$F_{JK} = BJ(J + 1) + (A - B)K^2, \tag{2.9}$$

where $A = h/8\pi^2 cI_A$, $B = h/8\pi^2 cI_B$ (I_A = unequal moment of inertia; I_B = equal moment of inertia) and $K = 0, \pm 1, \pm 2, \ldots, \pm J$ is the quantum number which determines the component of the angular moment in the direction of the axis of the top.

Taking account of the non-rigidity of actual symmetric tops introduces additional terms to eqn. 2.9, which however involve only exceedingly small (D_J, D_{JK} and D_K) coefficients operating on $J^2(J + 1)^2$, $J(J + 1)K^2$ and K^4.

If the two equal moments of inertia of a symmetric top are bigger than its third moment it is called a *prolate symmetric top* while it is an *oblate symmetric top* in the opposite case.

Symmetric tops possessing permanent dipole moments absorb electromagnetic radiation to go over to higher rotational levels if the conditions

$$\Delta J = 1 \quad \text{and, usually,} \quad \Delta K = 0 \tag{2.10}$$

as well as other less limiting selection rules are fulfilled. This absorption gives rise to bands having frequencies (cm^{-1})

$$\nu = 2B(J + 1) - 2D_{JK}K^2(J + 1) - 4D_J(J + 1)^3. \tag{2.11}$$

If in a symmetric top $A = B$ the molecule is called a *spherical top*. Since such molecules, usually, do not possess a permanent dipole moment no spherical tops exhibiting a pure rotation spectrum are known.

2.4 Asymmetric Top Molecules

In the case a molecule possesses three different principal moments of inertia I_A, I_B, I_C we have an *asymmetric top*.

The expressions calculated for its rotational levels are quite complex and yield the following values (cm^{-1}) for the lowest levels of a rigid top:

$$F(0_0) = 0; \quad F(1_{-1}) = B + C, \quad F(1_0) = A + C, \quad F(1_1) = A + B;$$
$$F(2_{-1}) = A + B + 4C \qquad (2.12)$$

(the first index of F gives its J number, the second distinguishing between the $2J + 1$ possible sub-levels with equal J; $B = h/8\pi^2 cI_B$; $C = h/8\pi^2 cI_C$).

The main selection rule for the absorption of asymmetric tops having a permanent dipole moment is

$$\Delta J = 0, 1, \qquad (2.13)$$

allowing also transitions between different sub-levels of the same J number.

2.5 Vibrational Spectra of Diatomic Molecules

The interatomic potential of a diatomic molecule cannot be described accurately by any known function. However, to a first approximation it can be represented by the harmonic potential model which is governed by the equation:

$$V = k(r - r_e)^2/2 \qquad (2.14)$$

where V = the molecular potential energy, k = the harmonic force constant, r = the interatomic distance and r_e = the equilibrium interatomic distance of the molecule.

This model obeys Hooke's law for harmonic oscillators according to which

$$-dV/dr = -k(r - r_e) = \mu \, d^2(r - r_e)/dt^2, \qquad (2.15)$$

where μ is the *reduced mass* of the molecule, being equal to

$$m_1 m_2/(m_1 + m_2) \qquad (2.16)$$

(m_1 = mass of atom 1, m_2 mass of atom 2).

The vibration frequency ν (sec^{-1}) of such an oscillator is given by:

$$\nu = (k/\mu)^{1/2}/2\pi. \qquad (2.17)$$

The solution of the quantum mechanical equation of motion for the harmonic oscillator gives its possible vibrational levels as:

$$E_v = h\nu(v + \tfrac{1}{2}), \qquad (2.18)$$

ν having the value given by (2.17) and v being the vibrational quantum number.

It can be seen from eqn. 2.18 that even for $v = 0$, i.e. in the vibrational ground state, the molecule still has some vibrational energy and is not at rest at the equilibrium position.

Expressed in cm^{-1} units the vibrational levels are given by (2.19),

$$G(v) = \frac{E_v}{hc} = \frac{h\nu(v + \frac{1}{2})}{hc} = \omega(v + \tfrac{1}{2})\ \mathrm{cm}^{-1}, \qquad (2.19)$$

where G is called the *vibrational term* and ω is the stretching frequency, in cm^{-1}.

In an actual molecule, as distinct from a harmonic oscillator, however, the stretching frequency is somewhat dependent upon the amplitude of the vibration; ω_e is defined in this case as the stretching frequency for very small displacements from the equilibrium interatomic distance. The vibrational term of an actual molecule is given by the expression:

$$G(v) = \omega_e(v + \tfrac{1}{2}) - \omega_e x_e(v + \tfrac{1}{2})^2 + \omega_e y_e(v + \tfrac{1}{2})^3 + \cdots, \qquad (2.20)$$

where $\omega_e x_e$ and $\omega_e y_e (y_e \ll x_e)$ are small *anharmonicity* constants.

For a harmonic oscillator only vibrational transitions obeying the selection rule $\Delta v = 1$ are possible. For actual molecules, however, transitions of $\Delta v = 2$ or 3 are also possible although they are of a much lower probability; hence overtone bands also appear in their absorption spectrum but, usually, only with low absorption intensities.

The frequency of the fundamental absorption bands of diatomic molecules is, according to (2.20), equal to:

$$v = \omega_e - 2\omega_e x_e + \omega_e y_e[(1 + \tfrac{1}{2})^3 - \tfrac{1}{8}] + \cdots. \qquad (2.21)$$

2.6 The Polyatomic Molecule Vibrations

The number of Cartesian co-ordinates needed to determine completely the positions of the N atoms constituting the polyatomic molecule is $3N$. This molecule thus possesses $3N$ degrees of freedom. Three co-ordinates out of these will determine the position of its centre of gravity, i.e. will describe the translation of the molecule as a whole. If the molecule is not linear it possesses also three rotational degrees of freedom while a linear molecule has only two such degrees of freedom (see Section 2.2). This means that a non-linear molecule has only $3N - 3 - 3 = 3N - 6$ and a linear one only $3N - 3 - 2 = 3N - 5$ degrees of freedom which bring about its vibration but do not effect the translation or rotation of this molecule.

The polyatomic molecular vibrational motion can always be represented as made up of one, or a superposition of several, of its so-called *normal vibrations* taken with suitable amplitudes. Their number is equal to that of its vibrational degrees of freedom and they are characterized by the fact that during these vibrations all the atoms of the molecule move in one phase.

In order to find out the frequency and form of all the normal modes of vibration of a molecule it is necessary to analyze the motion of its atoms during its vibration using some system of co-ordinates.

If displacement co-ordinates (i.e. Cartesian co-ordinates having their origin in the equilibrium position of the relevant atom) are used and it is assumed that the force field of the molecule is harmonic (i.e. such that the restoring forces acting upon each displaced atom are equal to sums of $-k_i s_j$ terms, where k_i is a *force constant* peculiar to this atom under the field of this molecule and s_j is an atomic displacement in a certain direction) one gets, for the restoring force in the x direction, F_x, acting upon *one* displaced atom (No. 1):

$$F_x^1 = -k_{xx}^{11}s_x^1 - k_{xy}^{11}s_y^1 - k_{xz}^{11}s_z^1, \tag{2.22}$$

and similarly:

$$F_y^1 = -k_{yx}^{11}s_x^1 - k_{yy}^{11}s_y^1 - k_{yz}^{11}s_z^1 \tag{2.23}$$

$$F_z^1 = -k_{zx}^{11}s_x^1 - k_{zy}^{11}s_y^1 - k_{zz}^{11}s_z^1. \tag{2.24}$$

If, as is usually the case for normal modes, atom No. 2 is also displaced, new forces arise affecting *also atom No. 1* and we get

$$\begin{aligned} F_x^1 &= -k_{xx}^{11}s_x^1 - k_{xy}^{11}s_y^1 - k_{xz}^{11}s_z^1 - k_{xx}^{12}s_x^2 - k_{xy}^{12}s_y^2 - k_{xz}^{12}s_z^2, \\ &\vdots \\ F_x^2 &= -k_{xx}^{21}s_x^1 - k_{xy}^{21}s_y^1 - k_{xz}^{21}s_z^1 - k_{xx}^{22}s_x^2 - k_{xy}^{22}s_y^2 - k_{xz}^{22}s_z^2, \\ &\vdots \\ F_z^2 &= -k_{zx}^{21}s_x^1 - k_{zy}^{21}s_y^1 - k_{zz}^{21}s_z^1 - k_{zx}^{22}s_x^2 - k_{zy}^{22}s_y^2 - k_{zz}^{22}s_z^2. \end{aligned} \tag{2.25}$$

And since in fact, usually, *all* the atoms are then displaced we finally get:

$$\begin{aligned} F_x^1 &= -k_{xx}^{11}s_x^1 - \cdots - k_{xx}^{12}s_x^2 - \cdots - k_{xx}^{13}s_x^3 - \cdots - k_{xz}^{1N}s_z^N, \\ &\vdots \\ F_x^2 &= -k_{xx}^{21}s_x^1 - \cdots - k_{xx}^{22}s_x^2 - \cdots - k_{xx}^{23}s_x^3 - \cdots - k_{xz}^{2N}s_z^N, \\ &\vdots \\ F_x^3 &= -k_{xx}^{31}s_x^1 - \cdots - k_{xx}^{32}s_x^2 - \cdots - k_{xx}^{33}s_x^3 - \cdots - k_{xz}^{3N}s_z^N, \\ &\vdots \\ F_z^N &= -k_{zx}^{N1}s_x^1 - \cdots - k_{zx}^{N2}s_x^2 - \cdots - k_{zx}^{N3}s_x^3 - \cdots - k_{zz}^{NN}s_z^N. \end{aligned} \tag{2.26}$$

Since the restoring force acting upon a particle carrying out a harmonic motion of frequency ν is given by eqn. 2.27,

$$F = -4\pi^2\nu^2 m_i s_i, \tag{2.27}$$

where m_i is its mass and s_i is its displacement, it follows from the assumed harmonicity of the molecular force field and from the in-phase character of the normal vibrations that this expression can be substituted for the various F forces of eqn. 2.26.

One thus gets:

$$\begin{aligned}
4\pi^2\nu^2 m_1 s_x^1 &= k_{xx}^{11}s_x^1 + k_{xy}^{11}s_y^1 + k_{xz}^{11}s_z^1 + k_{xx}^{12}s_x^2 + \cdots + k_{xz}^{1N}s_z^N,\\
4\pi^2\nu^2 m_1 s_y^1 &= k_{yx}^{11}s_x^1 + k_{yy}^{11}s_y^1 + k_{yz}^{11}s_z^1 + k_{yx}^{12}s_x^2 + \cdots + k_{yz}^{1N}s_z^N,\\
4\pi^2\nu^2 m_1 s_z^1 &= k_{zx}^{11}s_x^1 + k_{zy}^{11}s_y^1 + k_{zz}^{11}s_z^1 + k_{zx}^{12}s_x^2 + \cdots + k_{zz}^{1N}s_z^N,\\
4\pi^2\nu^2 m_2 s_x^2 &= k_{xx}^{21}s_x^1 + k_{xy}^{21}s_y^1 + k_{xz}^{21}s_z^1 + k_{xx}^{22}s_x^2 + \cdots + k_{xz}^{2N}s_z^N,\\
&\vdots\\
4\pi^2\nu^2 m_N s_z^N &= k_{zx}^{N1}s_x^1 + k_{zy}^{N1}s_y^1 + k_{zz}^{N1}s_z^1 + k_{zx}^{N2}s_x^2 + \cdots + k_{zz}^{NN}s_z^N.
\end{aligned} \tag{2.28}$$

This system of linear equations for the various displacements s can be solved only if the determinant of their coefficients is equal to zero. This condition gives a $3N$th grade determinantal equation for ν which as such must have $3N$ solutions. Out of these, 6 in the case of non-linear, and 5 in the case of linear molecules, correspond to *non-genuine normal vibrations* (translations and rotations) and are equal to zero.

Knowing the values of the different k (force) constants of the molecule thus allows the calculation of the frequency of its vibrational (genuine) normal modes as well as the ratios $s_x^1 : s_y^1 : s_z^1 : s_x^2 : \ldots : s_z^N$ which determine the distribution of the potential energy of the normal vibration between the various bonds of the molecule.

Usually, however, the force constants of an investigated molecule are unknown although its normal vibrations can be found out from its infrared and Raman spectra. The force constants are then "calculated" from (in fact: modified until they fit) the vibrational data. Since the number of the normal vibrations of a molecule is as a rule much smaller than that of its constants, many of the latter are neglected in this process in order to be able to derive the approximate value of its more important constants. More vibrational data are also generally sought from isotopic molecules (see Section 3.6).

A highly needed simplification of the very complicated calculations outlined above is obtained by introducing other types of co-ordinates such as *mass weighted co-ordinates*, *normal co-ordinates*, *internal co-ordinates*, etc.

Mass weighted co-ordinates q_i of atoms Nos. 1 and 2 are defined by the equations:

$$q_1 = (m_1)^{1/2}\,\Delta x_1; \quad q_2 = (m_1)^{1/2}\,\Delta y_1; \quad q_3 = (m_1)^{1/2}\,\Delta z_1;$$
$$q_4 = (m_2)^{1/2}\,\Delta x_2, \ldots \tag{2.29}$$

The normal co-ordinate Q_k (k designates the *normal vibration* to which a specific Q belongs) is defined by the expression:

$$Q_k = \sum_{i=1}^{3N} \left(A_{ik} \Big/ \sqrt{\sum_{i=1}^{3N} A_{ik}^2} \right) q_i, \tag{2.30}$$

where A_{ik} is an atomic displacement in one direction during the kth mode.

Internal co-ordinates η_i measure directly the changes in the interatomic distances or angles during vibrations. In case, e.g. of the non-linear molecule BAC (see Fig. 2.1), the three internal co-ordinates are $\eta_1 = \Delta r_{AB}$, $\eta_2 = \Delta r_{AC}$ and $\eta_3 = \Delta 2\phi$, respectively.

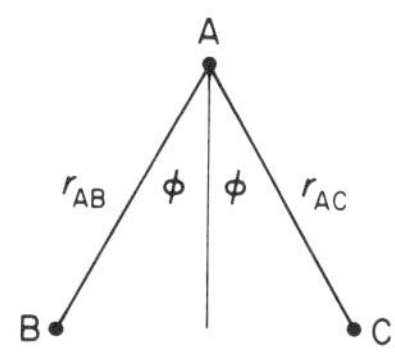

Fig. 2.1 The internal co-ordinates of a BAC molecule.

A further simplification of the calculation of the normal vibrations is possible for molecules which possess some elements of symmetry.

2.7 Symmetry of Polyatomic Molecules

The main symmetry elements which polyatomic molecules may have are the following:

(*a*) *A plane of symmetry*, designated by σ; a reflection of the molecule at this plane produces a molecule indistinguishable from the original.

(*b*) *A centre of symmetry*, designated by i; an inversion of the molecule at this point reproduces it exactly in the same form.

(*c*) *A* p-*fold axis of symmetry*, designated by C_p ($p = 2, 3, 4, \ldots$); a rotation by an angle of $360°/p$ about this axis gives the same configuration as before the rotation.

(*d*) *A* p-*fold rotation–reflection axis*, designated by S_p; a rotation by an angle of $360°/p$ about it followed by a reflection in a plane perpendicular to it results in an identical configuration with the original one.

The symmetry elements which a molecule possesses determine its *point group* (i.e. the combination of symmetry operations that leave at least one point unchanged, possible for it).

Table 2.1 gives the main symmetry elements of the more important point groups.

TABLE 2.1 The Main Symmetry Elements of Various Point Groups

Point group	Symmetry elements
C_2	One two-fold axis
C_3	One three-fold axis
C_i	Centre of symmetry
S_6	One six-fold rotation–reflection axis; one three-fold axis
C_s	One symmetry plane
C_{2v}	One two-fold axis; two symmetry planes (through axis)
C_{3v}	One three-fold axis; three symmetry planes (through axis)
$C_{\infty v}$	One infinitely-fold axis; infinite number of planes
C_{2h}	One two-fold axis; one plane (perpendicular to C_2); centre of symmetry
C_{3h}	One three-fold axis; one plane (perpendicular to C_3); one S_3
D_4	One four-fold axis; four C_2-axes (perpendicular)
D_{2d} (V_d)	Three two-fold axes; one S_4; two planes (through S_4)
D_{3d}	One three-fold axis; three C_2-axes; S_6; i; three planes
D_{2h} (V_h)	Three two-fold axes; three planes; centre of symmetry
D_{3h}	One three-fold axis; three two-fold axes; 3 + 1 planes
D_{4h}	One C_4-axis; four C_2-axes; 4 + 1 planes; S_4; centre of symmetry
D_{5h}	One five-fold axis; five C_2-axes; 5 + 1 planes
D_{6h}	One six-fold axis; six C_2-axes; 6 + 1 planes; S_6; centre of symmetry
$D_{\infty h}$	One infinitely-fold axis; infinite number of C_2-axes and planes; centre of symmetry
T_d	Three two-fold and four three-fold axes; six planes; three S_4 axes

2.8 SYMMETRY TYPES OF NORMAL VIBRATIONS

The normal vibrations of molecules that possess some degree of symmetry either maintain this symmetry, or part of it, or else are totally unsymmetrical. Of the three normal vibrations of the water molecule, e.g. ν_1, its in-phase stretching vibration and ν_2, its bending fundamental, do not destroy the original symmetry of the water molecule which belonging to the C_{2v}-point group has both a C_2-axis and two symmetry planes (see Table 2.1). This can be seen from Fig. 2.2.

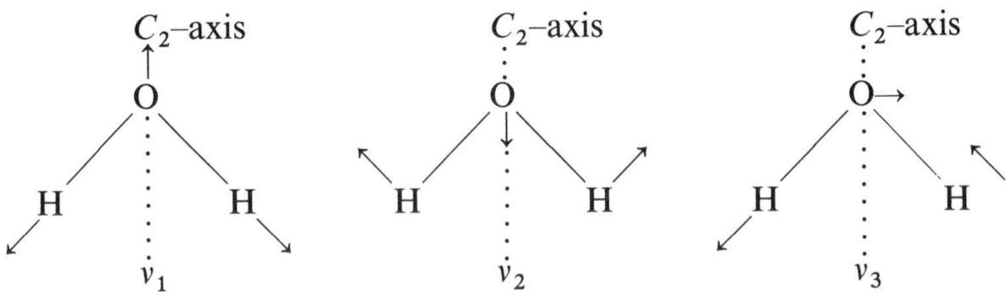

Fig. 2.2 The normal vibrations of the water molecule.

When, however, H_2O vibrates in the ν_3, out-of-phase stretching mode, it does not possess any more a C_2 symmetry-axis.

The normal vibrations are therefore divided into different *symmetry types* or *species* according to their symmetry. Each point group has a limited number of possible species characterized by their typical combinations of symmetry properties. Table 2.2 gives the properties of the symmetry types of a number of point groups. Symmetry in respect to a symmetry element is designated by +1 and no symmetry by −1.

TABLE 2.2 Symmetry Properties of the Various Species of Point Groups C_2, C_i, C_s, C_{2v}, C_{2h} and D_{2h}

Point group	C_2	C_i	C_s	C_{2v}			C_{2h}			D_{2h}						
Species	C_2	i	σ	$C_2(z)$	$\sigma(xz)$	$\sigma(yz)$	C_2	σ	i	σ_{xy}	σ_{xz}	σ_{yz}	i	C_z	C_y	C_x
A	+1															
B	−1															
A_g		+1					+1	+1	+1	+1	+1	+1	+1	+1	+1	+1
A_u		−1					−1	+1	−1	−1	−1	−1	−1	+1	+1	+1
A'			+1													
A''				−1												
A_1				+1	+1	+1										
A_2				+1	−1	−1										
B_1				−1	+1	−1										
B_2				−1	−1	+1										
B_g							−1	−1	+1							
B_u							−1	+1	−1							
B_{1g}										+1	−1	−1	+1	+1	−1	−1
B_{1u}										−1	+1	+1	−1	+1	−1	−1
B_{2g}										−1	+1	−1	+1	−1	+1	−1
B_{2u}										+1	−1	+1	−1	−1	+1	−1
B_{3g}										−1	−1	+1	+1	−1	−1	+1
B_{3u}										+1	+1	−1	−1	−1	−1	+1

Each species of a point group also includes the different possible molecular translations and rotations which share the symmetrical properties characteristic of this species. In the case of the C_{2v} point group, e.g. the rotation about the C_2-axis (in the z-direction), R_z, also belongs to the A_2 species since this rotation does not affect the C_2-symmetry but is not symmetrical with respect to the $\sigma(xz)$ and $\sigma(yz)$ planes.

The +1 or −1 values given in Table 2.2 for each symmetry type also indicate whether the normal co-ordinates corresponding to normal vibrations belonging to this type must be multiplied by +1 or by −1 as a result of carrying out a symmetry operation in respect to the indicated element. Such a value or, in case it is a matrix, the sum of its diagonal elements, is called the *character* of the species. In case a molecule possesses two or more normal vibrations with the same frequency these vibrations are said to be *degenerate*.

The ν_{2a} and ν_{2b} normal modes of the CO_2 molecule are examples of doubly degenerate vibrations (see Fig. 2.3).

The ν_{2a}, in-plane bending, mode.

The ν_{2b}, out-of-plane bending, mode (+ = above the paper; − = below it).

Fig. 2.3 The doubly degenerate modes of the CO_2 molecule.

Double degenerate modes constitute the various E (and other) symmetry types while triply degenerate modes form the F species. Table 2.3 gives the characters for a number of point groups which contain such species.

TABLE 2.3 Symmetry Types and Characters for Point Groups C_{3v}, D_{2d} and T_d

Point group	C_{3v}		D_{2d}				T_d			
Species	C_3	3σ	$2S_4$	C_2''	$2C_2$	2σ	$8C_3$	6σ	$6S_4$	$3C_2$
A_1	+1	+1	+1	+1	+1	+1	+1	+1	+1	+1
A_2	+1	−1	+1	+1	−1	−1	+1	−1	−1	+1
E	−1	0	0	−2	0	0	−1	0	0	+2
B_1			−1	+1	+1	−1				
B_2			−1	+1	−1	+1				
F_1							0	−1	+1	−1
F_2							0	+1	−1	−1

The number of normal vibrations belonging to each symmetry type of a molecule of a certain point group depends upon its structure and can be determined easily.†

In order to calculate the form and frequency of the various normal vibrations of symmetrical molecules it is advantageous to introduce *symmetry co-ordinates*. These co-ordinates are combinations of internal co-ordinates (Section 2.6) defined in equal numbers to those of the normal vibrations of each symmetry type of the molecular fundamentals and having the same symmetrical properties as the corresponding symmetry types. They are chosen in such a way that they never bring about a motion of the molecular centre of mass or its rotation as a whole. Figure 2.4 shows the symmetry co-ordinates defined for the water molecule.

†See e.g. Herzberg, G. (1945). "Infrared and Raman Spectra of Polyatomic Molecules", pp. 132, 134, 136–139. D. Van Nostrand Company, Princeton, N.J. for tables to calculate this number in each case.

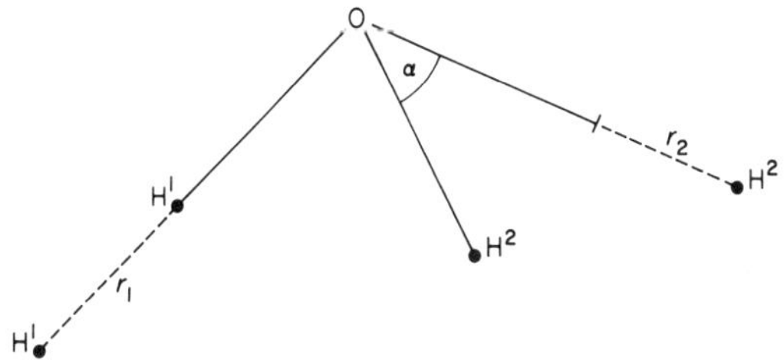

$S_1 = \alpha$ (increase in bond angle)
$S_2 = (r_1 + r_2)/2^{1/2}$
$S_3 = (r_1 - r_2)/2^{1/2}$

r_1 = extension of the first OH bond on vibration
r_2 = extension of the second OH bond on vibration

Fig. 2.4 The symmetrical co-ordinates of the water molecules.

Using these co-ordinates it is possible to obtain a secular determinant (analogous to that of Section 2.6 but) that can be written as a product of factors corresponding to the various symmetry types of the molecule in question. Each of these factors being put equal to zero yields the fundamentals of this type and it is thus necessary to solve determinantal equations of a much lower degree than that of the original equation.

2.9 The Infrared Absorption of Polyatomic Molecules

Solving the Schroedinger equation for the polyatomic harmonic oscillator yields the expression

$$E_v = \sum_{k=1}^{3N-6} (v_k + \tfrac{1}{2})h\nu_k \tag{2.31}$$

(v_k = the vibrational quantum number of the kth normal vibration) for the allowed energy levels of this oscillator.

The corresponding term value will thus be

$$\begin{aligned} G(v_1, v_2, v_3, \ldots) &= \frac{E(v_1, v_2, v_3, \ldots)}{hc} \\ &= \omega_1(v_1 + \tfrac{1}{2}) + \omega_2(v_2 + \tfrac{1}{2}) \\ &\qquad + \omega_3(v_3 + \tfrac{1}{2}) + \cdots \ (\text{cm}^{-1}) \end{aligned} \tag{2.32}$$

where $\omega_1 = \nu_1/c$, $\omega_2 = \nu_2/c$, $\omega_3 = \nu_3/c, \ldots$

Since actual polyatomic molecules also are not really harmonic oscillators their exact energy term values are reproduced more accurately by the expression:

$$\begin{aligned} G(v_1, v_2, v_3, \ldots) &= \omega_1(v_1 + \tfrac{1}{2}) + \omega_2(v_2 + \tfrac{1}{2}) + \omega_3(v_3 + \tfrac{1}{2}) + \cdots \\ &\quad + x_{11}(v_1 + \tfrac{1}{2})^2 + x_{22}(v_2 + \tfrac{1}{2})^2 + x_{33}(v_3 + \tfrac{1}{2})^2 + \cdots \\ &\quad + x_{12}(v_1 + \tfrac{1}{2})(v_2 + \tfrac{1}{2}) + x_{13}(v_1 + \tfrac{1}{2})(v_3 + \tfrac{1}{2}) + \cdots \\ &\quad + x_{23}(v_2 + \tfrac{1}{2})(v_3 + \tfrac{1}{2}) + \cdots, \\ &= \sum_i \omega_i(v_i + \tfrac{1}{2}) + \sum_i \sum_{k \geqslant i} x_{ik}(v_i + \tfrac{1}{2})(v_k + \tfrac{1}{2}), \end{aligned} \quad (2.33)$$

where the various x_{ik}'s are (small) *anharmonicity constants.*

In this case again the various ω's are only the *zero-order frequencies* for infinitesimal amplitudes of the normal vibrations while the molecular vibration frequencies at the lowest *possible* vibrational level are given by:

$$\omega_k^\circ = \omega_k + x_{kk} + \tfrac{1}{2} \sum_{i \neq k} x_{ik} + \cdots \quad \text{(higher terms).} \qquad (2.34)$$

If one or more of the molecular vibrations is degenerate a further term must be added to eqn. 2.33 which brings about a splitting of the term values of levels having at least one doubly excited degenerate mode (*l splitting*).

An actual polyatomic molecule can absorb light to go over from a $G(v_1'', v_2'', \ldots)$ level to a $G(v_1', v_2', \ldots)$ level both if one of its vibrational quantum numbers is increased thereby by 1 and if it is increased by 2, 3, ... (to yield *overtone bands*) or more than one of these are changed, by 1 or more (to yield *combination bands*), although the absorption intensity in the latter case is usually appreciably weaker than that in the first case.

The frequency of its *fundamental* bands, i.e. those brought about by its transitions from the vibrational ground state to levels having only one $v_k = 1$, the rest remaining equal to zero, is given by the equation:

$$\bar{\nu}_k = \omega_k^\circ + x_{kk} = \omega_k + 2x_{kk} + \tfrac{1}{2} \sum_{i \neq k} x_{ik} + \cdots (\text{cm}^{-1}). \qquad (2.35)$$

In order that a transition between two existing vibrational energy levels $G(v)$ be allowed to take place under the effect of an electromagnetic field its *transition moment* defined as:

$$[\mathbf{M}]^{v', v''} = \int \psi_{v'} \psi_{v''}^* \mathbf{M} \, d\tau, \qquad (2.36)$$

where $\psi_{v'}$ and $\psi_{v''}$ are the vibrational wave functions of the upper and lower level, respectively, and **M**, is the molecular dipole moment vector, must always be different from zero. This means that only such vibrations which do affect the molecular dipole moment will usually be excited by the electromagnetic field of light, but homogeneous diatomic molecules under normal conditions, for example, will not show a vibrational absorption spectrum.

It is easily seen that certain normal vibration species of various symmetrical polyatomic molecules do not fulfil this requirement since, due to their high symmetry properties, they do not change the molecular moment. They can therefore be expected to be *infrared inactive* (although, in most cases, Raman active). It is thus possible to calculate theoretically the number of fundamentals to be observed in the infrared absorption spectrum of a compound belonging to a certain point group by finding out the number of normal modes in each of its symmetry types (see Section 2.8) and making clear whether this species is infrared active or non-active.

Table 2.4. collects some data about infrared activity of the various vibration species of a number of important point groups.

TABLE 2.4 Infrared Activity of the Symmetry Types of Various Point Groups (+ = active; − = inactive)

Point group	A	B	A_g	A_u	A'	A''	A_1	A_2	B_1	B_2	B_g	B_u	E	F_1	F_2
C_2	+	+													
C_i			−	+											
C_s					+	+									
C_{2v}							+	−	+	+					
C_{3v}							+	−					+		
C_{2h}			−	+							−	+			
D_{2d}							−	−	−	+			+		
T_d							−	−					−	−	+

Data on the infrared activity of symmetry types of all the point groups not listed in Table 2.4 are also readily available.†

The activity of transitions from the ground state to levels having $\sum_k v_k > 1$ (i.e. of *binary* or higher vibrations) depends upon the symmetry type of the higher levels which is obtained by finding out their characters in respect to the various symmetry elements of the molecule in question. These characters are in each case the product of the respective characters of the component vibrations (taken to the v_kth power). If the resultant symmetry type is infrared active for the point group of this molecule (as given, e.g. in Table 2.4) this transition will give rise to a weak overtone or combination band in the absorption spectrum of this compound.

In the case of double or multiple excitation of degenerate vibrations the higher levels usually split between a couple of symmetry types.

The frequency of overtone or combination bands can be obtained from

†See, e.g. pp. 252, 253 of the book referred to in footnote on p. 16 above.

eqn. 2.33 and is very nearly the multiple or sum of the respective fundamentals.

2.10 Fermi Resonance

If it so happens that two molecular vibrations (including their combinations) belonging to the same species have originally very close frequencies, these vibrations interact to yield one vibrational mode of a *higher* frequency than the original ones and one mode of a *lower* frequency. Each of the new vibrational levels is a "mixture" of the two original levels, which are said to be in *Fermi resonance*, and their energies E are given by the equation

$$E = \bar{E} \pm (4w_{ni}^2 + \delta^2)^{1/2}/2 \tag{2.37}$$

where $\bar{E}$ is the mean energy of the original levels, w_{ni} is the absolute value of the integral $\int \psi W \psi_i^* \, d\tau$ (ψ_z, ψ_i = wave functions of the original states; W = perturbation factor depending upon the anharmonicity of the molecule) and δ is the difference between the original energy levels.

Such a resonance was first observed in the case of CO_2, the first excited v_1 level of which interacts with the lower $2v_2$ sub-level to yield shifted levels at 1388·3 and 1285·5 cm^{-1}.

Fermi resonance brings about a redistribution of the absorption intensity to be expected for the original bands and in the case of an interaction between a fundamental and an overtone band it usually results in the appearance of two strong absorption bands instead of the one (fundamental band) which would have appeared if Fermi resonance did not occur.

2.11 Simplified Force Fields

The calculation of the force constants of a molecule containing more than 4 atoms, on the basis of equations such as (2.28), is still very difficult even if it possesses some degree of symmetry and symmetry co-ordinates are applied. If the assumed force field of the molecule is the "general harmonic force field" ("G.F.F" for which eqn. 2.28 holds) usually there also exist too many force constants to calculate from the observed normal frequencies. A number of simplified force fields have therefore been used in such calculations, the more important of these being:

(*a*) *The Central Forces Field*; the underlying assumption of this field is that the forces acting upon the atoms in a molecule depend only upon the changes made in the distances between them and lie in the directions of the lines connecting these atoms.

In the case of a non-linear YXY molecule this assumption leads to the following equations for its fundamental frequencies:

$$4\pi^2\nu_3^2 = (1 + 2m_Y \sin^2 \alpha/m_X)\alpha_{11}/m_Y, \qquad (2.38)$$

$$4\pi^2(\nu_1^2 + \nu_2^2) = (2a_{33} + a_{11} + 2a_{11}m_Y \cos^2 \alpha/m_X)/m_Y, \qquad (2.39)$$

$$16\pi^4\nu_1^2\nu_2^2 = 2(1 + 2m_Y/m_X) \cos^2 \alpha\, a_{11}a_{33}/m_Y^2, \qquad (2.40)$$

where m_Y and m_X are the masses of atoms Y and X, respectively, α is half the YXY angle and a_{11} and a_{33} are force constants.

In the case of a tetrahedral XY_4 molecule the equations obtained are:

$$4\pi^2\nu_1^2 = (k_2 + 4k_1)/m_Y \qquad (2.41)$$

$$4\pi^2\nu_2^2 = (k_1 - k')/m_Y, \qquad (2.42)$$

$$4\pi^2(\nu_3^2 + \nu_4^2) = 2k_1/m_Y + (4m_Yk_2 + 3m_Xk_2)/3m_Xm_Y - 2k'(3m_X + 16m_Y)/3m_Xm_Y, \qquad (2.43)$$

$$16\pi^4\nu_3^2\nu_4^2 = 2(4m_Y + m_X) \times (k_1k_2 - 8k'k_1 - 5k'k_2 - 8k'^2)/3m_Xm_Y^2, \qquad (2.44)$$

where the various k's are force constants.

(*b*) *The Valence Forces Field*; this field is produced by: (1) restoring forces which act along every valence bond the length of which is different from its equilibrium value, (2) forces opposing changes in the angles between valence bonds.

The equations for the normal frequencies of a bent YXY molecule are in this case:

$$4\pi^2\nu_3^2 = (k_1 + 2m_Y \sin^2 \alpha\, k_1/m_X)/m_Y, \qquad (2.45)$$

$$4\pi^2(\nu_1^2 + \nu_2^2) = (k_1 + 2m_Y \cos^2 \alpha\, k_1/m_X)/m_Y + 2k_\delta(1 + 2m_Y \sin^2 \alpha/m_X)/m_Y l^2, \qquad (2.46)$$

$$16\pi^4\nu_1^2\nu_2^2 = 2k_1k_\delta(1 + lm_Y/m_X)/m_Y^2l^2, \qquad (2.47)$$

where k_1 is the restoring force constant, k_δ is the angle deformation constant and l is the equilibrium X–Y distance.

The frequency of the normal vibrations of a planar XY_3 oscillator is given by the expressions:

$$4\pi^2\nu_1^2 = k_1/m_Y, \qquad (2.48)$$

$$4\pi^2\nu_2^2 = (1 + 3m_Y/m_X)k_\Delta/m_Y l^2, \qquad (2.49)$$

$$4\pi^2(\nu_3^2 + \nu_4^2) = (1 + 3m_Y/2m_X)(k_1/m_Y + 3k_\delta/m_Y l^2), \qquad (2.50)$$

$$16\pi^4\nu_3^2\nu_4^2 = 3(1 + 3m_Y/m_X)k_1k_\delta/m_Y^2 l^2, \qquad (2.51)$$

where k_Δ is the constant for the force opposing a deviation of the X–Y bond from the molecular plane.

In the case of the tetrahedral XY_4 molecule the fundamentals are given by the equations:

$$4\pi^2\nu_1^2 = k/m_Y, \tag{2.52}$$

$$4\pi^2\nu_2^2 = 3k_\delta/m_Y l^2, \tag{2.53}$$

$$4\pi^2(\nu_3^2 + \nu_4^2) = k(1 + 4m_Y/3m_X)/m_Y + 2k_\delta(1 + 8m_Y/3m_X)/m_Y l^2, \tag{2.54}$$

$$\nu_3\nu_4/\nu_1\nu_2 = [(2 + 8m_Y/m_X)/3]^{\frac{1}{2}}. \tag{2.55}$$

The Valence Forces Field ("V.F.F.") is usually a better approximation to the actual potential field than the Central Forces Field.

(*c*) *The Urey-Bradley Force Field*; this field is a refined V.F.F. which takes into account also Van der Waals interactions between non-bonded atoms. The potential energy of a bent XYZ molecule having such a field, e.g. is given by

$$2V = K_1' r_1 \,.\, \delta r_1 + K_1 \,.\, \delta r_1^2 + K_2' r_2 \,.\, \delta r_2 + K_2 \,.\, \delta r_2^2$$
$$+ H' \,.\, \delta\alpha + H \,.\, \delta\alpha^2 + F'q \,.\, \delta q + F \,.\, \delta q^2, \tag{2.56}$$

where K_1', K_1, K_2', K_2, H', H, F' and F are force constants, r_1 and r_2 are the equilibrium X–Y and Y–Z distances, α is the X–Y–Z angle and q is the X–Z distance.

2.12 Vibration–Rotation Spectra

Since gaseous molecules (in contradistinction to liquid molecules, which only rarely possess rotational freedom, and solid state molecules, which are even more hindered) can exist in many rotational levels, their light-induced transition to higher vibrational states can also be accompanied by a *change* in their rotational energy. Such molecules therefore show vibration–rotation absorption bands made up of very many lines, rather than one absorption line having the frequency ν_0 which corresponds to the pure vibrational transition. If these bands are recorded for low pressure samples (to eliminate *pressure broadening*) and are observed under high enough resolution (of the order of 0·02 cm^{-1}) their rotational fine structure can be resolved and analyzed to yield their rotational and vibrational constants.

For linear molecules undergoing transitions from the ground state (or other non-degenerate levels) to vibrational levels in which the molecular dipole moment changes along the molecular axis (giving rise to *parallel bands*) the selection rule for the accompanying change in rotational energy is $\Delta J = \pm 1$. If $\Delta J = +1$ the resulting absorption line belongs to the so-called *R branch* of the vibration–rotation band and if $\Delta J = -1$ it belongs to its *P branch*.

If the transition is from a degenerate vibrational level (such as, e.g. the level $v_1 = v_3 = 0$, $v_2 = 1$, of CO_2—see Fig. 2.3) the resulting parallel band will also contain a weak *Q branch* of lines, for which $\Delta J = 0$ (belonging to different J values).

Transitions increasing the change of the dipole moment in a direction perpendicular to the molecular axis result in (*perpendicular*) absorption bands with a strong Q branch and weaker P and R branches.

In the case of symmetric tops the selection rules for parallel bands are:

$$\text{if } K \neq 0 \qquad \Delta K = 0 \qquad \Delta J = 0, \pm 1, \tag{2.57}$$

$$\text{if } K = 0 \qquad \Delta K = 0 \qquad \Delta J = \pm 1, \tag{2.58}$$

and for perpendicular bands:

$$\Delta K = \pm 1 \qquad \Delta J = 0, \pm 1. \tag{2.59}$$

Asymmetric tops show PQR branches for all their bands.

The values of the B and D rotational constants depend slightly upon the vibrational state of the molecule since (as a result of anharmonicity) this state affects the mean distances between its constituent atoms, hence also its moments of inertia and the extent of their change by centrifugal forces. The constants are given by the equations:

$$B_{v_1 v_2 \ldots} = B_e - \alpha_1(v_1 + \tfrac{1}{2}) - \alpha_2(v_2 + \tfrac{1}{2}) - \cdots, \tag{2.60}$$

$$D_{v_1 v_2 \ldots} = D_e - \beta_1(v_1 + \tfrac{1}{2}) - \beta_2(v_2 + \tfrac{1}{2}) - \cdots, \tag{2.61}$$

where the subscript e refers to the equilibrium position and the various α's and β's are constants.

The α coefficients are also the result of: (*a*) the fact that even the *mean* moments of inertia of harmonic oscillators usually depend somewhat upon their oscillation amplitudes, (*b*) the *Coriolis interaction* between the vibrations and rotations of the polyatomic molecule. This interaction depends upon the apparent velocity of the vibrating atoms in respect to the coordinate system fixed in the rotating molecule (and upon its angular velocity) and excites to a certain degree apparent vibrations which belong to other normal modes (of various species). This leads to higher B values for the higher frequency interacting vibrational levels and lower B values for the lower ones.

The α_i values can be calculated from the respective molecular constants, e.g. the Coriolis component of the α_2 coefficient for a linear XY_2 molecule is given by:

$$\alpha_2^{(\text{Cor.})} = 4B_{\text{e}}^2\omega_2/(\omega_3^2 - \omega_2^2). \tag{2.62}$$

In the case of symmetric and spherical top molecules a Coriolis interaction between degenerate branches of the same vibration sometimes brings about a splitting of this level, the extent of which depends upon the *Coriolis zeta constant* ζ_i which measures the magnitude of the vibrational angular momentum when this vibration is singly excited. The following expressions give the energy values of the rotational sub-levels of the ν_3, triply degenerate, vibration of the spherical top, tetrahedral XY_4 molecule:

$$B_vJ(J+1) + 2B_v\zeta_i(J+1), \quad B_vJ(J+1), \quad B_vJ(J+1) - 2B_v\zeta_iJ. \qquad (2.63)$$

CHAPTER 3

Isotope Effects on Frequencies

I. LAULICHT

3.1 The Isotope Effect on the Pure Rotation Spectrum

Since the linear molecule rotational energy levels are given (Section 2.2) by: $F_J = BJ(J + 1) - DJ^2(J + 1)^2$ and $B = h/8\pi^2cI$ (I = molecular moment of inertia) it is evident that the rotational spectrum of such an isotopic molecule varies considerably from that of the respective normal molecule, as a result of its different mass which changes its I.

Neglecting the non-rigidity correction, and remembering that the internuclear distances in isotopic molecules in the vibrational ground state are practically the same as in the corresponding normal isotopomers, we get for the rotational levels of an isotopic diatomic molecule in the ground state

$$F_J^i = B^iJ(J + 1) = (h/8\pi^2cI^i)J(J + 1) = (h/8\pi^2c\mu^ir^2)J(J + 1), \quad (3.1)$$

where μ^i = the reduced mass of the isotopic molecule and r = the interatomic distance. Substituting $h/8\pi^2cr^2$ by μB gives

$$F_J^i = (\mu/\mu^i)BJ(J + 1), \quad (3.2)$$

and therefore,

$$B^i = (\mu/\mu^i)B. \quad (3.3)$$

The spacing of the pure rotation spectrum of an isotopic diatomic molecule ($= 2B^i$) is thus lower by a factor of μ/μ^i from the corresponding normal spacing.

Rigorously, however, eqn. 3.3 holds only for B_e as even for the ground state the different zero-point vibration amplitudes of the various isotopomers bring about small changes between their mean internuclear distances, as a result of their anharmonicity.

The following relations were calculated for the other rotational constants of the isotopic diatomic molecules:[1]

$$\alpha_e^i = \rho^3\alpha_e, \qquad D_e^i = \rho^4D_e, \qquad \beta^i = \rho^5\beta \quad \text{etc.}, \quad (3.4)$$

where $\rho = (\mu/\mu^i)^{1/2}$.

To a good approximation the isotope shift of a rotational line of such a molecule increases as the expression $J(J + 1)$, J referring to the respective upper rotational level.

In the case of non-linear molecules also a strong isotope effect on their pure rotation spectra can be expected as a result of the changes brought about in the principal moments of inertia (A, B, C) of the labelled isotopomers by the isotopic masses.

The I^i values of linear isotopic molecules, which can be obtained from their pure rotation or vibration–rotation spectra, can be used to supply the additional equations needed for the calculation of the internuclear distances of the less simple linear molecules[2] as well as to yield the mass of the constituent isotopic atoms, in case these distances are known.

3.2 The Effect on the Diatomic Molecular Vibration

The potential function of an isotopic molecule is generally, to a high order of approximation, the same as that of the respective normal modification.† The vibration frequency of an *harmonic* isotopic diatomic molecule can therefore be represented, according to eqn. 2.17, by:

$$\nu^i = (k/\mu^i)^{\frac{1}{2}}/2\pi, \tag{3.5}$$

as compared with the expression: $\nu = (k/\mu)^{\frac{1}{2}}2\pi$, holding for the respective normal molecule. This gives:

$$\nu^i/\nu = (\mu/\mu^i)^{\frac{1}{2}}. \tag{3.6}$$

Equation 3.6 is also a very good approximation for the ratio of the isotopic fundamental stretching frequencies observed for real (anharmonic) diatomic molecules since their anharmonicity is usually not appreciable and it changes only moderately with isotopic substitution. In fact, the anharmonicity constants $\omega_e^i x_e^i$, $\omega_e^i y_e^i$ of the isotopic diatomic molecule are approximated quite precisely by the expressions

$$\omega_e^i x_e^i = \rho^2 \omega_e x_e \quad \text{and} \quad \omega_e^i y_e^i = \rho^3 \omega_e y_e. \tag{3.7}$$

Even in the case of polyatomic molecules eqn. 3.6 is very useful in predicting the correct —R—X^i stretching frequency if most of the energy of this vibration is centered in the RX group which then behaves as an almost independent oscillator. This is usually the situation when X has a low atomic weight, e.g. when X = hydrogen, or, to a lesser degree, when the RX bond is exceptionally strong, as in the case of R═X or R≡X bonds (e.g. C═O or C≡N linkages). Whenever the observed ν^i/ν ratio for corresponding —R—X stretching frequencies of such polyatomic molecules differs widely from the

†Slight differences between these functions are theoretically expected in case there exist low-lying excited electronic states or when the ground electronic state is degenerate.

value calculated for this ratio according to eqn. 3.6 a special effect can safely be assumed to operate. Coupling of the —R—X stretching fundamental with another mode of vibration and intermolecular interactions affecting the —RX group (as, e.g. when hydrogen bonds are formed) are responsible for most of the unexpectedly high v^i/v ratios observed (v^i is here always taken to refer to the *heavier* isotopomer). In these cases a considerable part of the vibration energy is localized in other parts of the polyatomic molecule and the diatomic R—X model can no longer provide the correct —R—X^i stretching frequency which, as a rule, is higher than that of the model the less this energy is centred in the —R—X group.

The bending fundamental frequencies of R—X^i groups belonging to polyatomic molecules can also be predicted satisfactorily on basis of eqn. 3.6 if the kinetic energy of these vibrations is localized preponderately in these groups. This requirement, however, is fulfilled less often in this case than in the case of the —R—X stretching frequencies which are more isolated from the rest of the polyatomic vibrations.

3.3 The Normal Frequencies of Isotopic Triatomic Molecules

It can be shown[3] that for harmonic, non-linear, symmetrically labelled Y—X—Y molecules, the following relations hold:

$$\left(\frac{\omega_3^i}{\omega_3}\right)^2 = \frac{m_X m_Y (m_X^i + 2m_Y^i \sin^2 \alpha)}{m_X^i m_Y^i (m_X + 2m_Y \sin^2 \alpha)}, \tag{3.8}$$

where ω_3 is the normal asymmetric stretching frequency of the Y—X—Y molecule, m_X and m_Y are the atomic weights of its X and Y atoms, α is half the YXY angle and the superscript i denotes quantities belonging to the respective labelled molecule; and

$$\left(\frac{\omega_1^i \omega_2^i}{\omega_1 \omega_2}\right)^2 = \frac{(2m_Y^i + m_X^i) m_Y^2 m_X}{(2m_Y + m_X) m_Y^{i2} m_X^i}, \tag{3.9}$$

where ω_1 and ω_2 are the YXY symmetric stretching and deformation frequencies, respectively.

In the case of real (anharmonic) molecules eqns. 3.8 and 3.9 can strictly be applied only to their (harmonic) zero-order normal frequencies (ω_e's) and not to their *observed* fundamentals. Substituting the latter frequencies for the respective ω_e's results, however, usually in a quite satisfactory agreement (within 3%) with the calculated frequency ratios.[4] These equations may thus: (*a*) serve in checking assignments of YXY molecules absorption bands, and (*b*) be used to calculate part of the normal frequencies of their labelled isotopomers.

In order to calculate the apex angle of such molecules from their ω_3's and ω_3^i's according to eqn. 3.8 one must use, however, only zero-order frequencies since this angle is very sensitive to the correct frequency ratio. The x_{ik}^i anharmonicity coefficients of the isotopic molecules, needed for the calculation of their ω_3^i's from the observed (anharmonic) ν_3^i fundamentals, can be approximated quite satisfactorily by the expression[5]

$$x_{ik}^i = \frac{\omega_i^i \omega_k^i}{\omega_i \omega_k} x_{ik}. \tag{3.10}$$

For harmonic linear YXY molecules there also exist separate expressions[4] for ω_1^i and ω_2^i:

$$\left(\frac{\omega_1^i}{\omega_1}\right)^2 = \frac{m_Y}{m_Y^i}, \tag{3.11}$$

$$\left(\frac{\omega_2^i}{\omega_2}\right)^2 = \left(\frac{\omega_3^i}{\omega_3}\right)^2 = \frac{m_Y(1 + 2m_Y^i/m_X^i)}{m_Y^i(1 + 2m_Y/m_X)}. \tag{3.12}$$

If an YXY molecule is labelled only at one Y position, i.e. asymmetrically, the resulting YXY^i molecule no longer retains the symmetry of the normal modification and the relations between the normal frequencies of these isotopomers are complicated.[6]

In case the $m_Y^i - m_Y$ difference is small, compared with m_Y, the isotope shifts for the YXY^i fundamentals may, however, be taken as being practically half the respective shifts calculated for the Y^iXY^i molecule.[7] These shifts are therefore given approximately by the equations[7]

$$\frac{\Delta\omega_3}{\omega_3} = \frac{m_X \,.\, \Delta m_Y}{4(m_X + 2m_Y \sin^2 \alpha)(m_Y + \Delta m_Y)} \tag{3.13}$$

where $\Delta\omega_3 = \omega_3$ of YXY $- \ \omega_3$ of YXY^i, and

$$\frac{\Delta\omega_1}{\omega_1} + \frac{\Delta\omega_2}{\omega_2} = \frac{(m_X + m_Y)\,\Delta m_Y}{2(m_X + 2m_Y)(m_Y + 2\,\Delta m_Y)}. \tag{3.14}$$

If YXY^i is linear the following relations also hold:[7]

$$\frac{\Delta\omega_1}{\omega_1} = \frac{\Delta m_Y}{4(m_Y + \Delta m_Y)}, \tag{3.15}$$

$$\frac{\Delta\omega_2}{\omega_2} = \frac{\Delta\omega_3}{\omega_3} = \frac{m_X \,.\, \Delta m_Y}{4(m_X + 2m_Y)(m_Y + \Delta m_Y)}. \tag{3.16}$$

Expressions for the fundamental frequencies of unsymmetrical triatomic molecules labelled by an isotope for which $m^i - m \ll m$ have also been derived.[8]

3.4 THE TELLER–REDLICH PRODUCT RULE

It has been shown[9, 10] by Teller and Redlich (independently) that since the force fields of isotopic molecules are practically identical, the product of all their ω^i/ω ratios for normal vibrations belonging to any one symmetry type depends only on the masses of their constituent atoms and their geometrical structure. It is thus possible to write for any pair of isotopic molecules:

$$\frac{\omega_1^i}{\omega_1} \cdot \frac{\omega_2^i}{\omega_2} \cdots \frac{\omega_f^i}{\omega_f} = \sqrt{\left(\frac{m_1}{m_1^i}\right)^{\alpha} \left(\frac{m_2}{m_2^i}\right)^{\beta} \cdots \left(\frac{M^i}{M}\right)^{t} \left(\frac{I_x^i}{I_x}\right)^{\delta x} \left(\frac{I_y^i}{I_y}\right)^{\delta y} \left(\frac{I_z^i}{I_z}\right)^{\delta z}}, \quad (3.17)$$

where $\omega_1, \omega_2, \ldots, \omega_f$ are the harmonic frequencies of all the f normal vibrations of a given species, m_1, $m_2, \ldots$ are the atom masses of each of the various existing sets of identical nuclei (i.e. reproduced by the allowed symmetry operations) and α, $\beta, \ldots$ are the numbers of normal vibrations, including the non-genuine ones, contributed by each of these sets to the species in question (these numbers can be found in Tables 35 and 36 of ref. 3). M is the molecular weight and t is the number of translations which belong to the species considered. I_x, I_y and I_z are the molecular moments of inertia about the x, y and z axes, passing through the centre of mass, and δx, δy and δz are equal in each case to 1 only when a rotation about the x, y or z axis, respectively, belongs to this species, otherwise being equal to 0. In all these calculations a degenerate vibration is counted once only.

As before (Sections 3.2 and 3.3), substituting the observed, anharmonic, fundamentals of a (polyatomic) molecule for its harmonic frequencies $\omega_1, \omega_2, \ldots$ also, usually, gives satisfactory results so that the *Teller–Redlich product rule* (eqn. 3.17) is very useful in predicting the normal frequencies of labelled isotopomers and in checking the point group and absorption band assignment of normal and isotopic molecules. Evidently, a good agreement between the product of a number of ω^i/ω ratios (which according to some assumptions regarding the point group and band assignment of certain isotopic molecules are all the respective ω^i/ω ratios of one of their species) and the value calculated for this product on basis of eqn. 3.17 supports the assignment on which this calculation is strongly based.

If the isotopic substitution is not complete and leaves a number of certain atoms besides their isotopic ones, the symmetry of the resulting labelled molecule is, in many cases, lower than that of the homogeneous normal modification. In this case, the product rule (eqn. 3.17) can be applied only to the symmetry types of the labelled molecule, the ω's taken for the normal modification in this case being always those corresponding to all the ω^i's included in the calculation.

In the case of labelling CH_4 by one deuterium atom, e.g. the molecular symmetry drops from that of the $\boldsymbol{T_d}$ point group to that of the $\boldsymbol{C_{3v}}$ group. The CH_4 normal vibrations ω_1 and ω_2 belong to the A_1 and E species, respectively, while its other ω_3 and ω_4 frequencies are of the F_2 symmetry type. Of the six CH_3D normal vibrations, ω_1^i, ω_{3a}^i and ω_{4a}^i belong to the A_1 species while ω_2^i, ω_{3bc}^i and ω_{4bc}^i are in that of E. The correlations between the CH_4 frequencies and those of CH_3D are shown in Fig. 3.1.

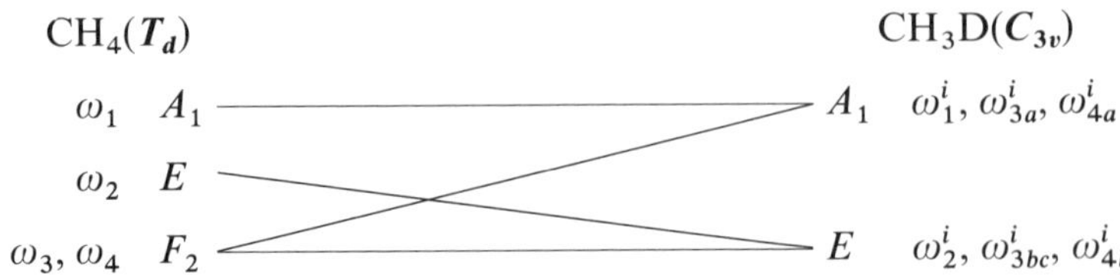

Fig. 3.1 The correlations between the CH_4 and the CH_3D normal vibrations.

The 3.17 equation as applied to the A_1 symmetry type of CH_3D will therefore be:

$$\frac{\omega_1^i \cdot \omega_{3a}^i \cdot \omega_{4a}^i}{\omega_1 \cdot \omega_3 \cdot \omega_4} = \sqrt{\left(\frac{m_{\mathrm{H}}}{m_{\mathrm{D}}}\right)\left(\frac{M^i}{M}\right)}. \tag{3.18}$$

When the correlations between the lower symmetry point group ω^i frequencies and the corresponding normal frequencies are not straightforward they can be found out easily by applying group theory or else read out directly from the appropriate correlation table (see, e.g. ref. 3, p. 237).

The correct assignment of the individual ω^i frequencies of a lower symmetry isotopic molecule within a certain species (in cases of great mass differences where the large isotopic shifts observed make it difficult) is assisted appreciably by the rule that *mass differences cannot bring about a crossing over of frequencies within the same species.*

Applying the product rule to the tetrahedral XY_4^i oscillator yields the following relations:

$$\frac{\omega_1^i}{\omega_1} = \frac{\omega_2^i}{\omega_2} = \left(\frac{m_{\mathrm{Y}}}{m_{\mathrm{Y}}^i}\right)^{1/2}, \tag{3.19}$$

$$\frac{\omega_3^i\omega_4^i}{\omega_3\omega_4} = \frac{m_{\mathrm{Y}}}{m_{\mathrm{Y}}^i}\left(\frac{M^i}{M}\right)^{1/2}. \tag{3.20}$$

In the case of the $X^iY_4^i$ molecule, eqn. 3.19 remains the same (since the X atom does not move at all during the ω_1 or ω_2 vibration) while the right side of eqn. 3.20 must be multiplied by the factor $(m_x/m_x^i)^{1/2}$.

Again, for small mass differences between Y and Y^i the isotope shifts of the ν_1 and ν_2 XYY_3^i frequencies are very nearly equal to $\frac{3}{4}$ of the corresponding XY_4^i shifts while for $XY_2Y_2^i$ and XY_3Y^i they are about $\frac{1}{2}$ and $\frac{1}{4}$ of these shifts, respectively.

3.5 The Decius–Wilson–Sverdlov Sum Rule

The sums of the squares of all the normal frequencies, ν_k, of each of a number of isotopic molecules which when superimposed geometrically with appropriate signs vanish at all the atomic positions, obey the Decius–Wilson–Sverdlov sum rule[11, 12]. In the case of HOH, HOD and DOD, e.g. this rule states:

$$\sigma(\text{HOD}) + \sigma(\text{DOH}) - \sigma(\text{HOH}) - \sigma(\text{DOD}) = 0$$

or (3.21)

$$\sigma(\text{HOD}) + \sigma(\text{DOH}) = 2\sigma(\text{HOD}) = \sigma(\text{HOH}) + \sigma(\text{DOD})$$

where $\sigma = 4\pi^2 \sum_k \nu_k^2$.

This rule is based on the fact that the sum of the squared normal frequencies of a molecule is a linear function of its reciprocal atomic masses; if, therefore, these masses are made to vanish by superimposing them appropriately these sums must also vanish.

In the case of $^{14}NH_3$, $^{15}NH_3$, $^{14}ND_3$ and $^{15}ND_3$ this rule gives

$$\sigma(^{14}\text{NH}_3) + \sigma(^{15}\text{ND}_3) - \sigma(^{14}\text{ND}_3) - \sigma(^{15}\text{NH}_3) = 0. \qquad (3.22)$$

Since all these modifications have the same symmetry, eqn. 3.22 also holds for the sums over any of the symmetry species of ammonia, separately. When this is not the case (e.g. for the eqn. 3.21 group) factoring of the summed frequencies is complicated and sometimes even impossible.[13]

3.6 The Bigeleisen–Sverdlov Second-Order Sum Rule and Other Sum Rules

The following rule was derived for the $\sum_k \lambda_k^2 = \sum_k (4\pi^2\nu_k^2)^2$ sums of the various isotopic XY_z molecules:[14, 15] *In a molecule having at least two equivalent positions for an* Y *atom, the difference in the* $\sum_k \lambda_k^2$ *sum of a molecule containing both* Y *and* Y^i *atoms brought about by exchanging* X^i *for* X *is the arithmetic mean of the corresponding differences brought about in the* $\sum_k \lambda_k^2$ *sums of the respective pure isotopic species.* Mathematically stated this rule has the form:

$$z\left[\sum_k^{3n-6} \lambda_k^2(X^iY_{z-r}Y_r^i) - \lambda_k^2(XY_{z-r}Y_r^i)\right]$$
$$= r\left[\sum_k^{3n-6} \lambda_k^2(X^iY_z^i) - \lambda_k^2(XY_z^i)\right] + (z-r)\left[\sum_k^{3n-6} \lambda_k^2(X^iY_z) - \lambda_k^2(XY_z)\right] \qquad (3.23)$$

Equation 3.23 also holds for sums taken over *single* symmetry types when these are common to all the relevant isotopomers.

Special consideration was given to the case where X and X^i are in fact Y and Y^i in positions equivalent to those of the z Y's in XY_z.[14]

It was shown that a similar second-order sum rule exists not only for λ_k^2 sums but also for the respective $\sum_{i \neq j}^{3n-6} \sum_{j}^{3n-6} \lambda_i \lambda_j$ sums.[14]

The product rule (eqn. 3.17) together with the sum rules make it possible in simple cases (such as, e.g. that of XYY^i) to calculate the fundamentals of one isotopomer from those of the corresponding symmetrical molecules without evaluating their force constants.

Sverdlov has proved the existence of sum rules of still higher orders for the fundamental frequencies of isotopic molecules.[16,17]

For the $\sum_k^{3n-6} \lambda_k^3$ sums, e.g. the rule is:

$$\sum_{i=0}^{3+1} \sum_{j} (-1)^i C_{ij} \sum_{k}^{3n-6} \lambda_{ijk}^3 = 0 \tag{3.24}$$

where C_{ij} is the coefficient for the molecule with i substituted atoms in its jth isomeric modification. The C_{ij} coefficients are chosen in such a way that multiplying the respective isomers by them and superimposing the products of all the isotopomers with the $(-1)^i$ sign bring about the vanishing of all the atomic masses.

For the vibrations of deuterated ethylenes this equation becomes[16b]

$$\left[\sum_k \lambda_k^3(C_2H_4) + \sum_k \lambda_k^3(C_2D_4)\right]$$

$$+ 2\left[\sum_k \lambda_k^3(trans\text{-}C_2H_2D_2) + \sum_k \lambda_k^3(cis\text{-}C_2H_2D_2) + \sum_k \lambda_k^3(CH_2{=}CD_2)\right]$$

$$- 4\left[\sum_k \lambda_k^3(C_2HD_3) + \sum_k \lambda_k^3(C_2H_3D)\right] = 0. \tag{3.25}$$

Substituting experimental data in eqn. 3.25 gives:[16b] 2·9299 units − 2·9308 units = 0.

The "complete isotopic rule" described by Brodersen and Langseth[18–21] gives in certain cases a sufficient number of relations which allows the calculation of all the fundamental frequencies of one isotopomer from those of two other respective isotopic modifications.

New sum rules for the fundamentals of isotopic molecules were also derived by Heicklen[22] who also stressed the importance of knowing the frequencies of enough labelled molecules, in respect to *all* their constituent atoms, for the complete determination of their force constants. This information is so valuable since the number of constants which must be determined in order to reproduce satisfactorily a given molecular force field is usually much larger than the number of fundamentals observed for it (Section 2.11). The normal frequencies measured for the respective labelled species thus constitute necessary added vibrational data (i.e. added equations) for evaluating these force constants† which remain practically the same as in the normal molecule (see Section 3.2).

3.7 Determination of Isotopic Frequencies by other Techniques

The individual fundamental frequencies of isotopic molecules can be determined from the normal isotopomer frequencies, the masses of the isotopic atoms and the molecular geometry by means of the so-called *Green's function*. This function, G (inverse of the secular matrix), is defined by eqn. 3.26:

$$G(\lambda) = (\lambda \boldsymbol{E} - \boldsymbol{D})^{-1}, \qquad (3.26)$$

where $\boldsymbol{E}$ is the unit matrix and $\boldsymbol{D}$ is the force constant matrix in terms of the q co-ordinates, and the procedure is based upon the assumption that all the dynamics of an isotopic molecule is present in this function for the respective normal modification.[28, 29] Since, however, the involved calculations are tedious this method has so far been applied only to simple molecules.[29]

If the mass of the isotopic atoms is only, relatively, little different from that of the normal ones a perturbation treatment[30, 31] yields the following important inequality for the changes brought about in the fundamental vibration frequencies of molecules as a result of an isotopic substitution:

$$\frac{\delta(\omega_k^a)^2}{\delta \mu_\beta} \geqslant 0, \qquad (3.27)$$

where ω_k^a is the kth *angular* fundamental frequency, equal to $2\pi\nu_k$, and $\mu_\beta = 1/m_\beta$, is the reciprocal of the mass of atom β. This inequality means that

† The question of the *uniqueness* of the force constants obtained on the basis of labelled molecule spectra only (without being supplemented by vibration–rotation constants) is, however, yet open to discussion.[23–26] In any case it seems that, generally, the partly labelled, unsymmetrical isotopomers, such, e.g. as HDS and HC≡CD, are more useful in determining the potential constants of the respective molecules than the fully labelled modifications.[27]

no original† fundamental vibration of a molecule can rise in frequency upon labelling it with isotopic atoms of a higher atomic weight.

A first-order perturbation treatment of the vibration frequencies of such labelled molecules[32] also yields the equation:

$$\Delta\lambda_k/\lambda_k = [(L^x_{nk})^2 + (L^z_{nk})^2 + (L^z_{nk})^2]m_n^2\,\Delta(1/m_n), \tag{3.28}$$

where $\Delta\lambda_k$ is the difference in the kth characteristic vibrational value $\lambda_k = 4\pi^2\nu_k^2$ of a labelled molecule, as a result of a change in the m_n mass of its nth atom and L^x_{nk}, L^y_{nk} and L^z_{nk} are the components of the L^X_k vector associated with the nth atom and defined by

$$X = L^X_k Q_k, \tag{3.29}$$

where X = the cartesian displacement vector and Q_k = the kth normal co-ordinate (see Section 2.6).

Equation 3.28 was found to be useful for estimating satisfactorily the isotope shifts in the case of various labelled boron and nitrogen compounds.[32]

For individual modes of vibration centered in homogeneously fully labelled hydrogen-containing groups (such as XD_3) in complex molecules an approximate isotopic frequency rule was derived[33] which is given by

$$\lambda^i/\lambda = 1 - \sum_j \Delta T_j/\rho T, \tag{3.30}$$

where $\lambda = 4\pi^2\nu^2$, $\sum_j \Delta T_j = \sum_j \Delta m q_j^2$ is the *differential* kinetic energy of this vibration due to all the substituted hydrogen atoms, T = the total kinetic energy of this vibration and $\rho = m^i/m$ (the isotopic mass ratio).

When applied to the zero-order vibrations of D_2O this rule gives results within 0·2% of the correct values while the frequencies calculated for its anharmonic (i.e. observed) fundamentals are within 1·2% of the observed values.[33] The application of eqn. 3.30 to the various CD_2 and CD_3 vibration modes gives predicted ν^i/ν ratios which agree well with the respective ratios observed for various high polymeric compounds.[34] This equation can therefore be made use of in the assignment of the characteristic XD_n bands.

3.8 Isotopic Frequency Ratios for Valence Forces Models

In the case of oscillators which can be taken to be approximated satisfactorily by valence forces models (see Section 2.11(b)) further relations between the various respective isotopic fundamentals can be shown to hold.

† *Observed* fundamentals can sometimes disobey this inequality as a result of Fermi resonance (Section 2.10) occurring for one isotopomer but not for the other, due to the isotope shift.

For a linear X^iY^iZ molecule, e.g. the following can be written[34]

$$\frac{\nu_1^{(i)2}\nu_3^{(i)2}}{\nu_1^2\nu_3^2} = \frac{m_X m_Y \,.\, M^i}{m_X^i m_Y^i \,.\, M}, \tag{3.31}$$

$$\frac{\nu_2^{(i)2}}{\nu_2^2} = \frac{l_1^2/m_Z + l_2^2/m_X^i + (l_1 + l_2)^2/m_Y^i}{l_1^2/m_Z + l_2^2/m_X + (l_1 + l_2)^2/m_Y}, \tag{3.32}$$

where ν_1 and ν_3 are the two XYZ stretching frequencies, ν_2 is the bending vibration, m_X, m_Y, m_Z are the atomic masses of X, Y, Z, respectively, M is the molecular weight, and l_1 and l_2 are the XY and YZ distances, correspondingly.

In the case of a planar $X^iY_3^i$ oscillator one gets[35]

$$\nu_1^i/\nu_1 = (m_Y/m_Y^i)^{1/2}, \tag{3.33}$$

$$\nu_2^i/\nu_2 = (1 + 3m_Y^i/m_X^i)^{1/2}(1 + 3m_Y/m_x^i)^{-1/2}(m_Y/m_Y^i)^{1/2}, \tag{3.34}$$

$$\frac{\nu_3^{(i)2} + \nu_4^{(i)2}}{\nu_3^2 + \nu_4^2} = \frac{(2m_X^i + 3m_Y^i)m_X \,.\, m_Y}{(2m_X + 3m_Y)m_X^i m_Y^i}, \tag{3.35}$$

$$\frac{\nu_3^i\nu_4^i}{\nu_3\nu_4} = \frac{m_Y}{m_Y^i}\sqrt{\frac{m_X(3m_X^i + 9m_Y^i)}{m_X^i(3m_X + 9m_Y)}}, \tag{3.36}$$

where ν_1 = the in-phase stretching fundamental, ν_2 = the out-of-plane bending vibration, ν_3 = the out-of-phase stretching frequency and ν_4 = the in-plane bending mode of vibration of this oscillator.

For planar $X^iY^iZ_2^i$ molecules and for linear, symmetric $X_2^iY_2^i$ oscillators, relatively simple relations between the isotopic fundamentals and the respective normal frequencies can be derived.[36]

If the valence forces non-linear, symmetric XY_2 model (Section 2.11) is refined by taking into account also the force, k_{12}, acting on one Y atom as a result of changing the distance between the other Y atom and the X atom, the following formulae for its $\lambda_1 + \lambda_2$ sum and $\lambda_1\lambda_2$ product are obtained:[37]

$$\lambda_1 + \lambda_2 = [1 + (2m_Y/m_X)\cos^2\alpha](k_1 + k_{12})/m_Y + 2\,[1 + (2m_Y/m_X)\sin^2\alpha]k_\delta/m_Y l^2, \tag{3.37}$$

$$\lambda_1\lambda_2 = 2(1 + 2m_Y/m_X)(k_1 + k_{12})k_\delta/m_Y^2 l^2. \tag{3.38}$$

These equations allow the ready and satisfactory estimation of ν_1^i and ν_2^i of such a $X^iY_2^i$ molecule, separately, from its force constants, k_1, k_{12} and k_δ and its geometry.†

† Its ν_3 can be calculated from eqn. 3.8. In the case of the equilateral triangle Y_3^i molecule the isotopic relations are:[38] $\nu_1^i/\nu_1 = \nu_2^i/\nu_2 = \nu_3^i/\nu_3 = (m/m^i)^{1/2}$.

For the pyramidal $X^iY^i_3$ oscillator fundamentals, on basis of either the simple or the refined valence forces model, one gets[39]

$$\frac{\nu_1^i\nu_2^i}{\nu_1\nu_2} = \frac{m_Y}{m_Y^i}\sqrt{\frac{(m_X^i + 3m_Y^i)m_X}{(m_X + 3m_Y)m_X^i}}, \tag{3.39}$$

and

$$\frac{\nu_3^i\nu_4^i}{\nu_3\nu_3} = \frac{m_Y}{m_Y^i}\sqrt{\frac{1 + \cos^2\beta + 3m_Y^i \sin^2\beta/m_X^i}{1 + \cos^2\beta + 3m_Y \sin^2\beta/m_X}}, \tag{3.40}$$

where β is the angle between an XY bond and the symmetry axis.

The effect of the isotopic masses on the Coriolis[40, 41] (Section 2.12) and anharmonicity constants[16d] of various polyatomic molecules has also been studied.

References

1. Dunham, J. L. (1932). *Phys. Rev.* **41**, 721.
2. Herzberg, G., Patat, F. and Spinks, J. W. T. (1934). *Z. Physik* **92**, 87.
3. Herzberg, G. (1945). "Infrared and Raman Spectra of Polyatomic Molecules", p. 228. D. Van Nostrand Company, Princeton, N.J.
4. Ref. 3, pp. 229, 230.
5. Darling, B. T. and Dennison, D. M. (1940). *Phys. Rev.* **57**, 128.
6. Rosenthal, J. E. (1934). *Phys. Rev.* **45**, 426.
7. Ref. 3, p. 231.
8. Adel, A. (1934). *Phys. Rev.* **45**, 56; **46**, 222.
9. Angus, W. R. *et. al.* (1936). *J. Chem. Soc.* 966.
10. Redlich, O. (1935). *Z. Physik Chem. B.* **28**, 371.
11. Decius, J. C. and Wilson, E. B., Jr. (1951). *J. Chem. Phys.* **19**, 1409.
12. Sverdlov, L. M. (1951). *Doklady Akad. Nauk S.S.S.R.* **78**, 1115.
13. Wilson, E. B., Jr., Decius, J. C. and Cross, P. C. (1955). "Molecular Vibrations", pp. 187, 188. McGraw-Hill Book Company, Inc., New York.
14. Bigeleisen, J. (1958). *J. Chem. Phys.* **28**, 694.
15. Sverdlov, L. M. (1952). *Doklady Akad. Nauk S.S.S.R.* **86**, 513.
16. Sverdlov, L. M. (1951). *Doklady Akad. Nauk S.S.S.R.* (a) **78**, 1115; (b) (1952) **86**, 513; (c) (1953) **88**, 249; (d) (1954) **94**, 451.
17. Sverdlov, L. M. (1960). *Optics and Spectroscopy* **8**, 17.
18. Brodersen, S. and Langseth, A. (1958). *Kgl. danske Videnskab. Selskab. Mat.-fys. Skrifter* **1**, No. 5.
19. Brodersen, S. and Langseth, A. (1959). *J. Mol. Spectroscopy* **3**, 114.
20. Brodersen, S. (1959). *J. Mol. Spectroscopy* **3**, 450.
21. Brodersen, S. and Johannesen, T. (1967). *J. Mol. Spectroscopy* **23**, 293.
22. Heicklen, J. (1962). *J. Chem. Phys.* **36**, 721; *J. Phys. Chem.* (1966) **70**, 989.
23. Duncan, J. L. and Mills, I. M. (1964). *Spectrochim. Acta* **20**, 523, 1089.
24. Aldous, J. and Mills, I. M. (1962). *Spectrochim. Acta* **18**, 1073.
25. Chalmers, A. A. and McKean, D. C. (1966). *Spectrochim. Acta* **22**, 251.
26. McKean, D. C. (1966). *Spectrochim. Acta* **22**, 269.
27. Nibler, J. W. and Pimentel, G. C. (1968). *J. Mol. Spectroscopy* **26**, 294.
28. DeWames, R. E. and Wolfram, T. (1964). *J. Chem. Phys.* **40**, 853.

29. Cyvin, S. J. (1966). *J. Phys. Chem.* **70**, 458.
30. Teller, E. (1934). *Hand und Jahrb. Chem. Physik* **9** (II), 141.
31. Edgel, W. (1945). *J. Chem. Phys.* **13**, 539.
32. Miyazawa, T. (1964). *J Mol. Spectroscopy* **13**, 321
33. Krimm, S. (1960). *J. Chem. Phys.* **32**, 1780.
34. Ref. 3, p. 173.
35. Ref. 3, p. 178.
36. Ref. 3, pp. 180, 181.
37. Ref. 3, p. 187.
38. Ref. 3, p. 172.
39. Ref. 3, p. 188.
40. Cyvin, S. J., Brunvoll, J., Cyvin, B. N., Kristiansen, L. A. and Meisingseth, E. (1964). *J. Chem. Phys.* **40**, 96.
41. Cyvin, S. J., Brunvoll, J., Cyvin, B. N. and Meisingseth, E. (1964). *Z. Naturforsch.* **19a**, 780.

CHAPTER 4

Deuterated and Tritiated Compounds

4.1 INTRODUCTION

By far the largest part of the labelled compounds hitherto investigated in respect of their characteristic absorption in the infrared region is no doubt that of their various deuterated molecules. The relatively long period that has elapsed since D_2O first became available for deuteration purposes (over 30 years), its moderate price and the ease of carrying out deuteration of many important chemical compounds, as well as the great significance of the spectra of their deuterated products for the correct assignment of their characteristic absorption bands, have all contributed to the extensive amount of infrared spectra of deuterated substances, measured so far. While their number can be estimated at well over 3000, the number of similar spectra of compounds labelled with the other isotopic modification of hydrogen, tritium, is probably below 20; this is the result of its strong radioactivity and very high price, which usually hinder their preparation in high enough enrichments.

The very large relative change in the mass of the hydrogen atom, which occurs on changing from protium to deuterium or tritium, has a very strong effect upon the frequency of the vibrational modes of an oscillator containing such atoms in as far as their kinetic energy is located in these atoms. In order to get a rough estimate of the shifts to be expected in such a case, one can make use of the fact that (see Section 3.2) the frequency of —R—H stretching modes is generally not very far from that of the corresponding diatomic R—H oscillators.

The corresponding isotopic diatomic frequencies ratio being equal to $(\mu/\mu^i)^{1/2}$ (eqn. 3.6) one gets the following equations for the ratio between the diatomic R—D or R—T stretching frequency and that of the respective R—H oscillator:

$$\frac{\nu^D}{\nu^H} = \left(\frac{M + 2}{2M + 2}\right)^{1/2} \tag{4.1}$$

$$\frac{\nu^T}{\nu^H} = \left(\frac{M + 3}{3M + 3}\right)^{1/2} \tag{4.2}$$

where M = mass of the R atom. Table 4.1 gives some values for these ratios as calculated with the help of these equations.

TABLE 4.1 Calculated Isotopic Frequency Ratios for Diatomic Hydrides

Compound	ν^D/ν^H	ν^T/ν^H
Hydrogen molecules (protium hydride)	0·866	0·82
Beryllium hydride	0·74	0·63
Boron hydride	0·74	0·625
Carbon hydride	0·73	0·62
Nitrogen hydride	0·73	0·615
Hydroxylic radicals	0·73	0·61
Hydrogen fluoride	0·725	0·605
Silicon hydride	0·72	0·60
Phosphorus hydride	0·72	0·595
Sulphydrilyc radicals	0·72	0·595

Since the diatomic R—H stretching frequencies are usually in the region of 1900–4200 cm^{-1} it is evident from Table 4.1 that the D-induced isotopic shifts of these frequencies can be expected to amount to about $3000 \times (1 - 0{\cdot}73) \approx 800$ cm^{-1}, while the corresponding T-induced shifts may be taken as *ca.* $3000 \times (1 - 0{\cdot}62) \approx 1100$ cm^{-1}. The usual —R—D stretching vibrations of polyatomic molecules can thus be expected to absorb not very far from 800 cm^{-1} below the corresponding —R—H-bands. In the case of —R—D bending modes also the frequency of an —RD centred absorption band can generally be arrived at quite satisfactorily by multiplying the frequency of its —RH analogous band by a factor of 0·7.

When the ν^D/ν^H ratio of a pair of corresponding isotopic bands is much higher than 0·7 it is clear that these bands belong to a vibrational mode which affects seriously also the positions of other atoms of the absorbing molecule. The nearer this ratio approaches unity the lower must then be the relative weight of the kinetic energy of the hydrogen atom(s) in the total vibration energy.

Labelled Inorganic Compounds

4.2 HYDROGEN MOLECULES

The isotopically homogeneous diatomic hydrogen molecules possess no permanent electric dipole moment, hence their stretching vibration is usually infrared inactive (Section 2.9). Only when subjected to the influence of either a strong static or alternating electric field[1] or else, pressures of the order of

200 atmospheres[2, 3] do these molecules show the fundamental vibration–rotation band of H_2 (centred at about 4160 cm^{-1}) or D_2 (2990 cm^{-1}), in the infrared region.

HD on the other hand does possess a certain dipole moment[4] (as a result of the anharmonicity of the potential field of the hydrogen molecule and the inequal displacements of the different atoms from their equilibrium locations) and as a result its fundamental frequency gives rise to a very weak vibration–rotation band even under normal conditions.[5, 6] Since the ω_e value for H_2 is 4395 cm^{-1} and the anticipated v^D/v^H ratio in this case is 0·866 (Table 4.1) the expected ω_e value for this band is 3806 cm^{-1}. This value agrees well with the value of 3812·293 cm^{-1} obtained for this constant from the HD absorption spectrum analysis.[5, 6]

This analysis also gave precise values for the rotational constants of HD, such as its B_e (45·663 cm^{-1}), its D_1 etc.

4.3 Isotopic Hydrogen Fluoride

The infrared absorption of deuterium fluoride has been measured both for a liquid sample and for a gas at room temperature or at very high temperatures.[7]

Since hydrogen fluoride exists in the liquid phase as a polymer held together by strong hydrogen bonds, the D—F stretching frequency of liquid deuterium fluoride is appreciably lower than that of the monomeric, gaseous DF. It was observed[7a] as a broad absorption band at about 2430 cm^{-1}, as compared with the 3375 cm^{-1} value observed for the $(HF)_n$ band (3962 cm^{-1} for gaseous HF). The molecular extinction coefficient for these bands is 84 and 83 1 mole^{-1} cm^{-1}, respectively.

The centre of the gaseous D—F vibrational rotational absorption band, on the other hand, was found at[7c] 2906·669 $\pm$ 0·005 cm^{-1} (in vacuum). Its ω_e was found to have the value of 2998·19 cm^{-1}, the first overtone of the stretching frequency appearing[7c] at 5721·815 $\pm$ 0·005 cm^{-1}.

The pure rotation spectrum of the DF vapour in the 40–340 cm^{-1} region[7d] yielded the following rotational constants for this molecule: B_0 = 10·871 $\pm$ 0·004 cm^{-1}, D_0 = (6·15 $\pm$ 0·18) $\times$ 10^{-4} cm^{-1}, as compared with the more accurate values of 10·8600 $\pm$ 0·0005 and (5·88 $\pm$ 0·005) $\times$ 10^{-4} cm^{-1}, respectively, obtained from analysing the vibration–rotation band.[7c]

The high temperature (4000°K) spectrum of gaseous DF[7b] was measured only in the pure rotation region of 230–850 cm^{-1}.

The absorption of TF was measured in the gaseous phase[8] and both the vibrational–rotational fundamental band and its first overtone absorption were observed and analysed. Figure 4.1. shows the fundamental band of TF. The following values were thus obtained for the main vibrational and rotational constants of TF (cm^{-1}): $\nu_{1\leftarrow 0}$ = 2443·86, $\nu_{2\leftarrow 0}$ = 4823·8, ω_e = 2508·5,

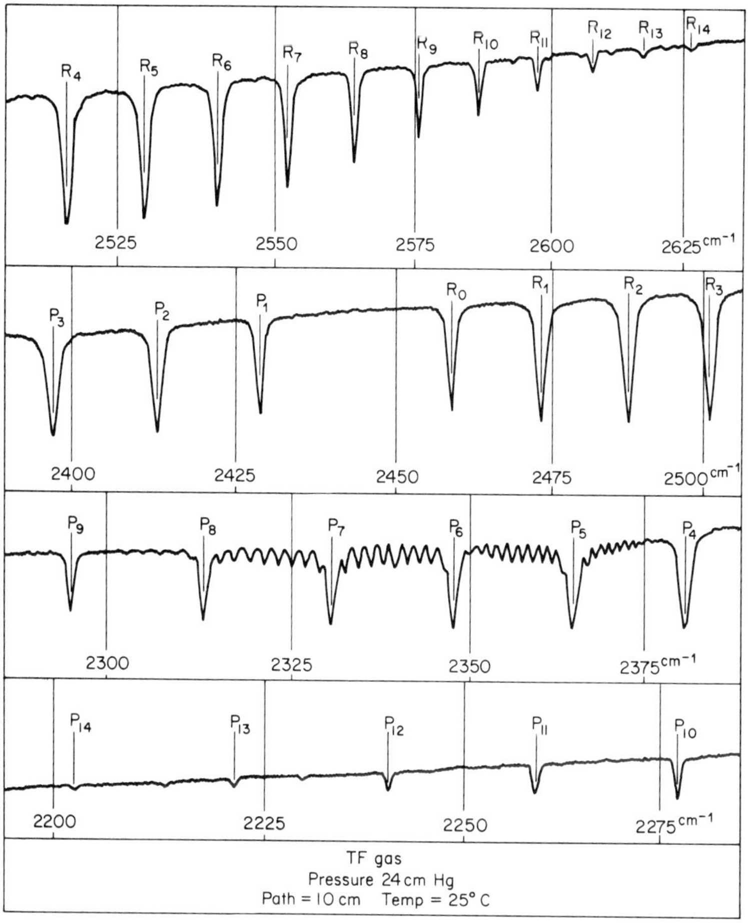

Fig. 4.1 Vibration–rotation fundamental of TF.[8]

$B_e = 7{\cdot}69$, $D_e = 2{\cdot}5{\cdot}10^{-4}$. A comparison of the corresponding values for the $(DF)_n$ and $(HF)_n$ band frequencies on one hand and of those for TF and HF on the other gives the ratios of 0·721 and 0·617, respectively, in good agreement with the values of Table 4.1.

4.4 Labelled Hydrogen Chloride

The infrared spectrum of both DCl[9,10] and TCl[11] was investigated thoroughly on gaseous samples. The highly precise values for their main molecular constants that have been obtained from the analysis of this spectrum are given in Table 4.2.

TABLE 4.2 Values of the Main Vibrational–Rotational Constants of Isotopic Hydrogen Chloride Molecules

Constant (cm^{-1})[a]	$H^{35}Cl$	$D^{35}Cl$	$T^{35}Cl$	$T^{37}Cl$
ν_0	2885	2091	1739·10	1735·51
ω_e	2990·946	2145·150	1775·86	1771·99
B_e	10·5934	5·4488	3·746	3·723
D_e	$53{\cdot}19 \times 10^{-5}$	[b]	$7{\cdot}7 \times 10^{-5}$	$6{\cdot}8 \times 10^{-5}$

[a] In vacuum.
[b] Not measured.

The infrared absorption fundamental band of HCl and DCl solutions in many solvents shows some splitting into three rotation components[12] similar to the PQR branches of a gaseous absorption band (see Section 2.12). The spacing between these components increases with the solubility of the hydrogen chloride in the solvent in question, being always higher than that for the gas phase. This effect of the solvent on the rotational energy of the dissolved DCl (and similar) molecules is also evident in the pure rotation region of the infrared spectrum where, e.g. a carbon tetrachloride solution of DCl shows a broad rotational band centred at *ca.* 100 cm^{-1} while the position of the respective gaseous bands is at appreciably lower frequencies.[13]

Since even gaseous hydrogen chloride, especially at low temperatures, shows some tendency to dimerize, an HCl–DCl mixture at -195°K usually contains a number of various isotopic dimers, each of which possessing its own set of vibration–rotation lines. The ν_0 H—Cl···· fundamentals of these dimers were observed in the null gap of the monomeric H—Cl molecules absorption spectrum.[14]

4.5 OTHER LABELLED HYDROHALIDES

The vibration spectrum of gaseous DBr[15] and TBr[11] was measured accurately and analysed carefully. In the case of DBr it was even possible to resolve the doublet structure of the vibration–rotation lines, due to the bromine isotopes of masses 79 and 81, respectively. The pure rotation spectrum of DBr, measured in the 60–220 cm^{-1} region,[16] is in agreement with the results of the analysis of its fundamental vibration–rotation band.

The infrared absorption of DI seems to have been investigated only in the pure rotation region.[16]

The values obtained thus for the more important vibration–rotation constants of these compounds are collected in Table 4.3.

TABLE 4.3 Vibration–Rotation Constants of HBr, DBr, TBr, HI and DI

Constant (cm^{-1})	HBr	$D^{79}Br$	$D^{81}Br$	TBr	HI	DI
ν_0	2559·25	1839·82	1839·29	1519·26		
ω_e	2649·67	1885·33	1884·75	1550·06		
B_e	8·473	4·291	4·288	2·899		
B_0					6·428	3·2535
D_e		$9·6 \times 10^{-5}$	$9·6 \times 10^{-5}$	$4·5 \times 10^{-5}$		
D_0					$20·5 \times 10^{-5}$	$5·3 \times 10^{-5}$

4.6 Isotopic Water Molecules

All water molecules, both those isotopically symmetrical and the other ones, show three fundamental bands, ν_1, ν_2 and ν_3, in their infrared absorption spectrum. The rotation–vibration spectrum of originally over 99 % D_2O vapour (which, however, contained appreciable amounts of HDO as a result of an exchange with the water film adsorbed on the walls of the cell-system) was measured under high resolution in the 2400–8000 cm^{-1} region[17,18] and the neighbourhood of[19,20] 1200 cm^{-1}. A thorough analysis of the *ca.* 5000 line frequencies observed resulted in accurate values for all the D_2O and HDO vibrational and rotational constants involved. These values are in excellent agreement with the calculated ones[21–25] and are summarized in Table 4.4.

TABLE 4.4 The Main Vibration–Rotation Constants of H_2O, D_2O and HDO

Constant (cm^{-1})[a]	$D_2{}^{16}O$	$H_2{}^{16}O$	$HD^{16}O$
ν_1	2671·46	3656·65	2726·73
ν_2	1178·45	1594·59	1403·42
ν_3	2788·05	3755·79	3707·47
ω_1	2783·80	3832·17	2824 32
ω_2	1206·39	1648·47	1441·43
ω_3	2888·78	3942·53	3889·84
C_0	4·8458	9·2869	6·417
B_0	7·2716	14·5092	9·102
A_0	15·3846	27·8778	23·379

[a] In vacuum.

The pure rotation spectrum of D_2O vapour in the 7–115 cm^{-1} region was also observed and analysed.[26]

The intensity of the liquid phase D_2O absorption bands was measured in the 0·7–40 μ region[27,28] (see also Section 12.2) while the liquid HDO bands were investigated down to[27] 10 μ. The broad bands (half-width of[27, 29] 200–400 cm^{-1}) of all the liquid water isotopic modifications become much sharper in the case of the corresponding ice samples;[28] Table 4.5 reports the frequency of the more important of these bands.

TABLE 4.5 The Main Absorption Bands of Liquid and Solid Isotopic Water (cm^{-1})

Symbol	Assignment[27, 28]	H_2O (liquid)	H_2O (ice)^b	D_2O(1)	D_2O(i)^b	HDO(1)
$\nu_L{}^a$	intermolecular	~500	—	~400	—	~600
ν_R	libration	710	850	530	630	
ν_2	bending	1645	1650	1215	1210	1460
$\nu_2 + \nu_R$	combination		2240^c		1620^c	
ν_1 } $2\nu_2$ }	symmetrical stretching + overtone	3280		2450		2500^d
ν_3	asymm. stretching	3490	3240^c	2540	2450	3410

^{a} Measured only in combinations.
^{b} Unless stated otherwise: at − 170°C.
^{c} At −55°C.
^{d} A pure ν_1 band.

It can be seen from Table 4.5 that a comparison of the liquid phase D_2O frequencies with the corresponding H_2O values still gives a ν^D/ν^H ratio of 0·73–0·74 as compared with a similar value for the water vapour molecules and the 0·73 ratio calculated for the diatomic O—H radical (Table 4.1).

The fundamental frequencies of D_2O and HDO solutions in organic and inorganic solvents were also measured[30, 31], D_2O solutions in carbon tetrachloride being likewise observed[13] (see Section 4.4) to show also a pure rotational band (in the vicinity of 140 cm^{-1}).

The interaction between (monovalent) ions and solvent molecules as well as the nature of the hydrogen bonding existing in aqueous solutions were investigated by studying the infrared absorption spectrum of HDO in water and various salt solutions and the near infrared (5600–12,500 cm^{-1}) absorption of D_2O, respectively.[32, 33]

DCl solutions in D_2O, which produce the heavy hydronium ion, D_3O^+, show its characteristic absorption bands[34, 35] at 960, 1400 and 2170 cm^{-1}.

The absorption of heavy water of crystallization was investigated in the cases of $CuCl_2.2D_2O$, $BaCl.2D_2O$, Kaolinite, Rochelle salt, etc.[36, 37]

The absorption of over 95% T_2O vapour was measured in the 700–7000 cm^{-1} region[38] under a resolution of *ca.* 1 cm^{-1}. The values obtained thus for the T_2O ν_2 and ν_3 fundamentals were 995·5 and 2370 cm^{-1}, respectively, with corresponding ω's of 1017·2 and 2438 cm^{-1}, accordingly. The ν_2 and ν_3 frequencies of HTO were observed at 1324 and 3720 cm^{-1}, respectively (ω_2 = 1358·8; ω_3 = 3895 cm^{-1}), while the ν_3 band of DTO appeared centred at 2735 cm^{-1} (ω_3 = 2829). These results are in satisfactory agreement with the corresponding calculated values.[21]

4.7 Heavy Hydrogen Hypochlorite

The absorption of DOCl vapour was measured in the 670–10,000 cm^{-1} region.[39] The observed fundamental frequencies, given in Table 4.6, were used to calculate the force constants of this molecule. Using these frequencies (instead of the corresponding ω's) the experimental Teller–Redlich ratio (see Section 3.4) obtained is 0·541 as compared with the theoretically predicted value of 0·531. This difference is attributed mainly to the effect of the anharmonicity of the hydroxylic bond.[39a]

Table 4.6 The Fundamental Bands (cm^{-1}) of HOCl and DOCl

		HOCl Frequency[a]		DOCl Frequency	
Band	Assignment	Observed	Calculated[b]	Observed	Calculated
ν_1	O—H(D) stretching	3626[39a] (3610[39b])	3627	2674[39a] (2665[39b])	2675
ν_2	bending	1242 (1239)	1240	911 (910)	916
ν_3	O—Cl stretching	739	742	739	733

[a] In vacuum.
[b] Using calculated force constants.

4.8 Metal Hydroxides

The infrared absorption of the crystalline ionic metal hydroxides is brought about by their OH^- stretching vibration and their lattice modes of vibration.[40]

The spectrum of crystalline LiOD, measured in the 450–4000 cm^{-1} region, shows a strong OD^- stretching band at 2713 cm^{-1} bracketed by nearly symmetrical sidebands of much lower intensity at 2671 and 2741, 2617 and 2780 cm^{-1}, etc. These bands have been shown to be the result of binary combinations of low lattice modes with the O—D$^-$ stretching vibration.[40] In the 450–2000 cm^{-1} vicinity a broad absorption, due to lattice vibration overtones, was observed. The ν^D/ν^H ratio for the main hydroxyl stretching band was found to be 0·738 as compared with the calculated value of 0·73 (Table 4.1).

The frequency of the sharp OD^- stretching band of the crystalline NaOD was observed at 2681·1 cm^{-1} (in vacuum) while the corresponding OH^-

band appears[41] at 3637·4 cm^{-1}. In the case of crystalline $Ca(OD)_2$ the OD^- stretching frequencies were found[42] at 2786 (ω_1) and 2773 (ω_2) cm^{-1}.

It is interesting to note that even for the covalent Sn—OH(D) hydroxyl group the O—H(D) stretching band shows quite similar frequencies, i.e. 3440 and 2570 cm^{-1}, accordingly.[43]

4.9 Deuterated Hydrogen Sulphide

The analysis of the ν_1 and ν_2 rotational–vibrational bands of gaseous D_2S has been carried out only recently[44–46] although the ν_3 absorption[47] and a number of combination bands were analysed much earlier.[47,48] This analysis yielded the centres of the ν_1 and ν_2 bands, the rotational constants A_0, B_0 and C_0 (5·472, 4·502 and 2·442 cm^{-1}, respectively), the D–S distance (1·345 Å) and the D–S–D angle[48] (92°). The spectra[49] of D_2S, HDS, T_2S, HTS and DTS, as well as the force constants (and mean amplitudes, heat content, free energy etc.) of HDS, were also calculated.[50,51]

It seems that of all the HDS bands only that due to the $\nu_2 + 2\nu_3$ combination was analysed[52] giving the following results: $\nu_0 = 6140{\cdot}0$, $A_0 = 9{\cdot}743$, $B_0 = 4{\cdot}922$, $C_0 = 3{\cdot}232$ cm^{-1}.

The frequency of the fundamental bands observed for gaseous D_2S and HDS[53] is reported in Table 4.7.

TABLE 4.7 The Fundamental Infrared Absorption Bands (cm^{-1})[a] of D_2S and HDS

Symbol	Assignment	D_2S	HDS
ν_1	in-phase stretching	1896·4	not measured
ν_2	bending	855·5	1090
ν_3	out-of-phase stretching	1906	2684

[a] In vacuum.

The absorption of D_2S and HDS in solution in organic solvents in the 1900–2600 cm^{-1} region was also measured[54a] as well as that of D_2S_2.[54b]

4.10 Deuterated Dihalogenoboranes

The dihaloboranes, DBX_2, being in equilibrium with diborane, B_2D_6, and the respective boron trihalides, have been investigated only in gaseous mixtures containing an intentional excess of these halides,[55–57] the absorption of which being compensated for. For these planar molecules which belong to the C_{2v} point group there exist six infrared active fundamental vibrations; these are shown in Fig. 4.2.

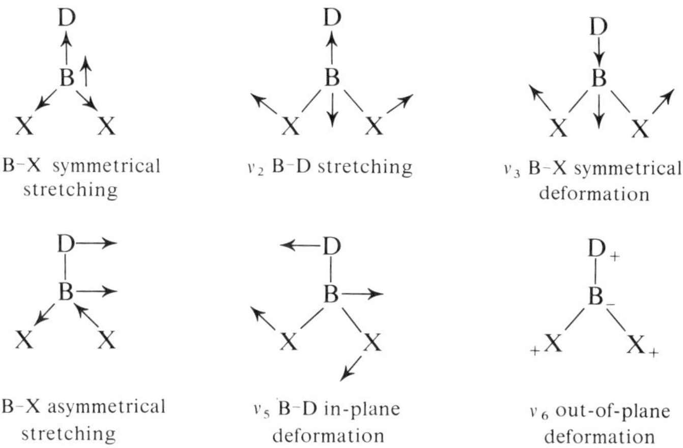

Fig. 4.2 The fundamental frequencies of dihaloboranes.

The frequencies of these vibrations in the case of the various deuterated dihaloboranes (which occur as isotopic mixtures of ^{11}B and ^{10}B boranes) are given in Table 4.8.

TABLE 4.8 The Infrared Absorption Bands (cm^{-1}) of Deuterated Dihaloboranes

Symbol	$H^{11}BF_2$[a]	$D^{11}BF_2$[a]	$D^{10}BF_2$[a]	$D^{11}BCl_2$[e]	$D^{10}BCl_2$[e]	$D^{11}BBr_2$[f]	$D^{10}BBr_2$[f]
ν_1	1162	1125	1147	[d]	[d]	585	[d]
ν_2	2619	1959	1959	1969	1969	1960	
ν_3	545[b]	540[b]	[d]	[d]	[d]	180–190[c]	180–190[c]
ν_4	1402	1394	[d]	[d]	[d]	870	901
ν_5	1348[c]	992[c]	991[c]	[d]	[d]	656	[d]
ν_6	924	787	806·5	645	661	625?	[d]

[a] Values from ref. 58 unless indicated otherwise.
[b] Values from ref. 59.
[c] Calculated.
[d] Not measured.
[e] Values of ref. 56.
[f] Values of ref. 60.

These frequencies are in good agreement with the respective values calculated from the corresponding BX_3 values by taking these boranes to be isotopic BX_3 compounds and using Green's function and matrix partitioning[55, 56] (see Section 3.7). In fact identification of the various borane absorption bands has been carried out with the help of such calculations.[55, 60]

Analysis of the high-resolution spectra of the deutero-dihaloboranes gave exact values for their inertial constants[57, 58, 61] as well as the B–D distance (1·20 Å) in deuterated difluoroborane.[58]

4.11 Labelled Ammonias

The isotopically homogeneous ammonia molecules, which belong to the C_{3v} point group, show four infrared active fundamental bands while their other modifications, belonging to the C_s point group, display six such bands. Since, however, ammonia molecules exchange their hydrogen atoms very quickly with each other the non-homogeneous ammonia species cannot be isolated and investigated in pure form.[62]

The analysis of a number of fundamental, combination and overtone bands of gaseous ND_3 (in the 1750–7100 cm^{-1} region) and NT_3 (530–2250 cm^{-1}), recorded under high resolution, has been carried out,[63,65] their vibrational, rotational and other molecular constants being determined.

A normal co-ordinate analysis yielded (all the most general, harmonic, ammonia force constants as well as) the rotational distortion constants for ND_3 and NT_3 in good agreement with these data.[66,67] The pure rotation spectrum of NT_3 is also in close agreement with the results of the vibration–rotation bands analysis.[68]

The characteristic absorption frequencies of ND_3, NHD_2 and NH_2D in the liquid phase and in solution in carbon tetrachloride were found to be not very different from the corresponding values of the gaseous phase.[62] The rotation spectra of gaseous NHD_2 and NH_2D in the 30–200 cm^{-1} region have also been measured and analysed.[69]

The known fundamental frequencies of the various isotopic ammonias in the gas phase are collected in Table 4.9.

The infrared absorption of ND_3 complexes formed with silicon tetrafluoride[70] and a number of Co^{++}, Pd^{++}, Ag^{+} and Pt^{++} salts[71–73] has similarly been investigated. This has shown that while the N—D stretching and asymmetrical N—D bending bands of such complexes appear near to their locations in the ND_3 spectrum the symmetrical N—D bending fundamental is generally much higher in frequency than it is for the deuterated ammonia itself. $Co(ND_3)_6Cl_2$, e.g. shows[72] its N—D stretching fundamental at 2430 cm^{-1} and its asymmetrical N—D bending absorption at 1021 cm^{-1} (strong) while its symmetrical N—D deformation band (weak) is at 1161 cm^{-1}. This complex also shows extra bands at about 663 (—ND_3 rocking) and 310 (Co—ND_3 stretching) cm^{-1}.

4.12 Deuterated Sodium Amides

The infrared absorption of Nujol (pure colourless paraffin oil) mulls of solid ($NaNH_2$)$NaND_2$ and NaNHD has been investigated in order to use the values obtained thus for the deuterated amide ions fundamentals to evaluate

TABLE 4.9 The Frequency (cm^{-1}) of the Fundamental Isotopic Ammonia Bands

Type of band	Symbol	NH_3	ND_3	NT_3	NHD_2	NH_2D
in-phase N—H(D)(T) stretching	ν_1	3336·2[a] 3337·2	2420	2014·1		
	ν_1'				2479	3366
symmetrical bending	ν_2	932·5[a] 968·3	745·7[a] 749·4	656·37[a] 657·19		
	ν_2'				808 817	874 894
out-of-phase stretching	ν_3	3443·6[a] 3443·9	2564	2184·76		
	ν_3'				3402	2484
	ν_3''				2557	3409
asymmetrical bending	ν_4	1626·1[a] 1627·4	1191	996·28		
	ν_4'				1177	1608
	ν_4''				1464	1380

[a] Splitting due to inversion of the pyramidal molecule.

the potential function of the NH_2^- group.[36a] The observed bands are reported in Table 4.10.

TABLE 4.10 The Isotopic Amide Ion Bands (cm^{-1})

Isotopic species	ν_1 (symmetrical stretching)	ν_2 (bending)	ν_3 (asymmetrical stretching)	$\nu_2 + \nu_3$
NH_2^-	3212·5	1539·5	3263	—
ND_2^-	2358	1130·5	2428	3551
NHD^-	2389	1350	3234·5	—

4.13 DEUTERATED HYDRAZOIC ACID

The spectrum of gaseous (and solid) N_3D, belonging to the C_s point group, was measured in the 400–3100 cm^{-1} region, where all its six fundamental bands appear.[74–77] This spectrum (among others) was used for the normal co-ordinate analysis of the HN_3 vibrations.

The frequency of its bands is reported in Table 4.11.

TABLE 4.11 The Frequency (cm^{-1}) of the Gaseous N_3H and N_3D Fundamentals

Band	N_3H	N_3D	Band	N_3H	N_3D
ν_1	3497·3	2478 (medium)	ν_4	1151	953·8 (strong)
ν_2	2139·8	2112 (in Fermi resonance)	ν_5	527	492·2 (weak)
ν_3	1265·2	1184·7 (medium)	ν_6	607	588·4

4.14 Labelled Nitrous Acids

The non-linear nitrous acid molecule, HO—N=O, exists only in an equilibrium mixture with its decomposition products.[78] Its absorption is complicated further by its existing concurrently both in the *cis* and in the *trans* forms. The final assignment of the two sets of its infrared bands has therefore been accomplished only after its D-, ^{15}N- and ^{18}O-labelled modifications have first been investigated. This enabled the calculation of the potential-energy minimum of the *cis* isomer (above that of the *trans* isomer) from the isotopic torsion fundamentals and a temperature study of the relative intensity of strong bands characteristic of the two forms.[78] Table 4.12 gives the band frequency of the isotopic gaseous nitrous acid fundamentals. The values calculated for these fundamentals on basis of a normal coordinate analysis agree reasonably well with the frequencies observed.

TABLE 4.12 Fundamentals (cm^{-1}) of Isotopic Nitrous Acids

Band	Type	HNO_2	DNO_2	$H^{15}NO_2$	$D^{15}NO_2$	$H^{16}ON^{18}O$	$H^{18}ON^{16}O$	$HN^{18}O_2$
					trans acids			
ν_1	O—H stretching	3588 (*m*)	2648	3588	2646	3588	3571	3571
ν_2	N=O stretching	1699 (*s*)	1692	1669	1661	1666	1702	1666
ν_3	N—O—H bending	1265 (*s*)	1014	1261	1008	1258	1251	1251
ν_4	N—OH stretching	791 (*s*)	737	774	723	785	779	779
ν_5	O—N=O bending	593 (*s*)	586	589	583	570	570	564
ν_6	torsion	540 (*m*)	416	538	413	540	537	537
					cis acids			
ν_1	O—H stretching	3424	2525	—	—	—	—	—
ν_2	N=O stretching	1640 (*m*)	1625	1611	1599	1606	1643	1606
ν_3	N—O—H bending	1261 (*w*)	1008	1258	1002	—	—	—
ν_4	N—OH stretching	853 (*s*)	814	836	802	846	840	840
ν_5	O—N=O bending	608 (*w*)	601	604	596	582	593	576
ν_6	torsion	638 (*m*)	504	—	582	638	631	631

m = medium *s* = strong *w* = weak — no band observed.

4.15 Deuterated Hydrogen Peroxide

The absorption of a mixture of (about 50%) D_2O_2, H_2O_2 and HDO_2 was measured in the 400–6600 cm^{-1} region in all its aggregation states and partly also in solution in carbon tetrachloride.[79] The D_2O_2 ν_1 and ν_2 fundamentals, however, have not yet been observed although their values have been estimated indirectly.[79] On the other hand analysis of its asymmetrical O—D stretching (ν_5) vibration–rotation band, centred at 2661 cm^{-1}, yielded values for the rotational constants of this molecule. The ν_3 and ν_4 bands of D_2O_2 have been observed only for its condensed phases (878 (*m*) and 538 cm^{-1}, (*m*), respectively, for the liquid) while its ν_6 absorption has been found both in this case (at about 1000 cm^{-1}) and for the D_2O_2 vapour (947 cm^{-1}).

The HDO_2 spectrum[80] has been investigated on a mixture of isotopic hydrogen peroxides containing *ca.* 40% of HDO_2. Its OH and OD stretching frequencies coincide exactly with the respective frequencies (ν_5) of H_2O_2 and D_2O_2.

4.16 Heavy Phosphine and Similar Hydrides

The phosphine, arsine and stibine molecules, being all (similarly to ammonia) pyramidal XY_3 oscillators, belong to the $\boldsymbol{C_{3v}}$ point group and show four infrared active fundamentals.

The absorption of gaseous PD_3 has been measured under high dispersion in the 600–5000 cm^{-1} region,[81] the analysis of its ν_1 vibration–rotation band yielding the value for its B_0 rotational constant (2·250 cm^{-1}) from which its geometry has been determined. Its pure rotation spectrum (between 55 and 101 cm^{-1}) was also reported.[82]

The absorption of AsD_3 was investigated in the 650–2500 cm^{-1} region[83] as well as in the (pure rotation) 50–90 cm^{-1} zone.[82]

In the case of SbD_3, both the ν_1, and the ν_3 bands were analysed thoroughly yielding the size and shape of this molecule.[84]

The frequency of the PD_3, AsD_3 and SbD_3 fundamentals is given in Table 4.13.

Normal co-ordinate analyses using the observed vibration–rotation data yielded values for the various force constants, harmonic frequencies etc. of PD_3, PT_3, AsD_3, AsT_3,[66] SbD_3 and SbT_3[86] and the partially deuterated and tritiated stibines.[87]

The rotational constants of AsD_3 have also been reported.[82]

The 400–4000 cm^{-1} region spectrum of YD_3, shows bands[88] at about 450, 650 and 920 cm^{-1} as compared with the respective bands of YH_3 at 640, 920 and 1295 cm^{-1}. The observed $1:\sqrt{2}$ ratio of corresponding YD_3/YH_3 frequencies is the one to be expected of a hydride of an element as heavy as yttrium (89).

TABLE 4.13 The Fundamentals (cm^{-1}) of the Isotopic Phosphines, Arsines and Stibines

Band	Type	PH_3[a]	PD_3[b]	AsH_3	AsD_3	SbH_3	SbD_3
ν_1	Symmetric stretching	2327	1682·7 (*w*)			1890·9	1358·8
ν_2	Symmetric bending	991	727·5 (*w*)[c]	906	652 (*m*)	781·5	561·1
ν_3	Asymmetric stretching	2421	1688·6 (*s*)[c]	2120	1531 (*vs*)[d]	1894·2	1362·0
ν_4	Asymmetric bending	1121	804·7 (*m*)	1003	716 (*s*)	830·9	592·5

[a] Data from ref. 53, p. 164

[b] The relative intensity is taken from ref. 85.

[c] The original assignment of ref. 81 seems to be erroneous as can be seen from a comparison with the respective PH_3 value.

[d] *vs* = very strong.

The infrared absorption of AlD_3[89] and GaD_3[90] has been investigated only in the case of their adducts with trimethylamine or tetrahydrofuran.

4.17 DO—Cr=O, DO—Co=O and DO—Ga=O

The absorption of the deuterated oxyhydroxide of trivalent chromium was measured in the solid phase in the 30–4000 cm^{-1} region in the course of a study of the very strong hydrogen bonding existing in it.[91,92] It was found that while HOCrO shows only *one* O—H stretching band, at about 1700 cm^{-1} (strong; very broad), DOCrO shows *two* such bands, at *ca.* 1920 (strong) and 1595 (strong) cm^{-1}. In the 300–800 cm^{-1} region also the heavy compound shows (at 80°K) more bands (6) than $HCrO_2$ (2). These results were taken to indicate that while the O····H····O band is *symmetrical* that of the O····D—O group is *asymmetrical*; hence, no centre of symmetry exists in the unit cell of the crystalline $DCrO_2$, which contains only one molecule.[91,93]. The O—D bending absorption appears at about 840 cm^{-1} (in $HCrO_2$: *ca.* 1200 cm^{-1}). The case of $HCoO_2$ and $DCoO_2$ seems to be analogous to that of $HCrO_2$ and $DCrO_2$.[91–93] Figures 4.3 and 4.4 show the infrared absorption spectra of $HCoO_2$ and $DCoO_2$, respectively.

The absorption of a potassium bromide pellet of α-DOGaO in the 290–3650 cm^{-1} region has been investigated in order to make possible the assignment of the α-HOGaO bands.[94] Its O—D stretching fundamental appears at 2185 cm^{-1} (O—H: 2915; 2840) while its O—D bending absorption is located at 691 and 780 cm^{-1} (OH: 954; 1025).

4.18 Deuterated Hydroxylamine

The spectrum of a crystalline film of fully deuterated hydroxylamine was investigated at −78°C in the 500–4000 cm^{-1} region in connection with the

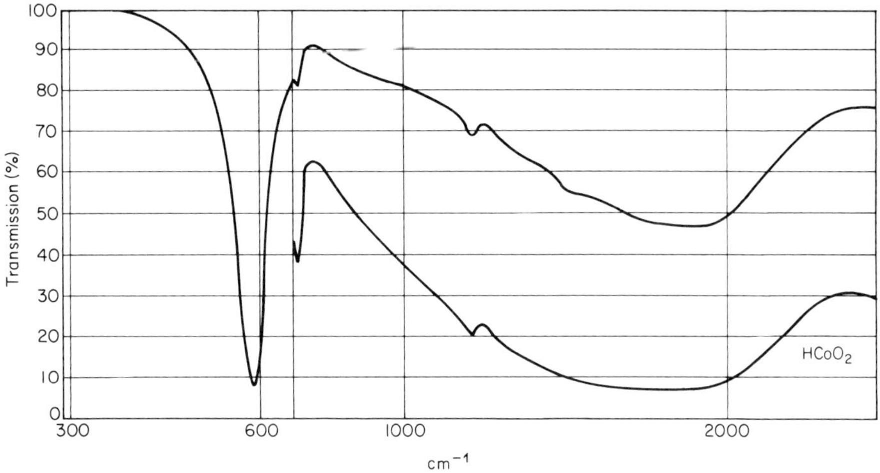

Fig. 4.3 The infrared absorption of $HCoO_2$.[93]

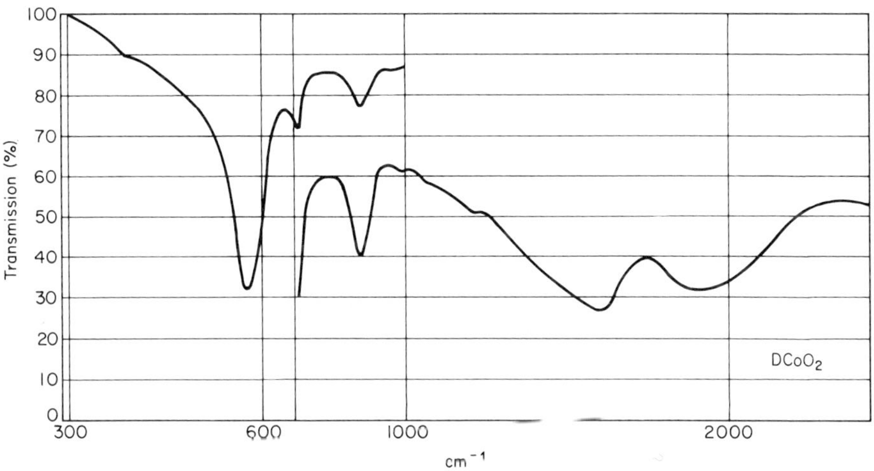

Fig. 4.4 The infrared absorption of $DCoO_2$.[93]

problem of its structure.[95] It was thus established that it exists in the ND_2OD form rather than in that of D_3NO. Table 4.14 reports the frequency of the crystalline NH_2OH and ND_2OD fundamentals.

4.19 Labelled Nitric Acid

$DONO_2$ belongs to the C_s point group and shows 9 fundamental infrared active bands. Its vapour absorption in the 300–2650 cm^{-1} region and its

TABLE 4.14 The Frequency (cm^{-1}) of the NH_2OH and ND_2OD Fundamentals at −78°C

Band	Type	NH_2OH	ND_2OD	Band	Type	NH_2OH	ND_2OD
ν_1	N—H(D) symmetrical stretching	3180	2330 (*m*)	ν_6	$NH_2(D_2)$ rocking	848	?
ν_2	O—H(D) stretching	2880	2195 (*s*)	ν_7	N—H(D) asymmetrical stretching	3305	2490 (*m*)
ν_3	H—N—H(DND) bending	1509	1130 (*w*)	ν_8	$NH_2(D_2)$ wagging	945	670 (*w*)
ν_4	O—H(D) bending	1186	942 (*m*)	ν_9	torsion	535	?
ν_5	N—OH(D) stretching	904	874 (*w*)				

hydrogen-bonded solid spectrum between 700 and 2300 cm^{-1} have both been reported.[96] The values observed for the DNO_3 fundamentals, together with other spectroscopic data, were used to carry out a normal co-ordinate vibration analysis for nitric acid molecules. These frequencies are collected in Table 4.15.

TABLE 4.15 The HNO_3 and DNO_3 Vapour Fundamentals (cm^{-1})

Band	Assignment	HNO_3	DNO_3	Band	Assignment	HNO_3	DNO_3
ν_1	OH(D) stretching	3550·0	2621·5 (*m*)	ν_6	N—OH(D) stretching	646·6 (*w*)	641·0
ν_2	NO antisymmetrical stretching	1708·2	1687·0 (*vs*)	ν_7	O—N—OH(D) bending	579·0 (*w*)	541·0
ν_3	H(D)ON bending	1330·7	1013·6 (*s*)	ν_8	NO_2 out-of-plane bending	762·2 (*s*)	762·8
ν_4	NO symmetrical stretching	1324·9	1308·4 (*vs*)				
ν_5	NO_2 deformation	878·6	888·0 (*s*)	ν_9	H(D)O torsion	455·8 (*m*)	342·0

4.20 DEUTERATED SILANES

The tetrahedral silane molecule belonging to the $\boldsymbol{T_d}$ point group can be expected to show, practically, only the two infrared active fundamentals (ν_3 and ν_4) due to the asymmetric Si—D stretching and Si—D bending vibrations, respectively.

The absorption of gaseous SiD_4 was measured in the 600–5000 cm^{-1} range, the fine structure of the Si—D stretching band being resolved.[97] The analysis of this structure yielded the value of the rotation constant of this molecule (1·44 cm^{-1}) as well as the respective Coriolis coupling constant (see Section 2.12).

$SiHD_3$ belongs to the $\boldsymbol{C_{3v}}$ point group hence all its six fundamentals are infrared active and these were in fact observed.[97] Their analysis yielded its two rotation constants (1·44 and 1·78 cm^{-1}).

The analysis of the highly resolved first overtone band of its symmetrical Si—D stretching vibration (ν_1), appearing at 4307·09 cm^{-1} (in vacuum), resulted in an appreciably greater value for the Si—H distance[98] than previously accepted.

All the eight infrared absorption bands to be expected of SiH_2D_2 (of the C_{2v} point group) were observed[97] as was also the Si—D stretching fundamental of SiH_3D, which appears[99] at 1594·4 cm^{-1}. The frequency of various isotopic silane fundamentals is given in Table 4.16.

TABLE 4.16 The Observed SiH_4, SiD_4, $SiHD_3$ and SiH_2D_2 Fundamental Bands (cm^{-1})

Band	Type	SiH_4	SiD_4	$SiHD_3$	SiH_2D_2
ν_1	Symmetrical stretch			1573	1587
ν_2	Symmetrical bend			683	ν_{2a} 944
ν_3	Asymmetrical stretch	2190·6	1597	ν_{3a} 2182	ν_{3a} 2189
				ν_{3b} 1598	ν_{3b} 2183
					ν_{3c} 1601
ν_4	Asymmetrical bend	914·2	681	ν_{4a} 683	ν_{4a} 862
				ν_{4b} 851	ν_{4b} 743
					ν_{4c} 682·5

The product of the ν_3 and ν_4 frequencies of SiH_4 divided over that of SiD_4 can be seen to be equal to 1·84 as compared with the value of 1·88 expected on basis of the product rule (Section 3.4). The sum of the squared frequencies of all the branches of the SiH_2D_2 ν_3 and ν_4 fundamentals can further be calculated to amount to $8{\cdot}65 \times 10^{-6}$ cm^{-2}, in excellent agreement with the value of $8{\cdot}63 \times 10^{-6}$ cm^{-2} obtained from the corresponding SiH_4 and SiD_4 frequencies on basis of the sum rule[97] (Section 3.5).

The force fields,[97, 100–102] normal vibrations[97, 103] and their mean amplitudes[104] as well as the Coriolis coupling and rotational distortion constants[103] of various deuterated silanes were all calculated. The fundamental frequencies of various tritiated silanes have also been calculated.[105]

The absorption of gaseous SiD_3F, SiD_3Cl,[106]† SiD_3I,[108] SiD_3CN[109, 110] and $SiDF_3$[106] has been measured from 300 to about 3500 cm^{-1} and almost all of their fundamentals have been observed and identified. The observed frequencies agree fairly well with the values calculated for them on the basis of the product rule[106, 109, 110] (Section 3.4) or the assumed force fields.[102, 108] In a number of cases, the rotational structure of such bands was analysed to yield rotational constants.[108, 109]

† For the spectra of adducts of SiD_2Cl_2(SiD_3I and SiD_2I_2) with pyridine, see ref. 107.

Various deuteration-sensitive fundamental frequencies of deuterated substituted silanes are reported in Table 4.17.

TABLE 4.17 Deuteration-Sensitive Fundamental Frequencies (cm^{-1}) of Deuterated Substituted Silanes

Band	SiH_3F	SiD_3F	SiD_3Cl	SiD_3I	SiD_3CN	$SiDF_3$
Symmetrical Si—D(H) stretch	2206	1577	1581	1575 (*s*)	1588 (*s*)	1688 (*m*)
Symmetrical (H)D—Si—D(H) bend	990	888	702	664(*vs*)	701(*vs*)	—
Antisymmetrical Si—D(H) stretch	2196	1615	1616	1607(*s*)	1624(*s*)	—
$-SiD_3(H_3)$ rocking	728	549	488	435(*w*)	542(*s*)	—

4.21 LABELLED GERMANE AND SIMILAR COMPOUNDS

Germane is fully similar to SiD_4 in its structure and should really show only its ν_3 and ν_4 infrared bands. However, as a result of the Coriolis interaction of its ν_4 vibration with its ν_2 mode, a weak band due to the latter was also observed when the absorption of gaseous GeD_4 was investigated in the 550–3200 cm^{-1} region.[111]

The analysis of the resolved rotational structure of its ν_3 band yielded the values of its rotation constants and a number of Coriolis coefficients (for the ν_3 and ν_4 bands).

The frequency of the GeD_4, $GeHD_3$, GeH_2D_2 and GeH_3D fundamentals[111,112] is given in Table 4.18.

TABLE 4.18 The Infrared Fundamental Bands (cm^{-1}) of Deuterated Germane

Type	GeH_4	GeD_4	$GeHD_3$	GeH_2D_2	GeH_3D
Symmetrical stretching	—	—	1504	1512	1520·4
Symmetrical bending	930·9	665	595	881	901
Antisymmetrical	2113·6	1522·2	2112·4	2112	2106
stretching			1522	1522	2112
Antisymmetrical	819·3	596	625	770	820
bending			792·3	657	706
				620	

The results reported in Table 4.18 are in good agreement with the fundamental frequencies calculated from the various force fields assumed to approximate satisfactorily the situation existing in the various germane molecules.[102, 103, 111]

The rotational constants of $GeHD_3$ and GeH_3D have also been reported.[113]

In the case of tritiated germanes, only calculated fundamental frequencies have so far been reported in the literature.[105]

$GeDCl_3$ shows[114] its Ge—D stretching band at 1552 cm^{-1} and its Ge—D bending absorption at 508 cm^{-1} while GeD_2Br_2 exhibits a Ge—D stretching absorption at 1521·9 (symmetrical) and 1543 (asymmetrical) cm^{-1}, a Ge—D deformation band at 607·1 cm^{-1} and GeD_2 rocking and GeD_2 wagging fundamentals at 362·6 and 543·1 cm^{-1}, respectively.[115]

The spectra of other substituted germanes have also been studied.[116, 117]

Stannane, being also tetrahedral, shows a very similar infrared spectrum to that of germane.[118] The ν_2, ν_3 and ν_4 fundamentals of SnD_4 appear at 539, 1367·5 and 487 cm^{-1}, respectively, in good agreement with the values expected on the basis of the product rule. The Coriolis coefficients obtained from the analysis of the rotational structure of the ν_3 band were used to determine uniquely a number of stannane force constants.[118]

4.22 Deuterated Borine Carbonyl

The $D_3BC{=}O$ molecule belonging to the C_{3v} point group shows eight infrared active fundamentals. These bands were assigned to the respective modes of vibration, as a result of an investigation of the absorption spectrum of a number of gaseous isotopic borine carbonyls in the 200–3300 cm^{-1} region.[119]

Table 4.19 gives the frequency of their main deuteration-sensitive fundamentals (in vacuum).

TABLE 4.19 Main Deuteration-Sensitive Borine Carbonyl Fundamentals (cm^{-1})

Band	Description	$D_3{}^{10}BCO$	$D_3{}^{11}BCO$	$H_3{}^{11}BCO$	$H_3{}^{10}BCO$
ν_1	B—D(H) stretching	1694·8	1679·0	2379	2385
ν_3	$BD_3(H_3)$ bending	~883	~862	1073·4	1083·2
ν_5	B—D(H) asymmetrical stretching	1852	1837	2441	2456
ν_6	BD_3(H) asymmetrical bending	801·5	801·3	1100	~1100

The force field of borine carbonyl has also been described.[120]

4.23 Labelled Ammonium Halides

The absorption of the crystalline, over 95% deuterated, ammonium halides was investigated in the 1000–3300 cm^{-1} region both at room temperature and at −190°C. These salts show an order-disorder transition point between −30° and −60°C and while their crystalline structure at low temperatures belongs to the T_dl symmetry, at room temperature the tetrahedral ND_4^+ ion is randomly distributed between the two possible orientations of the unit cell.[121] However, even the room temperature spectra of these halides are consistent with the existence of an "essentially" T_d symmetry in these salts.[122] Thus, they show only the ν_3 and ν_4 bands of the highly symmetrical isolated ND_4^+ ion (besides various combination and overtone bands) but not its ν_1 and ν_2 fundamentals. The ν_3 and ν_4 frequencies of ND_4F,[122] ND_4Cl,[123] ND_4Br[121] and various tritiated ammonium chlorides[124] are collected in Table 4.20. They are in good agreement with the calculated values.[121, 123, 124a]

Table 4.20 The Main Bands (cm^{-1}) of Labelled Ammonium Halides[a]

Band	NH_4F	ND_4F	ND_4Cl	NT_4Cl	ND_4Br	NHT_3Cl	NDT_3Cl
ν_3[b]	2815 (*vs*)	2148	2336	2022	2331	3084	2319
ν_4	1494 (*m*)	1120	1067	913	1065	1345	

[a] All the values for the non-tritiated halides are for *ca.* −190°C, the others being for room temperature.

[b] In cases of a split band, the frequency given refers to the main branch.

Hydroxylamine reacts with various mineral acids to form hydroxy-ammonium salts. The infrared absorption spectrum of solid $[ND_3OD]^+X^-$ (X = Cl, Br, NO_3, ClO_4, $\frac{1}{2}SO_4$) usually contains (among others) bands at about[125, 126] 840 (O—D bending), 890 (ND_3^+ rocking), 1110 (symmetrical ND_3^+ bending), 1130 (asymmetrical ND_3^+ bending), 2140 (symmetrical N—D stretching), 2250 (asymmetrical N—D stretching) and 2350 cm^{-1} (hydrogen-bonded O—D stretching).

The normal vibrations of labelled hydroxylamine have also been reported.[126]

4.24 Deuterated Hydrazine and Allied Compounds

The hydrazine molecules are assumed[127] to belong to the C_2 point group, hence are expected to show twelve infrared fundamentals. The vapour spectrum of N_2D_4 (absorption also measured in the other phases) shows, however, only eight fundamental bands in the 250–4000 cm^{-1} region, this being the result of an accidental degeneracy of two bands and low activity of others.

The absorption of the crystalline and liquid monohydrochloride and monohydrobromide of N_2D_4 in the 400–3400 cm^{-1} region has also been observed.[128] The main hydrazine bands are given in Table 4.21.

TABLE 4.21 The Main Infrared Absorption Bands (cm^{-1}) of Isotopic Hydrazines and their Hydrohalides

Description	N_2H_4 (In Matrix)	N_2D_4 (Vapour)	N_2H_5Cl (Solid)	N_2D_5Cl	N_2D_5Br
N—N stretching	1087	933	973	940 (*s*)	932
$NH_2(D_2)$ asymmetrical wagging	1265	985	1246	1000 (*m*)	998
$NH_2(D_2)$ symmetrical wagging	1312	1009	(1417	1100 (*m*)	1095)[a]
$NH_2(D_2)$ symmetrical bending	—	1140	(1500	1133 (*s*)	1135)[a]
$NH_2(D_2)$ asymmetrical bending	—	1197	1638	1190 (*m*)	1185
N—H(D) symmetrical stretching	3356 }	2431	{ 3150	2390 (*s*)	2370
N—H(D) asymmetrical stretching	3390 }		{ 3261	2440 (*s*)	2430

[a] Corresponding bands, although of lower symmetry.

4.25 Heavy Perchloric Acid

The absorption of vaporized and crystalline $DClO_4$ has been measured in the 300·5000 cm^{-1} region in connection with the study of the structure of absolute perchloric acid.[129] This absorption, which can be interpreted as belonging to a C_s point group oscillator, showed in the case of the solid phase 10 out of the 12 fundamentals to be expected for $DClO_4$ (one of which is below the 300 cm^{-1} limit). In the case of the vapour phase the observed fundamentals appear centred at about: 574 (O—Cl—O deformation), 585 (ditto), 1033 (^{35}Cl—O stretching), 1198 (O—D) deformation), 1268 (another Cl—O stretching), 1325 (ditto) and 2626 cm^{-1} (O—D stretching).

4.26 Deuterated Phosphonium Iodide

Crystalline PD_4I, the structure of which is similar to that of ammonium bromide at low temperatures, was investigated[130,131] both at room temperature and at $-190°C$. Its spectra in the 400–4000 cm^{-1} region showed fundamental bands, at about 683, 701, 1658 and 1740 cm^{-1} which were attributed to its ν'_4, ν_4, ν_3 and ν'_3, frequencies, respectively. These modes of vibration are due to a splitting of the original ν_3 and ν_4 frequencies of the isolated tetrahedral PD_4^+ ion as a result of the lower symmetry of the PD_4I crystal. This assignment is verified by a separate application of the product rule to the ν_3 and ν_4 bands of PD_4I and PH_4I (2280 and 954 cm^{-1}) and to the ν'_3 and ν'_4 absorptions of these compounds. One thus gets an agreement of 0·528 (calculated for both cases) versus 0·534 and 0·537 (observed).

The infrared spectrum of deuterated biphosphine, the phosphorous analogue of hydrazine, was investigated both for the gas phase and in the crystalline state.[85] While gaseous PD_2–PD_2 shows only one strong P—D stretching band at 1685 cm^{-1} crystalline P_2D_4 shows two such bands, at 1660 and 1674 cm^{-1}, besides other absorptions at 448, 466 (ν_5), 582 (ν_{12}) and 765 (ν_{10}) cm^{-1}.†

4.27 LABELLED SULPHURIC ACID

The absorption of D_2SO_4 and $HDSO_4$ vapours, at *ca.* 200°C, has been investigated in the 400–3800 cm^{-1} region[132] as was also that of the hydrogen-bonded condensed phases of D_2SO_4 and $HDSO_4$ in the 220–3600 cm^{-1} region.[132–134] The deuteration-sensitive bands of D_2SO_4 are reported in Table 4.22. $HDSO_4$ did not show any new bands which are not shown by H_2SO_4 or D_2SO_4.

TABLE 4.22 The Deuteration-Sensitive Absorption Bands (cm^{-1}) of Isotopic Sulphuric Acids

Band	Description	D_2SO_4 (Vapour)	H_2SO_4 (Vapour)	H_2SO_4 (Crystal)	D_2SO_4 (Crystal)	D_2SO_4 (Liquid)
ν_{12}	$S(OH(D))_2$ rocking			412	371	395
ν_{14}	OH(D) asymmetrical wagging			650	478	
ν_3	SOH(D) symmetrical bending	820	1159	1170		
ν_{10}	SOH(D) asymmetrical bending			1240	930	
ν_1	OH(D) symmetrical stretching			2450	1860	1860
ν_9	OH(D) asymmetrical stretching	2663	3610	2970	2280	2280

The observed value of 2663/3610 = 0·738 for the ν_{OD}/ν_{OH} ratio of sulphuric acid vapour is in agreement with the 0·73 value given in Table 4.1 for the OH radical as well as with the corresponding 2661/3602 ratio of fluorosulphonic acid, the 2667/3610 result for toluenesulphonic acid and the 2626/3560 value obtained for perchloric acid.[132]

† For the spectra of deuterated germyl phosphines and germyl arsines see: K. M. Mackay *et al.* (1969), *Spectrochim. Acta* **25A**, 925 and J. E. Drake *et al. Spectrochim. Acta* **25A**, 941.

4.28 Heavy Diboranes and Higher Borane Homologues

Diborane, the structure of which is described by Fig. 4.5, probably belongs to the $\boldsymbol{D_{2h}}$ point group.[135] The infrared absorption of its $^{10}B_2D_6$ isotopic

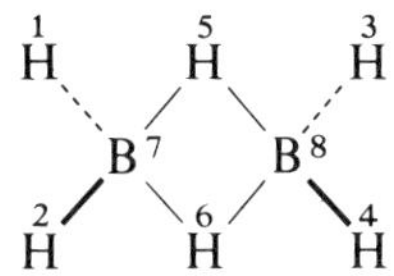

Fig. 4.5 The structure of diborane. (Atoms Nos. 1, 2, 3, 4 are in a plane perpendicular to that of atoms Nos. 5, 6, 7, 8.)

modification was investigated on a vapour sample in the 250–3800 cm^{-1} region, the bands observed thus being assigned to the vibrational modes responsible for them.[135]

The spectra of various vapour mixtures of over 20 isotopic species of diborane (obtained by mixing together B_2H_6 and B_2D_6 in various ratios), between 660 and 5000 cm^{-1}, were also measured yielding the characteristic bands of the B_2HD_5 and B_2H_5D groups of species by compensating spectroscopically for the absorption of the other isotopic groups present.[136]

The location of the main deuteration-sensitive fundamentals of B_2D_6, B_2HD_5 and B_2H_5D is reported in Table 4.23.

Table 4.23 The Main Deuteration-Sensitive Fundamentals (cm^{-1}) of B_2D_6, B_2HD_5 and B_2H_5D

Description	$^{10}B_2H_6$	$^{10}B_2D_6$	B_2HD_5	B_2H_5D
ν_{10}	368	262		
—$BD_2(H_2)$ wagging	977	728 (*s*)	~720, 730	
—BD_2 twisting			787	
ν_{18} (deformation)	1181	881 (*vs*)	(877, 898)[a]	
—BHD deformation			1073	1079
BD′(H′) asymmetrical stretching[b]	1606	1205 (*vvs*)	1200, 1232	1252
BD′(H′) symmetrical stretching	1887	1491 (*m*)	1408, 1481	1493
—$BD_2(H_2)$ symmetrical stretching	2528 (*vs*)	1857 (*vs*)	(1855, 1870)[c]	
—B—D(H) asymmetrical stretching	2625	1999 (*s*)	1910 (1930, 1965, 1985)[c]	1894 (1876, 1908)

[a] These bands correspond to the ν_{18} fundamental of the more symmetrical, homogeneous, species.

[b] D′ or H′ mean a bridging hydrogen atom.

[c] In resonance with other fundamentals.

The calculated anharmonicity factors and force constants of B_2D_6 have similarly been reported[137] as were also the calculated normal frequencies of 12 partially deuterated diboranes.[138]

The terminal and bridge B—D stretching bands of fully and di-deuterated tetraborane (10) (B_4H_{10}) were observed[139] at about 1826 and 1583 and *ca.* 1900 and 1605 cm^{-1}, respectively. The infrared spectrum of gaseous fully deuterated pentaborane has also been obtained, all its fundamentals[140] being assigned and found in agreement with the values calculated for them on basis of a normal co-ordinate analysis.[141]

The absorption of various 10–48 atom % deuterated decaborane ($B_{10}H_{14}$) samples in the B—D stretching range was investigated in a study of the course of the protium–deuterium exchange of decaborane[142] at 100°C.

4.29 Deuterated Disilane and Allied Compounds

Disilane, Si_2H_6, probably belongs to the D_{3d} point group, hence only five of its fundamentals can be infrared active.[143] The infrared absorption of a gaseous Si_2D_6 sample (which contained about 9% of Si_2HD_5 as an impurity) showed[143] all of its expected absorption bands. Their frequency is reported in Table 4.24.[143]

Table 4.24 The Observed Fundamental Bands (cm^{-1}) of the Disilanes

Symbol	Type	Vibration direction (relative to axis)	Si_2D_6	Si_2H_6
ν_9	$SiD_3(H_3)$ rocking	perpendicular	277·0 (*s*)	379·3
ν_6	$SiD_3(H_3)$ bending	parallel	624·9 (*vs*)	843·5
ν_8	$SiD_3(H_3)$ deformation	perpendicular	682·9 (*s*)	939·6
ν_5	Si—D(H) symmetrical stretching	parallel	1549·3 (*vs*)	2154·3
ν_7	Si—D(H) asymmetrical stretching	perpendicular	1585·0 (*vs*)	2178·6

The coarsely resolved rotational structure of the ν_8 and ν_9 bands yielded, after analysis, the corresponding Coriolis coefficients.[143]

The Si—D stretching bands of ethyldichlorosilane, methyldiethylsilane and triethylsilane also appear at about 1608, 1527 and 1530 cm^{-1}, respectively.[144] The infrared spectrum of a number of perdeuterated silyl compounds has been measured. These include: disiloxane,[145] disilylsulphide,[146] disilyl selenide,[146] trisilylamine,[147, 148] trisilylphosphine[149] and tetrasilylhydrazine.[150] These substances also, show both their symmetric and their asymmetric Si—D stretching fundamentals (as far as these are infrared active) as very intense bands between 1546 and 1592 cm^{-1}, their asymmetric SiD_3

bending vibrations absorbing strongly in the 674–741 cm^{-1} region. The very strong, symmetric SiD_3 deformation absorption of these molecules appears at 652–713 cm^{-1} while the rocking of this group gives rise in this case to a medium absorption at 550 ± 50 cm^{-1}.

4.30 Labelled Sulphamic Acid and Labelled Sulphamide

Solid sulphamic acid exists probably[151, 152] in the dipolar form, i.e. as $NH_3^+SO_3^-$. D_3-sulphamic acid shows its main absorption bands at 509 (*m*), 531 (*s*), 666 (*s*), 792 (*m*; ND_3^+ asymmetric rocking), 1057 (*m*), 1102 (*vs*; symmetric ND_3^+ bending), 1114 (*vs*; ND^+ bending), 1300 (*vs*), 1907 (*m*), 2299 (*m*; symmetric N—D stretching), 2370 (*s*; asymmetric N—D stretching) and 3176 (*m*) cm^{-1}. The spectra of its sodium and potassium salts were also reported.[151, 152]

The absorption of the mulled perdeuterated amide of the sulphamic acid, $ND_2SO_2ND_2$, in the 240–3700 cm^{-1} region[153] was investigated both in the case of ordinary radiation and in the case of polarized light. The dichroic ratios thus obtained, as well as the isotopic effects observed, were used for the vibrational assignment of the bands recorded. Table 4.25 gives the frequency of the main fundamentals of the isotopic sulphamides.

Table 4.25 The Main Fundamentals (cm^{-1}) of Isotopic Sulphamides

Type	$ND_2SO_2ND_2$	$NH_2SO_2NH_2$	NHD Compounds
$NH_2(D_2)$ torsion	322 (*s*)	420	
SO_2 rocking	420 (*w*)		
SO_2 twisting	437 (*w*)		
SO_2 bending	514 (*vs*)	532	
$NH_2(D_2)$ wagging	610 (*vs*)	~730 (*s*)	
N—S—N symmetric stretching	831 (*s*)	904	
N—S—N antisymmetric stretching	868 (*s*)	931	
$NH_2(D_2)$ twisting	902 (*m*)	1130	
$NH_2(D_2)$ bending	1140 (*s*)	1557	1410?
SO_2 symmetric stretching	1168 (*vs*)	1156	
SO_2 antisymmetric stretching	1349 (*vs*)	1358	
$NH_2(D_2)$ symmetric stretching	2385 (*s*)	3210	
$NH_2(D_2)$ asymmetric stretching	2505 (*vs*)	3220	2450, 3245, 3305

4.31 Potassium Deuterium Tetracarbonyl Ferrate

The infrared absorption of an aqueous solution of $K^+[DFe(CO)_4]^-$ in the 1750–2050 cm^{-1} region shows the 1897 (*vs*), 1937 (*sh*) and 2015 (*w*) cm^{-1}

bands characteristic of the acid anion, in contradistinction to the normal anion which absorbs at[154] 1786 (*vs*) cm^{-1}.

4.32 Manganese Carbonyl Deuteride

The absorption spectrum of gaseous $DMn(CO)_5$ has been recorded between 400 and 3000 cm^{-1} and was shown to display its Mn—D stretching band at[155] 1290 cm^{-1} as compared with the 1783 cm^{-1} value for the corresponding Mn—H absorption. Other deuteration-sensitive bands appear at 326, 518 and 674 cm^{-1} (and in the normal hydride at 361, 612 and 730 cm^{-1}).

4.33 Deuterated Borazoles

The infrared spectrum of almost pure N-trideuteroborazole (Fig. 4.6) has been recorded for its vapour phase as well as for its solutions in carbon

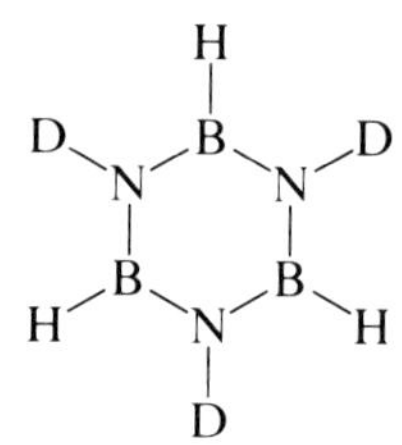

Fig. 4.6 The structure of N-trideuteroborazole.

disulphide and carbon tetrachloride.[156a, b] Similar measurements have also been carried out in the case of B-deuterated borazole and the N-deuterated B-trichloro-derivative. The results obtained thus are reported in Table 4.26.

4.34 The Ir—D and Pt—D Bands

The infrared absorption of various deuterated iridium complexes of the general structure as in Fig. 4.7 exhibits Ir—D stretching bands in the 1499–1608 cm^{-1} region and an Ir—D bending fundamental between[156c] 592 and 608 cm^{-1}. The ν_D/ν_H ratio in this case is about 0·72 for the stretching and 0·91 for the bending frequencies.

Deuterium gas chemisorbed on platinum metal gives rise to infrared absorption bands at *ca.* 1480 and 1520 cm^{-1}, as compared with the values of 2060 and 2110 cm^{-1} obtained for ordinary adsorbed hydrogen gas. These bands are due to the stretching of the Pt—D(H) bonds formed thus.[156d]

TABLE 4.26 The Main Infrared Absorption Bands (cm^{-1}) of Isotopic Borazoles

Type	$B_3N_3H_6$	$B_3N_3H_3D_3$ N-deuterated	$B_3N_3H_3D_3$ B-deuterated	$B_3N_3D_3Cl_3$
Ring distortion	393	383		
Ring deformation	519	509		
N—H(D) out-of-plane bending	718	547	713	503
N—H(D) in-plane bending	[a]	[a]		813
B—H(D) out-of-plane bending	918	899	808	
B—N stretching	1465, 1605	1436, 1589		1410, 1575
B—H stretching	2530	2527		
N—H(D) stretching	3490	2607		2570

[a] The literature assignment for this vibration[156a] does not seem correct.

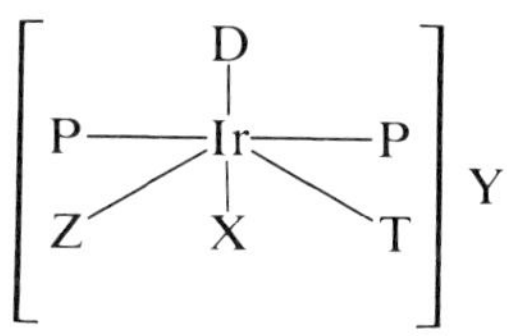

P = PPh_3, $\frac{1}{2}(Ph_2PCH_2)_2$
X = Cl,Br,I,D
Y = nil,BPh_4
Z = Cl, $\frac{1}{2}(Ph_2PCH_2)_2$
T = CO, $\frac{1}{2}(Ph_2PCH_2)_2$

Fig. 4.7 The structure of the investigated iridium complexes.

Labelled Saturated Hydrocarbons

4.35 DEUTERATED AND TRITIATED METHANE

Isotopically homogeneous methane molecules possessing tetrahedral symmetry belong to the T_d point group and hence should theoretically show only two infrared active fundamentals, ν_3 and ν_4. These bands have been observed also in the case of CD_4, the absorption spectrum of which was investigated[157] from 900 to 8500 cm^{-1}. Owing to a vibration–rotation interaction, however, the ν_2 fundamental of (CH_4 and)CD_4 also appears in the spectrum, although only as a weak infrared absorption band. After being resolved and analysed[158] its structure has yielded the value of 2·701 $\pm$ 0·002 cm^{-1} for the B_2 rotational constant of CD_4. All the four zero-point fundamental frequencies[157b, 159] of CD_4, as well as its force constants, Coriolis constants[159] and other vibration–rotation interaction constants,[160] have also been calculated, using the empirical relations existing between respective anharmonicity terms of corresponding isotopic molecules (eqn. 3.10).

The C—D stretching band of CD_4 solutions at about 2258 cm^{-1} was studied in the case of various solvents such as carbon tetrachloride, carbon disulphide, hexane, etc.[161] It was found that while the form of the C—H band did not change with the solvent the form of the C—D stretching absorption curve depended on the properties of the solvent of the deuterated methane.

Monodeuteromethane (similarly to CHD_3 and CHT_3) belongs to the C_{3v} point group, hence all of its six normal vibrations give rise to infrared absorption bands. Their frequency has been reported by several authors[157a, 162] together with the frequency of their overtone and combination bands[162] in the region up to 6000 cm^{-1}. All the fundamentals of CH_3D[162–166] as well as some of its combination bands[164] have been studied under high resolution to yield precise values for the centres of these bands and its rotational constants B_0 (=3·8800 cm^{-1}), D_0^J, etc.

Dideuteromethane belongs to the C_{2v} point group and therefore shows eight infrared active fundamentals, one of its normal vibrations being only Raman active. All these frequencies have been reported [157a] and the values of its rotational constants A_0, B_0 and C_0 have been determined as 4·303, 3·504 and 3·049 cm^{-1}, respectively, from an analysis of two of its combination bands which are centred at 2671·67 and 4425·61 cm^{-1}, correspondingly.[167] Somewhat different values for these constants were obtained from an analysis of the band centred at[168] 5998·4 cm^{-1}.

The absorption of CHD_3 in the 670–4000 cm^{-1} region [157a, 169] and the 5000–6000 cm^{-1} range[170] shows all of its fundamentals and a number of its overtone bands (such as the one centred at[170a] 5865·02 cm^{-1} and that at[164] 2592·6 cm^{-1}). The rotational analysis of these bands gave the value of 3·2795 $\pm$ 0·0005 cm^{-1} for the B_0 constant as well as the value of other rotational constants of CHD_3.[164, 170]

The infrared spectra of CT_4 and the less tritiated methane modifications, containing also deuterium or protium atoms, have been studied by adding T_2 gas to CH_4 or CD_4 and recording daily the spectrum of the exchanging methane gas sample.[159] This study has resulted in the identification of most of the fundamentals belonging to the various tritiated methanes. High resolution of parts of these spectra have made possible the calculation of the zero-point frequencies, force constants, etc. of CT_4 and this in turn has enabled the calculation of the harmonic frequencies of the other tritiated methanes. The various rotational constants of all of these tritiated compounds have also been reported.[159, 171, 172]

The frequency of the observed fundamental infrared absorption bands of various isotopic methanes is given in Table 4.27.

TABLE 4.27 The Infrared Absorption Bands (cm^{-1}) of Isotopic Methanes[a]

Band	Type	CH_4	CD_4	CH_3D	CH_2D_2	CHD_3	CT_4	CH_3T	CH_2T_2	CHT_3	CD_3T	CD_2T_2	CDT_3
ν_1	symmetrical stretching	—	—	2200[b]	2202	2142	—	1877[c]	—	1796	1865	1841	—
ν_2	symmetrical bending	1534	1092	1473[163]	1329	1291	—	1463	—	1236	1066	1049	—
					1436				1426			—	
ν_3	asymmetrical stretching	3019	2259	2970[d]	2234	2263	1937	—	1932	—	2165	1935	1939
				3017[e]	2976	2993		~3014[e]	—	2983	2260	2203	2230
					3013				3003			—	
ν_4	asymmetrical deformation	1306	996	1158[f]	1033	1003	858	1088	—	—	929	—	—
				1307[g]	1090	1036		1296	1005	875	990	903	873
					1234				1203			957	

[a] Unless otherwise stated, values are from ref. 157a.
[b] In Fermi resonance with another band; unperturbed value: 2210 cm^{-1} (ref. 162).
[c] All the values for the tritiated compounds are from ref. 159.
[d] In Fermi resonance with another band; unperturbed value: 2948 (ref. 162).
[e] Ref. 172.
[f] Ref. 162.
[g] Ref. 165: it is assigned to ν_3 there.

The agreement of the observed frequencies reported in Table 4.27 with the corresponding calculated values is generally very good.[159]

4.36 Substituted Heavy Methanes

The CDX_3 molecules (X = F, Cl, Br) all belong to the C_{3v} point group and can thus be expected to show three A_1 and three E infrared active fundamentals. The spectra of heavy fluoroform,[173,174] chloroform,[174,175] and bromoform[174,176,177] have indeed been observed to show all these fundamentals. Figure 4.8 reproduces the infrared absorption spectrum of liquid $CDCl_3$ as compared with that of $CHCl_3$. Table 4.28 gives the frequency of the various gaseous CDX_3 fundamentals.

Table 4.28 The Fundamental Frequencies (cm^{-1}) of Deuterated Fluoroform, Chloroform and Bromoform

Band		Type	CHF_3	CDF_3	$CDCl_3$	$CDBr_3$
A_1	ν_1	C—D(H) stretching	3031	2273 (*m*)	2245	2247
	ν_2	symmetrical C—X stretching	1140	1109 (*s*)	648	519
	ν_3	symmetrical bending	700	693 (*m*)	365	221
E	ν_4	out-of-phase C—X stretching	1152	1207 (*vs*)	735	628
	ν_5	C—D(H) bending	1372	974 (*m*)	910	848
	ν_6	asymmetrical C—X bending	507	504 (*m*)	261	153

These frequencies obey the product rule (eqn. 3.17) both in the case of the A_1 species and in the case of the E species.[173b] Using these values the potential energy function for these compounds has been calculated[174] on the basis of the hybrid orbital force field approximation which is the only one giving satisfactory results in this case.[174,178] The calculated Coriolis constants, rotation distortion constants and mean square amplitudes of vibration of the CDX_3 molecules have also been reported[179] as well as the observed anharmonicity terms of $CDCl_3$[188] (obtained from its overtone and combination bands in the 650–7400 cm^{-1} region).

The C—D stretching frequency of a CDX_3 solution in an inert solvent is about 12 cm^{-1} lower than the corresponding gaseous CDX_3 frequency and 1–7 cm^{-1} higher than the respective neat liquid value.[180] In the case of a CDI_3 solution in carbon tetrachloride the C—D stretching band appears at about 2240 cm^{-1} while the C—D bending absorption of its hexane solution is shown at *ca.* 790 cm^{-1} as compared with the value of over 905 cm^{-1} for $CDCl_3$ in hexane.[181]

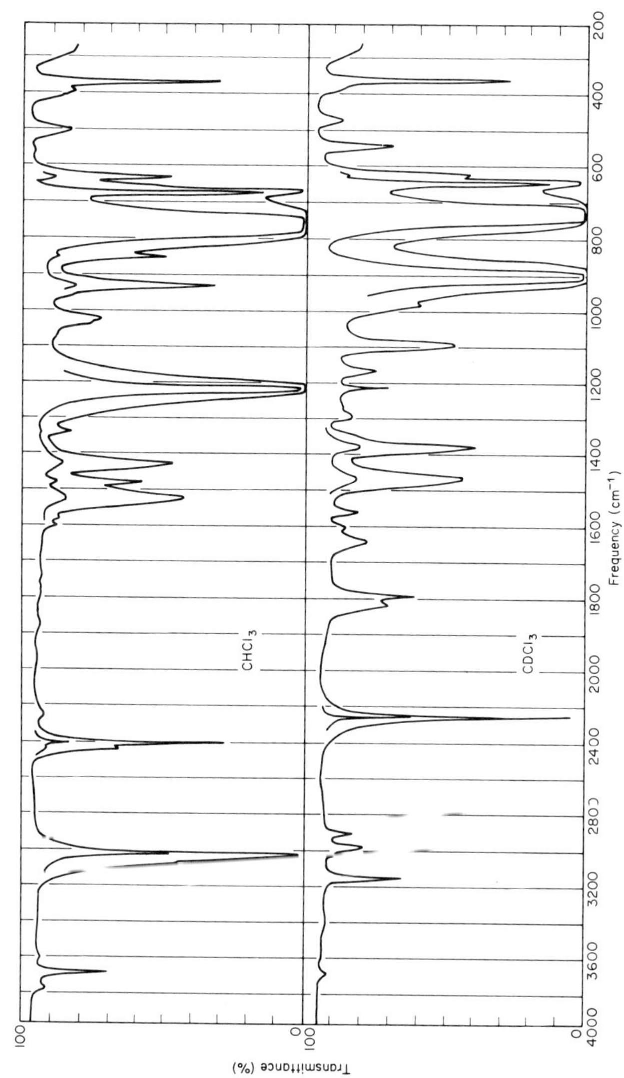

Fig. 4.8 The infrared spectra of $CHCl_3$ and $CDCl_3$. These spectra were run on a PE 521 grating spectrometer in a 0·180 mm shell with some regions run in a 0·023 mm cell.[175]

If a haloform is dissolved, however, in a solvent which can accept protons, such as, e.g. acetone, triethylamine or nitromethane, association between the CDX_3 molecules and those of the solvent occurs with formation of a complex of the type shown in Fig. 4.9.

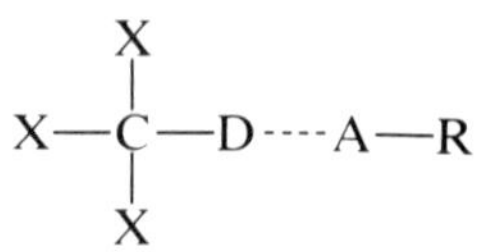

(*a*) A = O; R = $(CH_3)_2C$, CH_3NO
(*b*) A = N; R = $(C_2H_5)_3$

Fig. 4.9 A typical haloform-proton acceptor complex.

Since the formation of this complex affects very appreciably the length, polarity and force field of the C—D bond involved in it, the absorption intensity of the C—D stretching band of the complex is often increased ten-fold or more.[182,183]

The investigation of the C—D stretching band of deuterated chloroform and deuterated iodoform dissolved in various solvents has shown[181] that its intensity varies as a function of the solvent in parallel to the absorption intensity of a polar HX vibrator while its frequency changes differently, between *ca.* 2200 cm^{-1} (CDI_3 in pyridine) to about 2250 cm^{-1} ($CDCl_3$ in CH_3NO_2).

The observed increase in the C—D stretching band absorption of $CDCl_3$, $CDCl_2Br$ and $CDBr_3$ solutions in n-hexane, as the result of the addition of acetone, can be explained reasonably on basis of a charge-transfer mechanism.[183]

Studying the changes in optical density brought about in this case by various proportions of acetone and haloform has yielded values for the formation constants of these complexes while measuring the C—D band intensities as a function of temperature has given their heats of formation.[183]

The integrated intensity of the C—D bending absorptions of CDX_3 molecules in proton-accepting solvents remains, however, practically the same as in the case of inert solvents, although their frequencies are shifted by 15–30 cm^{-1} to higher values.[181]

The infrared absorption of $CDCl_2Br$ in the 200–4000 cm^{-1} region was measured both for the vapour and for the liquid phase.[184] Its nine, infrared active, fundamentals to be expected were observed for the liquid phase at about 215 (ν_9), 220 (ν_6), 329 (ν_5), 579 (ν_4), 699 (ν_3), 735 (ν_8), 872 (ν_2), 904 (ν_7) and 2253 (ν_1), cm^{-1}. Of these bands ν_1 belongs to the C—D stretching mode while ν_2 is due to the symmetrical, and ν_7 to the non-symmetrical, C—D bending vibration. These frequencies, together with the $CHCl_2Br$ vibrational data, have been used to calculate the force constants of this compound.[184]

The fundamental frequencies of $CDFCl_2$ and CDF_2Cl[185] as well as those of CDF_2Br[186] have all been calculated.

The fundamental bands appearing in the 400–4000 cm^{-1} spectrum of liquid $CHDCl_2$ are located at[187] 681 (*m*), 719 (*vs*), 778 (*m*), 885 (*vs*), 1218 (*vs*), 1275 (*s*), 2246 (*m*) and 3021 (*s*) cm^{-1}. They were assigned to their respective modes of vibration by comparing them with the assigned absorption frequencies of CH_2Cl_2.

The absorption bands of the liquid CHDBrCl derivative were obtained from the spectrum of a mixture of CH_2BrCl, CD_2BrCl and CHDBrCl by elimination of the bands due to the CH_2 and CD_2 modifications. The former were thus found[188] at 706, 743, 865, 1183, 1262, 2252 and 3031 cm^{-1}.

The infrared absorption of high isotopic purity gas samples of all the four CH_2DX (X = F, Cl, Br, I) compounds was measured in the 425–5000 cm^{-1} region,[189] all their (4 × 9) fundamentals being observed and assigned. An application of the product and sum rules (eqns. 3.17 and 3.21) to their frequencies resulted in a satisfactory agreement between the observed and the respective calculated values. Table 4.29 gives these frequencies.

TABLE 4.29 The Fundamental Frequencies (cm^{-1}) of the CH_2DX Compounds

Compound	Symmetrical (A') bands	Non-symmetrical (A'') bands
CH_2DF	938, 1052, 1364, 1466, 2197, 2950[a]	1133, 1299, 3016
CH_2DCl	713[b] (707[c]), 827, 1269, 1436, 2224, 2990	986, 1260, 3035
CH_2DBr	594, 769, 1225, 1425, 2244, 3001	930, 1240, 3052
CH_2DI	518, 710, 1172, 1415,[a] 2241, 3002	862, 1240, 3057

[a] Corrected for Fermi resonance.
[b] For the ^{35}Cl species.
[c] For the ^{37}Cl species.

The absorption of the various CHD_2X (X = F, Cl, Br, I) compounds in the gas phase has also been investigated thoroughly.[189, 190] Their symmetrical C—D stretching vibration results in a medium band at about 2195 cm^{-1} while the corresponding antisymmetrical C—D band appears between 2259 (for X = F) and 2336 (X = I) cm^{-1}. The symmetrical CD_2 bending fundamental gives in this case a strong band rising in frequency from 1018 (X = I) to 1092 (X = F) cm^{-1}, as compared with the corresponding CH_2 frequencies of 1415–1466 cm^{-1} reported in Table 4.29 for the analogous CH_2DX compounds.

Liquid CD_2Cl_2 shows these bands at[187] 2199 (symmetrical C—D stretching), 2305 (asymmetrical C—D stretching) and 1052 (CD_2 bending) cm^{-1}, the last band, however, being observed only in the Raman spectrum, since in the infrared region it is probably masked by the very strong absorption centred at 957 cm^{-1}. This latter absorption is due to the CD_2 rocking vibration which corresponds to the 1266 cm^{-1} frequency of[191] CH_2Cl_2.

Almost the same frequencies were reported[188] for these CD_2 vibrations in the case of CD_2BrCl (2195, 2302, 1041, 923).

All the 3 A_1 and 3 E fundamental vibration–rotation bands of gaseous perdeuteromethyl fluoride, as well as a number of its binary combination bands, have been measured on a 98 atom % sample and much of their fine structure has been resolved.[192] Its analysis yielded values for the various rotational constants of the CD_3F molecules, such as their A_0 (2·56 cm^{-1}), B_0 (0·6821 cm^{-1}), D''_J etc. Similar high resolution measurements have also been carried out on gaseous CD_3Cl[193, 194] CD_3Br[195, 196a] and CD_3I.[190, 196b–198]

The pure rotation absorption spectrum of CD_3F between 15 and 80 cm^{-1} (from $J = 10$ to $J = 49$) has likewise been reported.[199]

The fundamental frequencies of the various gaseous CD_3X halides are collected in Table 4.30.

TABLE 4.30 The Fundamental Frequencies (cm^{-1}) of CD_3F, $CD_3{}^{35}Cl$, $CD_3{}^{79}Br$ and CD_3I

Band	Species	Type	CH_3F[a]	CD_3F[b]	CD_3Cl[c]	CD_3Br[d]	CD_3I[e]
ν_1	A_1 (parallel)	symmetrical C—D(H) stretching	2964·5	2090·8	2159·6	2159·8	2155·4
ν_2	A_1	symmetrical C—D(H) bending	1475	1134·8	1029·6	992	949
ν_3	A_1	symmetrical C—X stretching	1048	992·2	700·7	576·7	501·2
ν_4	E (perpendicular)	asymmetrical C—D(H) stretching	2982	2258·4	2283·2	2297·3	2298·5
ν_5	E	asymmetrical C—D(H) bending	1471	1071[f]	1061·3	1055·6	1049·3
ν_6	E	asymmetrical deformation	1195·5	991·2	768·6	713	655·9

[a] Ref. 191, p. 315.
[b] Values are generally from ref. 192.
[c] Values are from ref. 193.
[d] Values are from ref. 196a.
[e] From ref. 196b.
[f] From ref. 200.

4.37 DEUTERATED ETHANES

C_2D_6 which must be assumed to exist in the stellar (staggered) configuration[201] (see Fig. 4.10), of point group $\boldsymbol{D_{3d}}$, can be expected to show only five infrared active fundamentals.[179]

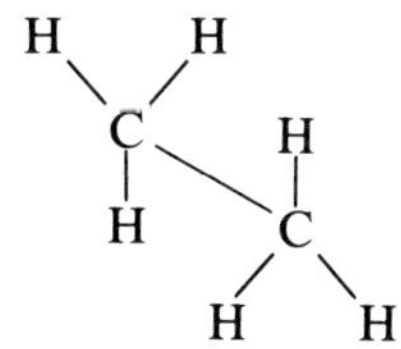

Fig. 4.10 The probable configuration of the ethane molecules.

All of these frequencies were observed in the infrared absorption spectra of gaseous and pure solid C_2D_6[201–204] (as well as for dilute solutions of various deuterated ethanes in one another)[205] two of them (ν_5 and ν_7) being involved in Fermi resonance with other modes of vibration. This brings about the splitting of the C—D stretching band at about 2100 cm^{-1} into two branches. The torsional mode of vibration of C_2D_6 and CH_3CD_3 was observed under 3–7 atm. pressure.[205]

The assignments of the C_2D_6 fundamentals were checked by an application of the sum rule (Section 3.5) and the Teller–Redlich rule (Section 3.4) and were always found to yield experimental values within 1·5% of the corresponding calculated values.[201]

The infrared spectra of gaseous C_2H_5D,[201] CH_2DCH_2D,[201] CH_3CD_3,[201,204] CHD_2CHD_2[201] and C_2HD_5[201] have also been measured, the observed absorption bands being assigned to the corresponding fundamentals. Table 4.31 reports the frequency of the main bands of these compounds.

The analysis of some of the near infrared absorption bands of CH_3CD_3 has recently been reported.[206] The absorption of CH_3CHD_2 and CH_2DCD_3 has similarly been measured.[201]

4.38 Substituted Deuterated Ethanes

The infrared spectra of gaseous 1-deutero- and 2-deutero-ethyl fluoride and gaseous and liquid 1-deutero- and 2-deutero-ethyl bromide (the former only in a mixture with the latter) were measured in the 400–4000 cm^{-1} range for gases and between 400 and 1500 cm^{-1} for liquids.[207]

The fact that the spectrum of liquid 2-deutero-ethyl bromide shows *two* strong bands in the C—Br stretching region, at 534 and 555 cm^{-1}, indicates that this compound exists as a mixture of two distinguishable rotational isomers: the *trans* isomer and the *skew* isomer (see Fig. 4.11).

Generally, rotational isomerism affects the whole infrared absorption spectrum of a compound very strongly since it changes both its symmetry and its force field. This can be seen from Table 4.32, which gives the observed infrared active fundamentals of the rotational isomers of liquid *meso-sym*-dideuterodibromomethane.[208]

TABLE 4.31 The Frequency (cm^{-1}) of the Main Bands of Deuterated Ethanes[201, 202]

Designation[a]	Type	C_2H_6	C_2D_6	C_2H_5D	$(CH_2D)_2$	$C_2H_3D_3$	$(CHD_2)_2$	C_2HD_5
ν_5	C—H(D) stretching	2954	2111 (*s*)	2954	2948[b]	2121[c]	2182[b]	2111
$\nu_8 + \nu_{11}$	combination	2894	2087 (*s*)					
ν_6	$CH_3(D_3)$[d] in-phase deformation[e]	1379	1072 (*s*)		1124	1120	1120	1073
ν_7	C—H(D) stretching	2995	2236 (*s*)	2980	2192	2975	2136	2237
				2995	2979		2948	2268
ν_8	$CH_3(D_3)$[d] out-of-phase	1472	1082 (*s*)	1312	1313	1469	1107	1134
	deformation			1470	1460		1309	1309
ν_9	C—C bending	821	594 (*s*)	715, 805	744, 791	685	635, 720	599, 631

[a] According to ref. 201.
[b] This frequency is not affected by Fermi resonance.
[c] This is the calculated original ν_5 frequency.
[d] Or CH_nD_{3-n}.
[e] "In phase" refers to the atoms of one group only.

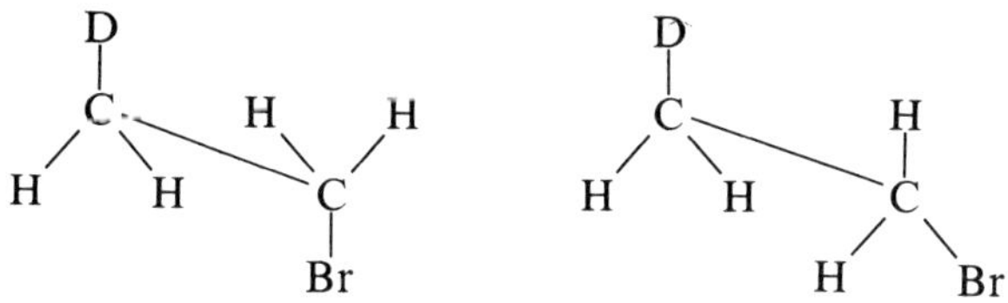

trans isomer (absorbs at 534 cm^{-1}) | *skew* isomer (absorbs at 555 cm^{-1})

Fig. 4.11 The structure of the rotational isomers of CH_2DCH_2Br.

TABLE 4.32 The Infrared-Active Fundamentals (cm^{-1}) of the Rotational Isomers of *meso* $(CHDBr)_2$

Trans isomer		*Skew* isomer	
Designation[208]	Frequency	Designation[208]	Frequency
ν_{17}	570	ω_5	527
ν_9	621	ω_{17}	556
ν_8	833	ω_{13}	721
ν_{16}	1144	ω_9	756
ν_{15}	1287	ω_8	849
ν_{14}	2235	ω_{12}	934
ν_7	3020	ω_3	1040
		ω_{16}	1207
		ω_{15}	1331
		ω_{14}	2177
		ω_7	3020

The spectrum of a mixture which contained the *trans* and *skew* rotamers of both the *meso*- and the racemic $(CHDBr)_2$ species was also measured, its absorption bands being tentatively assigned to the various species present in this mixture.[208]

The absorption of CH_2DCHDF and $CH_2DCHDBr$ was similarly investigated in the 400–4000 cm^{-1} region.[207]

The C—D stretching frequencies of CF_3CD_3 appear at 2172 (symmetrical) and 2284 cm^{-1} (corresponding to the 2975 and 3035 cm^{-1} values for CF_3CH_3) while its other isotope-sensitive bands are shown at 333, 798, 1043, 1069 and 1202 cm^{-1}, as compared with the respective frequencies of 365, 970, 1233, 1280 and 1443 cm^{-1}, in the case of CF_3CH_3.[209]

Besides the other trideuterated substituted ethanes which were studied, CHD_2CHDF and CHD_2CHDBr,[207] a number of tetradeuterated substituted ethanes were also examined.[210, 211]

Here again the absorption spectrum of the liquid (in contradistinction to the solid) phase is complicated by the co-existence of various rotamers of the same compound which absorb quite differently one from the other. Thus, e.g. while *trans* $(CD_2Br)_2$ shows[210] its C—D stretching bands at 2209 and 2290 cm^{-1}, its CD_2 bending vibration at 1061 cm^{-1} and its CD_2 rocking mode at 564 cm^{-1}, the *skew* $(CD_2Br)_2$ rotamer[210] shows its *three* C—D stretching bands at 2174, 2234 and 2272 cm^{-1}, its CD_2 bending absorption at 1050 *and* 1141 cm^{-1} and its CD_2 rocking vibrations both at 686 *and* at 711 cm^{-1}.

4.39 Deuterated Propane

The absorption of a gaseous sample containing about 90% of C_3D_8 was measured in the 230–10,000 cm^{-1} region.[212] Figure 4.12 shows this absorption between 650 and 3100 cm^{-1}.

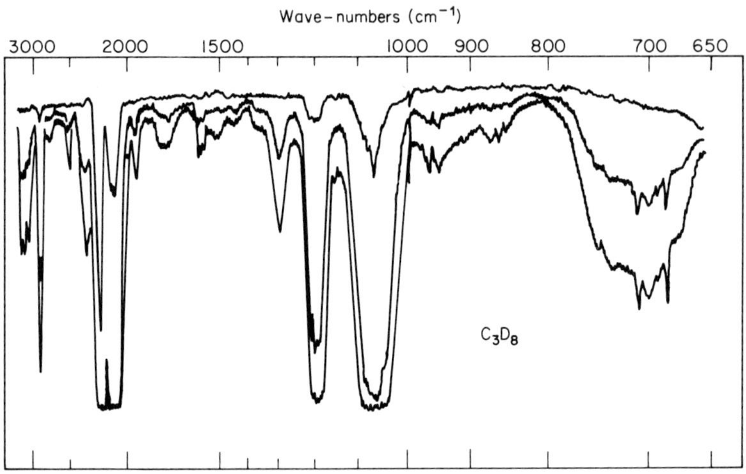

Fig. 4.12 Prism spectrum of C_3D_8.

Assuming the propane molecules to have a C_{2v} symmetry 22 infrared active fundamentals are expected for them. In practice, however, only about 14 separate bands have been observed in the region examined, this being due to band coincidences, vibrations absorbing below 230 cm^{-1} and weakness of other bands.

The absorption of gaseous 1-deutero-,[213] 2-deutero-,[213] 2,2-D_2-,[212, 214] 1,1,1-D_3-[212] and 1,1,1,3,3,3-D_6-propane[212, 214, 215] has also been reported. The spectra of all the symmetrically deuterated propanes have been used in establishing the correct assignment of the C_3D_8 fundamentals.[212]

Table 4.33 reports the main absorption bands observed so far for deuterated propanes.

TABLE 4.33 The Main Bands (cm^{-1}) of the Isotopic Propanes

C_3H_8[212]	C_3D_8[212]	1-C_3H_7D[213c]	2-C_3H_7D[213d]	$CH_3CD_2CH_3$[212]	$C_2H_5CD_3$[212]	$CD_3CH_2CD_3$[212]
369	—[a]	700 (*m*)[b]	670 (*m*)[b]	622	660	640
748	544	740 (*m*)	675 (*w*)	829	750	711
869	712	830 (*m*)	850 (*w*)	843	831	725
922	688	850 (*w*)	875 (*w*)	964	846	920
054	862	885 (*m*)	910 (*m*)	1064	999	962
158	959	900 (*m*)	915 (*w*)	1146	1062	1066
192	949	915 (*m*)	935 (*m*)	1203	1100	1086
338	1068	1030 (*w*)	1145 (*s*)	1207	1129	1131
378	1086	1040 (*m*)	1155 (*s*)	1374	1132	1331
392		1070 (*w*)	1315 (*m*)	1392	1285	1467
462		1130 (*m*)	1330 (*m*)	1458	1332	2091
464	1203	1165 (*w*)	1355 (*w*)	1461	1383	2226
472	1064	1285 (*s*)	1370 (*m*)	1476	1460	2883
476		1295 (*s*)	1385 (*s*)	2141	2075	2929
887	2080	1320 (*s*)	1465 (*vs*)	2182	2214	
962	2122	1330 (*s*)	1675 (*w*)	2883	2225	
968	2148	1360 (*s*)	1820 (*w*)	2962	2882	
973	2224	1385 (*w*)	2140 (*vs*)	2974	2934	
977		1395 (*s*)	2300 (*w*)		2966	
		1465 (*vs*)	2600 (*w*)			
		2140 (*s*)	2740 (*w*)			
		2740 (*w*)	2940 (*s*)			
		2940 (*vs*)	3140 (*m*)			
		3130 (*m*)				

[a] The values given in this column (only) correspond, row for row, to those cited for C_3H_8.
[b] The frequencies cited here were read from a curve, hence are usually only within 5–20 cm^{-1} of the rrect values.

The values observed for the absorption bands of the symmetric deuterated propanes are in good agreement with the frequencies calculated for them on basis of a normal co-ordinate analysis.[216]

The only substituted deuterated propane spectrum to be measured so far seems to be that of $CH_3CDClCH_3$.[217]

4.40 HEAVY BUTANES

Perdeuterated butane, C_4D_{10}, seems to have been investigated only in a mixture with deuterated *iso*butane.[218] This mixture showed strong bands at 2090 and 2200 cm^{-1} (C—D stretching), a medium band at 1065 cm^{-1} (C—D bending), weak absorption at 1150, 1180, 1210 and 2150 cm^{-1} and shoulders at 2050, 2115 and 2230 cm^{-1}.

The spectra of both *meso* and optically active $(CH_3CHD)_2$ have been observed in the 850–1150 cm^{-1} region for the gas phase[219] and it was found that they were characteristic enough to enable an easy differentiation between these compounds. Thus, while the *meso* modification shows strong bands at 973 and 1112 cm^{-1} its l-stereoisomer shows clear-cut doublets at 880, 962 and 1075 cm^{-1}.

$CH_3CDClC_2H_5$ was reported[217b] to show two C—D stretching bands, at 2188 and at 2198 cm^{-1}.

The absorption of perdeuterated *iso*butane was measured on its gas phase in the whole 290–3300 cm^{-1} region, all the bands observed being assigned to their parent modes of vibration.[220] *Iso*butane very probably belongs to the C_3 point group,[220] hence should show 12 symmetrical and 12 degenerate, infrared active fundamentals, almost all of which have in fact been observed. The appreciable change in mass brought about by the complete deuteration of the *iso*butane molecule affects considerably the frequency of *all* of its bands, i.e. even of those not conventionally assigned to any motion of the hydrogen atoms. The symmetrical C—C stretching band, e.g. which appears in the $CH(CH_3)_3$ spectrum at 966 cm^{-1} thus moves in the case of $CD(CD_3)_3$ to 789 cm^{-1}.

The tertiary C—D group stretching frequency of $CD(CD_3)_3$ appears at 2067 cm^{-1} (*s*) (the respective CH value being 2880 cm^{-1}) while its ν_1, out-of-phase CD_3 stretching, band is shown at 2216 cm^{-1} (*vs*) and the respective in-phase, ν_2, fundamental is situated at 2073 cm^{-1} (*s*). The ν_4 and ν_5 (asymmetrical and symmetrical, respectively) CD_3 bending vibrations of this *iso*-butane give rise to an absorption at 1118 (*w*) and 1068 cm^{-1} (*s*), respectively.

The infrared spectrum of $HC(CD_3)_3$ in the 290–3300 cm^{-1} region was also measured.[220] It shows the ν_1 and ν_2 C—D stretching bands at almost the same frequencies (2212 and 2076 cm^{-1}) as the perdeuterated *iso*butane and even in the case of the symmetrical C—C stretching fundamental ν_6 the difference between the $HC(CD_3)_3$ vibration and the $DC(CD_3)_3$ frequency is not much more than 1% of their values. Its C—H stretching frequency, which appears at 2896 cm^{-1}, is however different from the corresponding $DC(CD_3)_3$ analogue by a factor of 1·40, as can be expected.

The absorption of gaseous 2-deutero-methylpropane is analogously very similar to that of $HC(CH_3)_3$ almost in the whole infrared region.[220, 221] The C—D stretching and the C—D bending fundamentals, however, absorb at 2152 (*s*) and 952 cm^{-1} (*m*), respectively, as compared with the corresponding C—H values of 2880 and 1330 cm^{-1} in the case of $HC(CH_3)_3$.

Of the substituted butanes only the spectra of perdeuterated *t*-butylchloride and *t*-butylbromide appear to have been reported in the literature. The chloride was examined, both in the vapour phase and in solution in CCl_4, CS_2 or n-hexane, in the 100–3800 cm^{-1} region.[222]

The assignment of its absorption bands was facilitated by the results of a Urey–Bradley force field calculation, (2.11) which yielded the calculated frequencies of its various fundamentals together with their potential energy distribution.

Its CD_3 stretching bands appear for the vapour near 2120 (in-phase; medium) and 2240 cm^{-1} (out-of-phase, in respect to the D atoms of *one* CD_3 group; very strong) while the corresponding C—D bending absorptions are at 1016 (*vs*) and 1054 cm^{-1} (*s*), the CD_3 rocking vibration is seen at 824 cm^{-1} (*w*) and the symmetric C—CD_3 deformation band is observed at 330 cm^{-1} (*m*).

The spectrum of D_9-*t*-butylbromide was also measured[223] although in this case only under lower resolution and not farther down than 670 cm^{-1}.

4.41 Deuterated Tetramethylmethane

The absorption of perdeuterated tetramethylmethane (neopentane) in the gas phase was investigated above 400 cm^{-1} and the bands observed[224] were assigned to their respective modes of vibration on basis of a normal coordinate analysis and the isotopic product rule[225] (Section 3.4). The assignment of the infrared active fundamentals of $C(CH_3)_4$ and $C(CD_3)_4$ is reported in Table 4.34.

Table 4.34 The Infrared Active Fundamentals (cm^{-1}) of $C(CH_3)_4$ and $C(CD_3)_4$

Designation[201]	Frequency in $C(CH_3)_4$	Frequency in $C(CD_3)_4$	Type
ν_8	2959	2218 (*vs*)	C—H(D) asymmetrical stretch
ν_9	2876	2058 (*vs*)	C—H(D) symmetrical stretch
ν_{10}	1256	1218 (*s*)	C—C stretch
ν_{11}	1475	1066 (*s*)	C—H(D) out-of-phase deformation
ν_{12}	1372	1037 (*s*)	C—H(D) in-phase deformation
ν_{13}	925	765 (*w*)	$CH_3(D_3)$ rock
ν_{14}	418 (*w*)	(342)[a]	C—C—C deformation

[a] This value was obtained from the Raman spectrum since it is below the end of the measured infrared spectrum.

The spectrum of gaseous 99% $(CH_3)_3CCH_2D$, which was measured in the 670–3330 cm^{-1} range,[226] shows the following bands: 919 (*w*), 922 (*w*), 1217 (*m*), 1251 (*s*), 1298 (*s*), 1376 (*s*), 1476 (*s*), 2176 (*s*) and 2955 (*s*) cm^{-1}.

The C—D stretching bands of $C_2H_5CDClC_2H_5$ appear[217b] at 2179 and 2198 cm^{-1}.

4.42 Heavy Cyclopropane

The spectrum of fully deuterated cyclopropane, C_3D_6, was measured both for the vapour phase, in the 350–3700 cm^{-1} region, and in the case of the liquid state, in the 650–3700 cm^{-1} range.[227–230] All of its 6 ($2A_2'' + 4E'$)

infrared active fundamentals were thus observed, its other 8 fundamentals being infrared inactive since C_3D_6 very probably belongs to the $\boldsymbol{D_{3h}}$ point group. The frequencies of the main infrared bands of gaseous C_3D_6 are given in Table 4.35.

TABLE 4.35 The Main Infrared Bands (cm^{-1}) of C_3H_6 and C_3D_6

Designation		Type	C_3H_6	C_3D_6
A_2''	ν_6	C—H(D) stretching	3101·47[a] (*vs*)	2336·02[a]
	ν_7	$CH_2(CD_2)$ rocking	854[b]	614
E'	ν_8	C—H(D) stretching	3028 (*vs*)	2211
	ν_9	$CH_2(CD_2)$ bending	1442 (*s*)	1074
	ν_{10}	Ring bending	1029 (*s*)	887
	ν_{11}	$CH_2(D_2)$ wagging	868 (*vs*)	720

[a] In vacuum.
[b] Overlapped by the 868 cm^{-1} band.

The rotation–vibration 2336 cm^{-1} band was investigated under high resolution to yield the various rotation constants of C_3D_6 and the fundamental frequencies observed for this compound were used in deriving the force constants and eigenvectors of the cyclopropane molecule.[228]

4.43 DEUTERATED CYCLOBUTANE

The frequency of the main infrared absorption bands of gaseous D_8-cyclobutane is reported in Table 4.36.

TABLE 4.36 The Main Infrared Absorption Bands (cm^{-1}) of C_4H_8 and C_4D_8

Designation	Type	C_4H_8	C_4D_8
ν_1	C—H(D) symmetrical stretching	2895	2120
ν_2	$CH_2(D_2)$ deformation	1443	1160 } (coupled vibrations)
ν_3	a "breathing" mode	1001	882 } (coupled vibrations)
ν_6	C—H(D) asymmetric stretching	2987	2242
ν_7	$CH_2(D_2)$ rocking	627	483
ν_8	$CH_2(D_2)$ wagging	1219	1078
ν_9	ring deformation	926	746
ν_{14}	$CH_2(D_2)$ deformation	1443	1040
ν_{18}	$CH_2(D_2)$ twisting	1223	938
ν_{19}	$CH_2(D_2)$ rocking	749	556

The absence of a Q branch in the 556 cm^{-1}, ν_{19}, band, which (is free of overlapping from another band and) belongs to the E_g species, is taken to show that the Coriolis interaction constant (Section 2.12) for that vibration has a large positive value. Since calculations show that this is so only if cyclobutane exists in the puckered ring form but not if its ring is planar and hence its point group is $\boldsymbol{D_{4h}}$, the contour of this band indicates that C_4D_8 belongs to the $\boldsymbol{D_{2d}}$ point group.[231] The vibrational spectrum of monodeuterated cyclobutane has also been reported.[232]

The spectra of α-chloro- and α-bromo-deutero-cyclobutane and α,β,β,β′,β′-D_5-bromocyclobutane in the 250–4000 cm^{-1} region were measured both for the gaseous and for the liquid states;[233] the absorption of gaseous β,β,β′β′-D_4-Cl-cyclobutane[233a] and its bromo analogue,[233b] between 33 and 4000 cm^{-1}, and that of their liquid phase in the 250–4000 cm^{-1} range[233b] were also investigated. These spectra were interpreted on the assumption that these compounds belong to point group C_s and the assignment of their various bands was based on their position, respective isotope shift, band type, etc.

The (Cl)C—D stretching band of the monodeuterated chloride appears at 2223 cm^{-1} (medium) while its in-plane bending mode absorbs at 924 cm^{-1} (strong).

The asymmetric, ν_{19}, C—D stretching absorption of the D_4-chloride is shown at 2247 cm^{-1} (very strong) while its CD_2 scissoring vibration absorbs at 1074 (*m*) cm^{-1}, its CD_2 wagging mode appears at 1040 (*m*) cm^{-1} and the CD_2 twisting and rocking fundamentals are displayed at 924 (*vs*) and 634 (*m*) cm^{-1} respectively.[233a]

4.44 Labelled Cyclopentane

The observed perdeuterated cyclopentane fundamentals in the 400–2300 cm^{-1} region were compared with the respective calculated frequencies expected for this species on basis of a $\boldsymbol{D_{5h}}$ point group model in order to establish the symmetry of the cyclopentane molecule.[234] It was found, however, that although these frequencies are in a sufficiently good agreement the cyclopentane ring is still slightly puckered as evidenced, e.g. by the appearance of weak bands due to C_5D_{10} fundamentals which are infrared inactive for the $\boldsymbol{D_{5h}}$ planar model.[237]

The infrared absorption of D_5-cyclopentane shows bands (cm^{-1}) at 438 (C—C—D deformation), 738 (C—C stretching), 943 (C—D bending), 1064 (CD_2 scissoring), 1155 (C—D bending), 2125 (symmetrical C—D stretching) and 2220 cm^{-1} (asymmetrical C—D stretching).

The infrared spectrum of monodeuterated cyclopentane was also published[235] as was the location of the C—D stretching band of α-chloro-deutero-cyclopentane[217b] (2212 cm^{-1}).

4.45 Deuterated Cyclohexane

The perdeuterocyclohexane absorption in the 300–3450 cm^{-1} region was reported both for the vapour phase and for the liquid state.[236]

Figure 4.13 reproduces the spectra of gaseous C_6H_{12} and C_6D_{12} in most of this range.

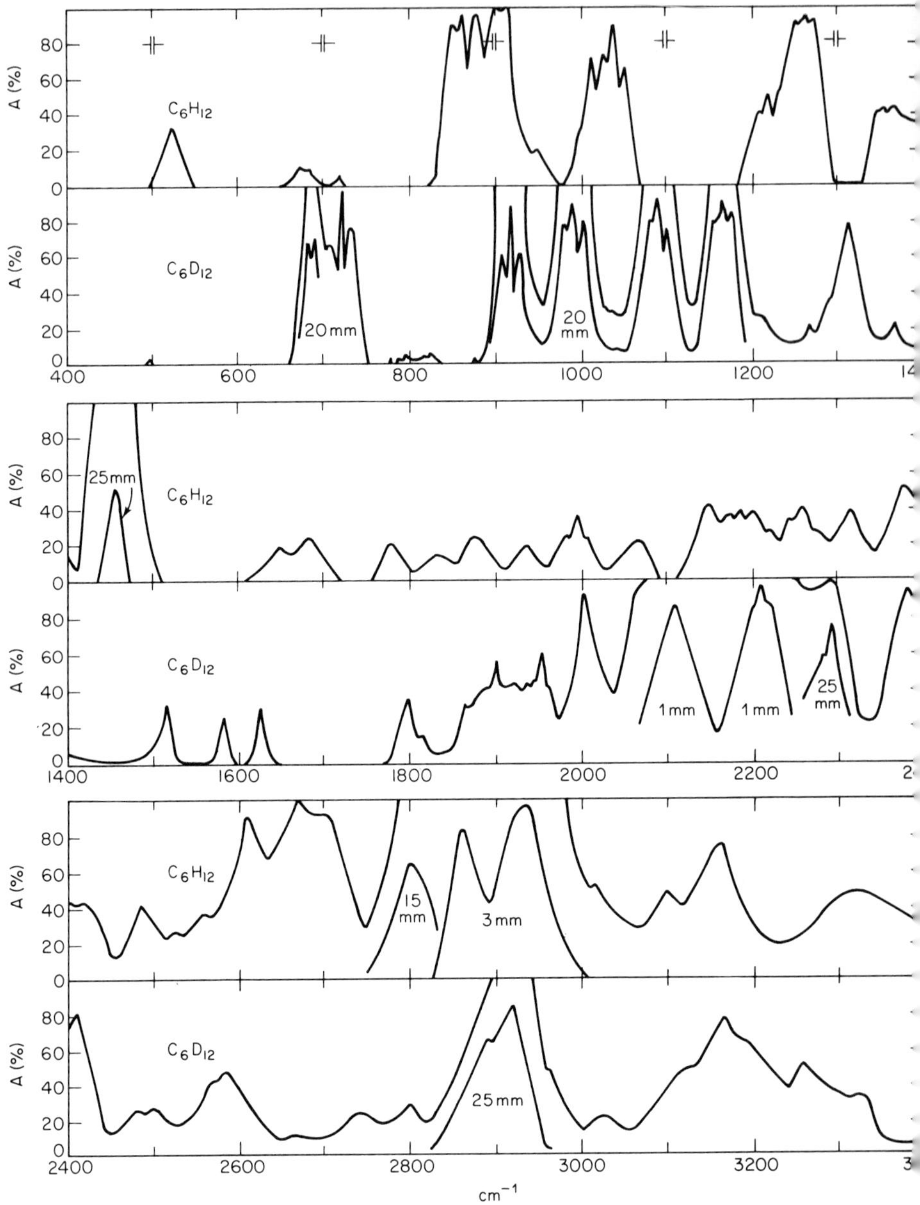

Fig. 4.13 Infrared spectra of gaseous cyclohexane and cyclohexane-D_{12}. 1 m cell, pressure *ca.* 85 mm, except where indicated otherwise.

The cyclohexane molecule which belongs to the D_{3d} point group is expected to show 13 infrared active fundamentals of which 9 should appear below 1500 cm^{-1}. C_6D_{12} shows, however, only six medium or strong bands in this range so that the assignment of a number of these fundamentals is by no means certain.[236a] Its parallel C—D stretching vibrations absorb at 2108 and 2289 cm^{-1} while its perpendicular C—D stretching modes are shown at about 2206 cm^{-1}.

The two CD_2 scissoring frequencies of C_6D_{12} are shown at 1090 and 1166 cm^{-1}, its CD_2 twisting mode is displayed at 990 cm^{-1} and its CD_2 wagging fundamental appears at 917 cm^{-1}. Two other strong absorptions are observed in its spectrum at 687 and 720 cm^{-1}.

The C—D stretching bands of 1-chlorodeuterocyclohexane are shown[217b] at 2174 and 2198 cm^{-1}.

4.46 D_2-Cyclododecane

The absorption of 1,1-D_2-cyclododecane, as a melt and in the solid phase, was investigated in the 50–4000 cm^{-1} region in an attempt to establish the correct conformation of the cyclododecane molecule.[237] However, the fact that this absorption shows in the case of the liquid phase *six* C—D stretching bands, at 2112 (*s*), 2122 (*w*), 2138 (*w*), 2187 (*s*), 2200 (*m*) and 2210 (*w*) cm^{-1}, proves that in this state this compound exists in an equilibrium between *three* different conformations, each of which absorbs quite separately. This makes it almost impossible to assign properly the enormous number of the observed absorption bands of this compound. The situation seems to be even more complex in the case of the solid state since here six *doublet* absorptions appear in the C—D stretching zone.

4.47 Labelled Bicyclic Hydrocarbons

The infrared absorption of a number of deuterated saturated bicyclic hydrocarbons has been reported in the literature.[238]

The C—D stretching region of the *exo*(axial)-2-D_1bicyclo[3.2.1]octane (Fig. 4.14) spectrum (as well as that of its OH derivative) shows its strongest band at about 2148 cm^{-1} while the corresponding *endo*(equatorial)-2-D_1-isomer exhibits it at 2167 cm^{-1} and the respective 2,2-D_2-modification absorbs at 2148, 2170 and 2187 (*s*) cm^{-1}. It is thus possible to determine the stereochemistry of the cyclohexane ring C—D bond from the location of its strongest stretching band.[238a]

The absorption of *trans*-9(D_3-methyl)decalin was measured[238b] in order to be able to identify the angular CH_3 rocking band which appears in the spectra of various steroids at 1160 cm^{-1} and in the case of the normal *trans* 9-methyldecalin at 1170 cm^{-1}.

The spectrum of its solution in carbon disulphide in the 700–1270 cm^{-1} region was indeed found to lack this band (since no CH_3 group is now present).

Some results concerning deuterated polyethylene are reported in Section 4.138 and the literature dealing with labelled steroids is reviewed in Section 4.136.

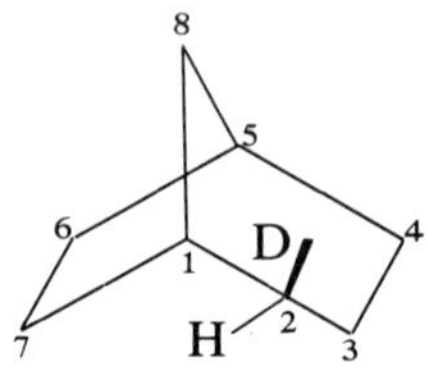

Fig. 4.14 The structure of *exo*(axial)-2-D_1-bicyclo[3.2.1]octane.

Deuterated Unsaturated Hydrocarbons

4.48 Deuterated Ethylene

The isotopically homogeneous ethylene molecules being planar and symmetrical belong to point group V_h and therefore possess 5 infrared active fundamentals out of a total number of 12 normal vibrations. All of these frequencies have been reported in the case of the $D_2C{=}CD_2$ species in the gas phase.[203, 239, 240] Their observed values agree well with those calculated for them on basis of a normal co-ordinate analysis of the fundamental vibrations of the ethylene molecule,[240] which yielded also the exact form of each of these modes of vibration.[241]

The infrared absorption of gaseous monodeuteroethylene,[239–242] $CH_2{=}CD_2$,[240, 241] *cis*-CHD=CHD,[240, 243] *trans*-CHD=CHD,[240, 241] and $CHD{=}CD_2$[239, 240] has also been measured, mainly in order to use the fundamentals of these additional isotopic modifications of ethylene, observed thus, in a calculation of all the planar force constants of the most general potential function of the ethylene molecule. The frequency of all the observed fundamental infrared absorption bands of these isotopic species of ethylene is given in Table 4.37.

The infrared spectrum of a $D_2C{=}CD_2$ complex with inorganic salts, $K[Cl_3Pt(C_2D_4)].H_2O$ (labelled Zeise's salt), was investigated between 33 and 4000 cm^{-1} in order to be able to identify the platinum-olefin stretching vibration and to estimate the strength of this bond.[244] This spectrum shows *all* the 12 fundamentals of C_2D_4 (all of which are now infrared active), its ν_7 and ν_9–ν_{12} modes absorbing here at 811, 2335, 536, 2185 and 1067 cm^{-1}, respectively.

TABLE 4.37 The Fundamental Bands (cm^{-1}) of Deuterated Ethylenes[240]

Band	Type	C_2H_4	C_2D_4	C_2H_3D	CH_2CD_2	*cis*-$(CHD)_2$	*trans*-$(CHD)_2$	C_2HD_3
ν_1	symmetrical C—H(D) stretching	—	—	3002[a]	3019	2300	—	2280
ν_2	C=C stretching	—	—	1605	1585	1567	—	1546
ν_3	$CH_2(D_2)$ symmetrical bending	—	—	1290	1032	1215	—	1047
ν_4	out-of-plane C—H(D) bending	—	—	1001	—	—	988	765
ν_5	C—H(D) stretching	—	—	3062	2335	3055	—	2232
ν_6	in-plane CH(D) bending	—	—	1128	1150	—	—	995
ν_7	out-of-plane C—H(D) bending	949	720 (*vs*)	807	752	842	727	724
ν_8	out-of-plane C—H(D) bending	—	—	943	944	—	—	919
ν_9	C—H(D) stretching	3106	2345 (*vs*)	3103	3095	3058	3065	3045
ν_{10}	C—H(D) bending	810	584	715 ?[b]	—	646	678	610
ν_{11}	C—H(D) stretching	2990	2200 (*m*)	2272	2231	2251	2270	2222
ν_{12}	C—H(D) bending	1444	1078 (*s*)	1403	1384	1342	1300	1290

[a] An average of the two resonating vibrations which have a ν_1 character.
[b] The corresponding Raman band at 730 is believed to be more reliable.[240]

4.49 Substituted Deuterated Ethylenes

The infrared absorption of gaseous deuterotrifluoroethylene, CDF=CF_2, has been measured in the whole 200–4000 cm^{-1} region.[245] The spectrum of a 91 atom % sample of the corresponding liquid chloro-ethylene, in the 600–4000 cm^{-1} range,[246] as well as that of the tribromo-derivative between 300 and 3000 cm^{-1}, have also been reported.[247] The main infrared bands observed for these compounds are collected in Table 4.38.

Table 4.38 The Main Infrared Bands (cm^{-1}) of the CDX=CX_2 Molecules

Band	Species	Type	C_2DF_3	C_2DCl_3	C_2HCl_3	C_2DBr_3
ν_1	A'	C—H(D) stretching	2347 (*s*)	2302	3085	2297 (*s*)
ν_2	A'	C=C stretching	1767 (*s*)	1563	1587[a]	1529 (*w*), 1574 (*w*)[b]
ν_3	A'	C—X stretching	1323 (*vs*)	1013[c]	1245	953 (*s*)
ν_4	A'	C—X stretching	1200 (*vs*)	873[c]	930	775 (*vs*)
ν_5	A'	(D)H—C—X bending	983 (*s*)	797[c]	840	678 (*s*)
ν_6	A'	C—X stretching	855 (*s*)			493 (*m*)
ν_7	A'	C—X rocking	620 (*m*)			
ν_8	A'	CX_2 deformation	480 (*s*)			
ν_9	A'	CX_2 rocking	228 (*w*)			
ν_{10}	A''	out-of-plane CX bending	612 (*s*)			609 (*s*)
ν_{11}	A''	out-of-plane CX_2 bending	528 (*m*)			385 (*m*)
ν_{12}	A''	torsion	291 (*w*)			

[a] In a CCl_4 solution.
[b] These bands are the result of a Fermi resonance between ν_2 and $2\nu_4$.
[c] In a CS_2 solution.

The values given in Table 4.38 are in agreement with those calculated for C_2DF_3,[248] C_2DCl_3[248] and C_2DBr_3[249] on basis of an analysis of their vibrational normal modes.

The infrared absorption of gaseous *cis*- and *trans*-FClC=CDCl[250] as well as of Cl_2C=CDF[251] has been measured down to the end of the far infrared region and the assignment of the 12 fundamentals of each of these isomers has been completed.[251]

The calculated vibration spectrum and potential energy constants of ClDC=CF_2 have also been reported.[248]

The spectrum of monodeuterated dihaloethylenes of the CHX=CDX type was examined in the case of X = F both for the *cis* and for the *trans* form.[252] While, however, the first isomer showed all its 12 fundamentals as absorption bands in the 250–4000 cm^{-1} region examined the *trans* isomer showed only 8 such bands here. The C—D stretching vibration of the *cis* form absorbs at 2364 cm^{-1} and its C—D bending mode appears at 889 cm^{-1}, as compared with the respective 2347 and 938 cm^{-1} values observed for the *trans* modification.

The planar vibrations of the *cis*-CHCl=CDCl derivative were also reported.[253]

The fundamentals of *cis*- and *trans*-CHBr=CDBr have been determined and assigned by studying the absorption of a sample in which the *cis* form predominated and a sample which was richer in the *trans* form.[254] The assignments made thus were checked by applying the sum rule (Section 3.5) to the vibrational data of $C_2H_2Br_2$, $C_2D_2Br_2$ etc. The C—D stretching and C—D bending vibrations of the *cis* isomer absorb at 2302 and 811 cm^{-1}, respectively, while the corresponding *trans* isomer values are 2316 and 827 cm^{-1}.

The spectrum of a 97 atom % sample of CF_2=CHD[255] showed its fundamental bands at 392 (*m*; CF_2 rocking), 544 (*s*; CF_2 bending), 617 (*s*; CF_2 wagging), 770 (*vs*; CHD wagging), 833 (*vs*; CHD rocking), 929 (*vs*; CF_2 stretching), 1224 (*vs*; CHD bending), 1298 (*vs*; CF_2 stretching), 1716 (*vs*; C=C stretching), 2325 (*s*; C—D stretching) and 3133 cm^{-1} (*m*; C—H stretching).

The absorption of the 1-deuterovinyl halides, CH_2=CDX, was measured in the case of X = F in the 400–3200 cm^{-1} region[256] while in the case of X = Cl it was examined[257] only down to 570 cm^{-1}. The labelled fluoride shows (all) its 12 expected fundamental bands at 477, 683, 791, 867, 902, 985, 1165, 1366, 1628, ~2310, ~3120 and ~3150 cm^{-1}. For the deuterovinyl chloride, however, only 10 fundamentals have been observed, these appearing at 588 (CH_2 wagging), 706 (C—Cl stretching), 808 (C—D wagging), 1598 (C=C stretching), 2305 (C—D stretching), 3045 (C—H symmetric stretching) and 3125 cm^{-1} (asymmetric C—H stretching).

The absorption of a mixture of *cis*- and *trans*-CHD=CHF in the 400–3200 cm^{-1} region yielded the values of most of their fundamentals.[256]

The infrared spectrum of the DHC=CHCl sample examined indicates that it probably contained only the *trans* isomer since its absorption is in agreement with the calculated spectrum of this isomer.[257] These calculations were made assuming a Urey–Bradley force field for the planar vibrations of vinyl chloride and yielded satisfactory values for the fundamentals of a couple of variously deuterated vinyl chlorides.[257] Calculated values for the normal frequencies of various other deuterated chloroethylenes were also published.[258, 259]

The infrared spectra of deuterated vinyl bromide have also been reported[260] as well as the absorption of deuterated vinyl chloride in the overtone region.[261]

A sample containing gaseous 93% CD_2=CF_2 showed its CD_2 stretching bands at 2274 (*s*; symmetric) and 2354 cm^{-1} (*w*; asymmetric), its CD_2 rocking absorption at 807 (*vs*) cm^{-1} and its CD_2 wagging fundamentals at 575 (*s*) and 660 (*vs*) cm^{-1}, while its CD_2 scissoring vibration, originally at *ca.* 1060 cm^{-1}, was observed to be in Fermi resonance with the 2 × 520 cm^{-1} overtone, giving rise to two strong bands, at 1021 and 1076 cm^{-1}, respectively.[255]

The corresponding fundamentals observed for the $CD_2{=}CCl_2$ derivative[262] have the respective values of 2230, 2330, 815—(not measured below 670 cm^{-1}) and 980 cm^{-1}.

The calculated vibration spectrum of $CD_2{=}CFCl$ was also reported[248] as were the observed fundamentals of $CD_2{=}CHCl$.[257]

The C—D stretching band of *cis*-CDF=CDF appears at 2320 cm^{-1} while its in-plane C—D bending vibrations absorb at about 847 (symmetrical mode) and 937 cm^{-1} (asymmetrical mode) and its out-of-plane C—D deformation frequency is observed at about[252] 597 cm^{-1}. The corresponding *trans* isomer, which contains a centre of symmetry, shows only 5 fundamentals, at 309 (torsion), 651 (out-of-plane deformation), 941 (C—D bending), 1175 (C—F stretching) and[252] 2315 cm^{-1} (C—D stretching).

The planar vibrations of *cis*-$C_2D_2Cl_2$[253] as well as the spectra of *cis*- and *trans*-$C_2D_2Br_2$[248, 254, 263] and *trans*-$C_2D_2I_2$[264] have all been described in the literature.

The absorption of gaseous *cis*- and *trans*-CHD=CDF shows their C—D stretching bands at about 2285 and 2325 cm^{-1} and their various C—D bending absorptions at 575, 726 and 974 cm^{-1} for the *cis* isomer and at 639, 695 and 969 cm^{-1} for the *trans* derivative.[256]

In the case of CHD=CDCl[257, 265] the *cis* and *trans* isomers could not be obtained separately.[237] Their mixture showed the C—D stretching bands at 2225 and 2285 cm^{-1} while their CD bending modes absorbed at 997, 1045, 1249 and 1269 cm^{-1}.

The spectrum of gaseous perdeuterovinyl fluoride[256] and its chloride analogue[257, 266] was measured in the 400–3200 cm^{-1} range. Table 4.39 gives the frequency of their fundamentals.

4.50 Labelled Propylenes

The infrared spectrum of perdeuterated propylene,[218, 267] C_3D_6, shows[268, 279] its fundamentals at 434 (*s*; out-of-plane C—D bending), 691 (in-plane C—D bending), 712 (*vs*; CD_3 rocking), 734 (*s*; out-of-plane C—D bending), 788 (*vw*; C—C stretching), 872 (*m*; C—D bending), 1022 (*m*; D—C=C bending), 1049 (*vs*; CD_2 scissoring), 1061 (asymmetrical CD_3 deformation), 1588 (*vs*; C=C stretching), 2130 (*m*; symmetrical CD_3 stretching), 2204 (*s*; asymmetrical CD_3 stretching), 2219 (*vs*; another asymmetrical CD_3 stretching), 2224 (*s*; symmetrical CD_2 stretching), 2240 (*s*; CD stretching) and 2331 cm^{-1} (*s*; asymmetrical CD_2 stretching). These frequencies are in satisfactory agreement with the values calculated for these fundamentals on basis of parameters taken from similar compounds.[268]

TABLE 4.39 The Fundamentals (cm^{-1}) of some Isotopically Homogeneous Vinyl Halides[256, 257]

Band[256]	Species	Type	CH_2=CHF	CD_2=CDF	CD_2=CDCl
ν_1	A'	C—X bending	483	~410	
ν_2	A''	twisting	711	563	490
ν_3	A'	C—X stretching	1156	1142	657
ν_4	A''	$CH_2(D_2)$ wagging	863	675	685
ν_5	A''	CH(D) wagging	940?	728	730
ν_6	A'	$CH_2(D_2)$ rocking	929	794	814
ν_7	A'	$CH_2(D_2)$ bending	1306	967	991
ν_8	A'	=C—H(D) bending	1380	1001	1044
ν_9	A'	C=C stretching	1654	1596	1537
ν_{10}	A'	$CH_2(D_2)$ symmetrical stretching	~3080	~2240	2225
ν_{11}	A'	CH(D) stretching	~3115	~2273	2300
ν_{12}	A'	$CH_2(D_2)$ asymmetrical stretching	~3150	~2325	2337

The infrared absorption of the complex compound K $[PtCl_3(C_3D_6)].H_2O$ was also reported in the literature.[269]

The spectra of a 92% sample of *cis*-CH_3CH=CHD and a 93% sample of *trans*-CH_3CH=CHD, as well as similar samples of CH_3CD=CH_2 and CH_3CH=CD_2, were measured in the 700–3200 cm^{-1} region.[270] The *cis* isomer shows its C—D stretching band at 2260 cm^{-1} while the *trans* isomer shows it at 2270 cm^{-1}. The out-of-phase, out-of-plane, C—H bending fundamental, which is practically inactive in infrared absorption in case of the light modification, gives rise to an appreciable band at about 820 cm^{-1} in case of the deuterated *trans* propylene.

The C—D stretching band of the 2-deuteropropylene appears at 2263 cm^{-1} while CH_3—CH=CD_2 shows two such bands, at about 2212 (*m*) and 2310 (*s*) cm^{-1}, respectively.[270]

4.51 DEUTERATED HALOGENOPROPYLENES

The spectra of the two gaseous 1-deutero-2-chloropropene geometrical isomers were measured in the whole 130–3500 cm^{-1} region[271] together with the spectra of other isotopic modifications of 2-chloropropene in order to evaluate its force field. The observed fundamentals were then assigned on basis of normal co-ordinate calculations. It was found that while the isomer in which the deuterium atom is *cis* to the methyl group shows its C—D stretching band at 2260 cm^{-1} the isomer in which this atom is *cis* to the chlorine atom displays it at 2300 cm^{-1}.

The absorption of the corresponding CHD=$CBrCH_3$ isomers was also reported.[272] The largest isotope shift observed in this case was for the torsion of the =CHD group which has a frequency of 553 (*s*) and 567 (*m*) cm^{-1}, respectively, as compared with the 680 cm^{-1} value for the corresponding $=CH_2$ torsion.

The C—D stretching absorption of $CD_2{=}CClCH_3$ appears[271] at 2335 (*w*) cm^{-1} (one band only) while the spectrum of the respective bromide[272] shows it at 2339 (*vw*) cm^{-1}.

The CD_3 stretching bands of $CH_2{=}CClCD_3$[271] are shown mainly at 2233 (*s*) and 2245 (*s*) cm^{-1}, its CD_3 bending bands appear at 1042 (*s*), 1051 (*s*) and 1071 (*w*) cm^{-1}, the CD_3 rocking absorption of this molecule is at about 823 (*s*) cm^{-1} and its torsion mode gives rise to a very weak band at 143 cm^{-1}. The corresponding values for $CH_2{=}CBrCD_3$[272] are 2150 (*m*), 2228 (*s*) and 2240 (*s*) for the C—D stretching bands, 1040 (*s*), 1053 (*s*) and 1071 (*s*) cm^{-1} for the deformation modes, and 818 (*s*) cm^{-1} for the rocking vibration.

The spectrum of both liquid and gaseous *cis*-$CDCl{=}CClCD_3$ was measured in the 400–2400 cm^{-1} region[273] as well as the absorption of the liquid and the gaseous *trans* isomer[273] between 580 and 2450 cm^{-1}. The characteristic C—D absorption bands of the two isomeric $CHD{=}CClCD_3$ species[271] are quite similar to the corresponding $CH_2{=}CClCD_3$ bands, the main difference between them being that most of the C—D bending frequencies of the former are about 20 cm^{-1} lower than the respective $CH_2{=}CClCD_3$ fundamentals. This is also more or less true for the $CHD{=}CBrCD_3$ isomers spectra[272] relative to the absorption of $CH_2{=}CBrCD_3$.

Gaseous perdeuterated 2-chloropropene,[271] C_3D_5Cl, shows five fundamental C—D stretching bands, at 2125 (*m*), 2210 (*w*), 2229 (*s*), 2242 (*s*) and 2375 (*w*) cm^{-1}. Its C—D bending modes absorb at 1022 (*vs*), 1035 (*vs*) and 1042 (*vs*) cm^{-1} while its CD_3 rocking and wagging vibrations give rise to absorptions at 809 (*w*) and 856 (*w*) cm^{-1}, respectively. The $=CD_2$ wagging vibration is shown at 701 (*vs*) cm^{-1}, its torsion mode is observed at 500 cm^{-1} and the CD_3 torsion fundamental is displayed at 142 cm^{-1}. The corresponding frequencies of the analogous bromide[272] are within 14 cm^{-1} of these values.†

4.52 Other Deuterated Olefins

The infrared absorption spectrum of 2-methyl-2-butene partly monodeuterated at the 2-position has been reported in the literature.[274,275]‡

The effect of deuteration on the absorption of 1- and 3-hexenes in the 300–3600 cm^{-1} range has been investigated[276,277] in the case of samples of 93% $CH_2{=}CDC_4H_9$(I), 97% *trans*-$CHD{=}CHC_4H_9$(II), 91% *cis*-$CHD{=}CDC_4H_9$(III), 98% *cis*-$C_2H_5CD{=}CHC_2H_5$(IV) and 88% *cis*- $C_2H_5CD{=}CDC_2H_5$(V) as compared with the spectrum of 1-D_2-n-hexane. A replacement of one protium atom of the $=CH_2$ group by a deuterium atom was

† For the spectrum of the respective iodide see ref. 278.

‡ For the vibrational spectrum of perdeuterated isobutene, see ref. 275.

found to result in the appearance of a band at about 2275 cm^{-1} (II-2271 cm^{-1}; III-2276 cm^{-1}) while a =CD-group was observed to absorb at about 2230 cm^{-1} (I-2227 cm^{-1}; III-2224 cm^{-1}; IV-2234 cm^{-1}; V-2220, 2242 cm^{-1}). The intensity and position of the out-of-plane C—H deformation band in this case are almost as if the protium atom were substituted by an alkyl group. The C=C stretching frequency decreases, however, with the progress of the deuteration near the double bond although it increases with alkylation at these positions.[276] This can be seen from Table 4.40 which gives the effect of deuteration on some fundamentals of a number of isotopic hexenes.[277]

TABLE 4.40 The Effect of Deuteration on Fundamental Frequencies (cm^{-1}) of 1- and 3-Hexenes

1-Hexenes				
Type	$H_2C{=}CHC_4H_9$	$H_2C{=}CDC_4H_9$	$HDC{=}CHC_4H_9$	$HDC{=}CDC_4H_9$
2-C—H(D) bending	1295	1107?	1298	1105?

3-Hexenes			
Type	$H(H_5C_2)C{=}C(H)C_2H_5$	$H(H_5C_2)C{=}C(D)C_2H_5$	$D(H_5C_2)C{=}C(D)C_2H_5$
C=C stretching	1654	1643	1632
C—C stretching	842	830	813
C—H(D) bending	1407	1361	979?
C—H(D) out-of-plane deformation	709	641	603

The reason for the different behaviour of the deuterated olefins is that while the alkyl groups are bonded to the ends of the olefinic bond relatively loosely, hence do not follow its vibration closely, the stretching frequency of the =C—H(D) bonds is very high. This means that during the C=C stretching vibration, which is appreciably lower, the hydrogen atoms to a large extent follow their carbon neighbours bringing about an increase in the effective masses of the vibrating double bond. A deuteration at the C=C bond has therefore the effect of a further increase in the masses of this oscillator with the inevitable result of lowering its stretching frequency.

4.53 LABELLED CYCLOBUTENE†

The infrared spectrum of (over 99·7% pure) perdeuterocyclobutene, C_4D_6, was measured in the 250–3800 cm^{-1} region both in the vapour and in the

† In the case of cyclopropene only the 250–4000 cm^{-1} absorption of the gaseous and solid (at −196°C and down to 600 cm^{-1}) 1,2-D_2-isotopomer seems to have been reported.[281]

liquid state.[280] All the fundamental bands observed thus were interpreted assuming the cyclobutene molecules to possess a C_{2v} symmetry. Table 4.41 reports the frequency of these bands.

TABLE 4.41 The Main Infrared Bands (cm^{-1}) of Cyclobutene and D_6-Cyclobutene

Band[280]	Type	C_4H_6	C_4D_6
ν_1	Symmetrical C—H(D) stretching	3058	2335
ν_2	Symmetrical $CH_2(D_2)$ stretching	2933	2164
ν_3	C=C stretching	1566	1504
ν_4	$CH_2(D_2)$ scissoring	1444	1054
ν_5	C—H(D) in-plane bending	1276	1031
ν_6	Ring expansion	1182	1136
ν_8	Symmetrical $CH_2(D_2)$ out-of-plane bending	986	723
ν_{14}	Antisymmetrical C—H(D) stretching	3126	2261
ν_{15}	Antisymmetrical $CH_2(D_2)$ stretching	2916	2129
ν_{19}	Ring deformation	875	810
ν_{20}	Antisymmetrical $CH_2(D_2)$ out-of-plane bending	850	702
ν_{21}	Another antisymmetrical $CH_2(D_2)$ stretching	2955	2228
ν_{23}	Out-of-plane C—H(D) bending	846	625
ν_{24}	$CH_2(D_2)$ rocking	635	462

4.54 PERDEUTEROCYCLOHEXENE

The absorption of completely deuterated cyclohexene in the vapour and liquid states was measured in the 200–4000 cm^{-1} region[282] in order to use the observed fundamental frequencies of C_6D_{10}, after their vibrational assignment, in the refinement of the force field of cyclohexene. Figure 4.15

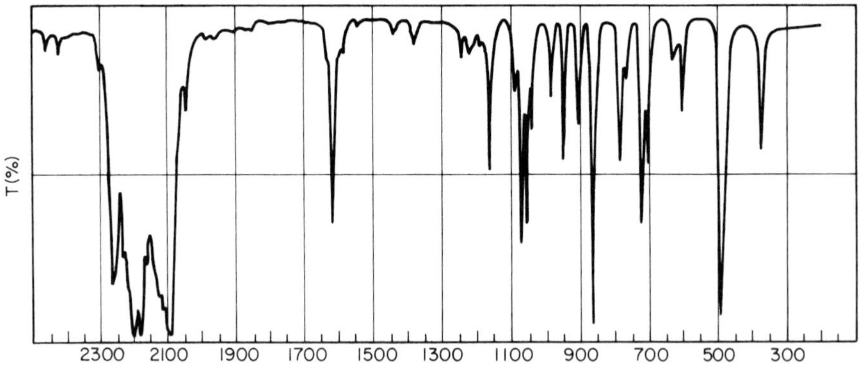

Fig. 4.15 Infrared spectrum of liquid D_{10}-cyclohexene.

shows the infrared spectrum of liquid D_{10}-cyclohexene in the 200 2500 cm^{-1} range.[282]

The infrared absorption bands of the C_6D_{10} vapour appear at 375 (*m*), 497 (*vs*), 608 (*m*), 712 (*m*), 731 (*m*), 795 (*m*), 872 (*vs*), 917 (*w*), 961 (*m*), 1052 (*w*), 1069 (*m*), 1083 (*vs*), 1174 (*m*), 1630 (*m*), 2102 (*vs*), 2110 (*vs*), 2191 (*vs*), 2211 (*vs*) and 2268 (*m*) cm^{-1}.

4.55 Labelled Allenes

The spectrum of gaseous $D_2C{=}C{=}CD_2$, the molecules of which belong to point group V_d[283] and possess 7 infrared active fundamentals, was measured in the 300–5000 cm^{-1} region.[284] All the observed absorption bands were assigned to their respective modes of vibration[284] and the rotational structure of several of these (the $\nu_8 - \nu_{11}$ fundamentals,[283, 284] the $2\nu_8$ overtone[285] and the $\nu_3 + \nu_6$ combination band[286]) was examined under high resolution and analysed to yield the vibrational and rotational constants of C_3D_4.[285, 286]

The absorption of 1,1-D_2-allene vapour between 260 and 6200 cm^{-1} was also investigated,[287–289] its bands being assigned on the assumption that its molecules belong to point group C_{2v}. Since the asymmetry brought about by the isotopic substitution is so slight, they can, however, be treated as if they contained a centre of symmetry as can be judged from the fact that the C=C symmetric stretching vibration, e.g. does not give rise to a clearly observable absorption band in the infrared spectrum.[288]

The rotational structure analysis of a number of fundamentals, such as the asymmetrical C=C stretching band of $CD_2{=}C{=}CH_2$, yielded the values of many of its rotational constants.[288, 289] The frequency of the observed fundamental infrared bands of D_4- and D_2-allene is given in Table 4.42.

TABLE 4.42 The Fundamental Infrared Bands (cm^{-1}) of Isotopic Allenes

Band	Type	C_3H_4	C_3D_4	$C_3H_2D_2$
ν_5	C—H(D) stretching[287]	3005 (*m*)[287]	2229[286]	2226[288]
ν_6	C=C=C stretching (asymmetrical)[287]	1957 (*s*)[287]	1920[286]	1940[289]
ν_7	$CH_2(D_2)$ scissoring[287]	1398 (*s*)[287]	1034[284]	1176[288]
ν_8	C—H(D) stretching (degenerate[a])[287]	3086[283]	2330[283]	2318[288]
ν_9	C—H(D) bending[283]	1033·5(*m*)[283]	836[283]	1022[288]
ν_{10}	C—H(D) out-of-plane bending[283]	841 (*vs*)[283]	667[283]	842[287]
ν_{11}	C=C=C bending[287]	356[283]	307[283]	317[287b]

[a] In case of C_3H_4 and C_3D_4.
[b] $C_3H_2D_2$ also shows bands at 1418 (ν_2), 3011 (ν_1) and 3084 (ν_{12}) cm^{-1}.

The normal vibrations of monodeutero-, 1,1-D_2-, 1,3-D_2- and D_3-allene have all been calculated.[290]

4.56 Deuterated Butadienes

The *trans*-buta-1,3-diene molecules which amount to 97% of the gaseous butadiene at room temperature[291] belong to the $\boldsymbol{C_{2h}}$ point group, hence can be expected to show $8B_u + 4A_u$ infrared active fundamentals. Perdeuterated butadiene, which was investigated as a gas, as a liquid and in the solid state,[292, 293] was indeed observed to show all these fundamentals. For the gaseous phase, however, only the following bands (cm^{-1}) were observed[293]: at 2335 (*vs*; ν_{10}), 2270 (*m*; ν_{11}), 2215 (*vs*; ν_{12}), 1520 (*vs*; ν_{13}), 1042 (*m*; ν_{14}), 1009 (*s*; ν_{15}), 260 (ν_{17}), 718 (*vw*; ν_{22}), ~390 (*m*; ν_{23}) and ~150 (ν_{24}). The ν_{16} and ν_{21} fundamentals were observed in the spectrum of the solid C_4D_6, at about 770 and at 731 cm^{-1}, respectively. All these values are in good agreement with the results of normal co-ordinate calculations, which also yielded the form of the normal vibrations of C_4D_6,[293] as well as the fundamental frequencies of $D_2C{=}CH{-}CH{=}CD_2$, $H_2C{=}CD{-}CD{=}CH_2$ and $H_2C{=}CD{-}CD{=}CD_2$.[291] The infrared absorption bands observed for gaseous, liquid and solid ~90% 2-deuterobuta-1,3-diene were assigned to their respective vibration modes.[294, 295] The liquid diene shows its C—D stretching band at 2235 (*m*) cm^{-1} and its =C—D bending absorption at 1218 (*s*) cm^{-1}.

The spectrum of a mixture of $D_2C{=}CD{-}CH{=}CH_2$ and $CH_2{=}CH{-}C{\equiv}CD$ has also been reported.[295]

4.57 Heavy Cyclodienes

The only case of deuterated cyclopentadienes investigated so far seems to be that of a mixture of 1-, 2- and 5-C_5H_5D[296] which showed its C—D stretching band at 2160 cm^{-1}.

The spectrum of gaseous (and liquid) 95 atom % D_8-cyclohexa-1,4-diene (containing about 25 mole per cent D_7-cyclohexadiene) in the 300(400)–3200 cm^{-1} range was measured in order to provide enough data for the assignment of the cyclohexadiene fundamentals.[297, 298] The assignment of the observed absorption bands was made assuming these molecules to belong to point group $\boldsymbol{D_{2h}}$. The =C—D stretching bands of liquid C_6D_8 appear at 2238 (*vs*) and 2279 cm^{-1} (*vs*) while its CD_2 stretching bands are at 2080 (*vs*) and 2144 cm^{-1} (*vs*), its in-plane CD=CD bending modes absorb at 918 (*s*) and 937 cm^{-1} (*vs*) and the respective out-of-plane vibrations are displayed at 464 cm^{-1} (*vs*). The CD_2 scissoring frequency of this diene is shown at 1058 cm^{-1} (*vs*) and another band which exhibits a high ν_H/ν_D ratio is observed at 740 cm^{-1} (*vs*; $\nu_H/\nu_D = 1{\cdot}28$) being due to the rocking of the CD_2 groups.

4.58 D_8-Cyclo-octatetraene

The absorption spectrum of liquid and gaseous perdeuterated cyclo-octatetraene was measured in the 240–3500 cm^{-1} region in order to assist in the spectroscopic determination of its steric structure.[299]

The interpretation of the infrared (and Raman) spectra of C_8H_8 and C_8D_8 led to the conclusion that this compound is in the "crown" structure (see Fig. 4.16) which has a D_4 symmetry. C_8D_8 must therefore show 14 infrared active fundamentals ($4A_2 + 10E$) out of a total of 32 normal vibrations. Table 4.43 gives the location of these bands in the case of the liquid compound.

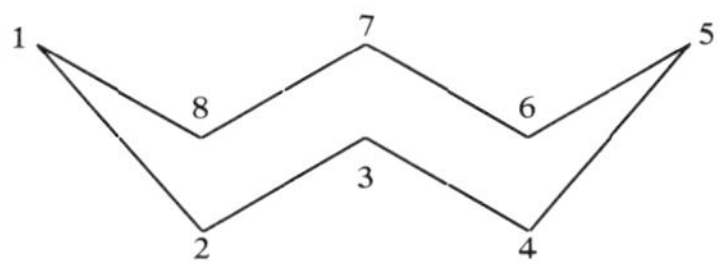

Fig. 4.16 The "crown" structure of 1,3,5,7-cyclo-octatetraene.

TABLE 4.43 The Infrared Fundamental Bands (cm^{-1}) of C_8H_8 and C_8D_8

Species	Band	Type	C_8H_8	C_8D_8
A_2	ν_7	C—H(D) stretching	3004	2236 (*vs*)
A_2	ν_8	C—H(D) perpendicular bending	1221	899 (*s*)
A_2	ν_9	Ring deformation	800	720 (*s*)
A_2	ν_{10}	C—H(D) parallel bending	648[a]	527 (*vs*)
E	ν_{23}	C—H(D) stretching	3004	2252 (*vs*)
E	ν_{24}	C—H(D) stretching	2955	2220 (*s*)
E	ν_{25}	C=C stretching	1635	1614 (*s*)
E	ν_{26}	Ring distortion (out-of-plane)	1399	1286 (*m*)
E	ν_{27}	C—H(D) perpendicular bending	1221	926 (*m*)
E	ν_{28}	C—H(D) perpendicular bending	1202	879 (*s*)
E	ν_{29}	C—H(D) parallel bending	966	(761)[b]
E	ν_{30}	C—C stretching	942	822 (*m*)
E	ν_{31}	C—H(D) parallel bending	366	313 (*m*)
E	ν_{32}	Ring puckering	290	265 (*m*)

[a] Corrected for the effect of Fermi resonance.
[b] Not directly observed.

The experimentally observed ratio between the product of the A_2 species fundamental frequencies of C_8H_8 and that of C_8D_8, evaluated on basis of the D_4 group assignment, was found to be 2.49 as compared with the respective calculated value of 2.53. The corresponding ratios in the case of the E species are 6·56 and 7·11. This agreement strongly supports the conclusion that cyclo-octatetraene really exists in the "crown" structure.

4.59 Labelled Azulenes

The spectra of 1,3-D_2-, 2,4,5,6,7,8-D_6- and D_8-azulene were measured in the 280–4000 cm^{-1} region both for solutions in CCl_4 or CS_2 and for the crystalline state.[300] On basis of these spectra, which also included polarized crystal spectra, as well as on basis of calculations of the out-of-plane normal vibration frequencies and modes of vibrations of D_1-, D_6- and D_8-azulene,[301] it was possible to assign most of the fundamentals of these isotopic species.

Since the agreement of the observed out-of-plane normal vibration frequencies with the values calculated for them on basis of a modified benzene force field is good, it is concluded that the force field of the azulene molecule is not very different from that of benzene.[301]

The main isotope-sensitive bands of the D_2-azulene solutions appear at[300] 500, 636, 893, 2299 and 2308 cm^{-1}. In the case of the D_6-species, these absorptions are observed at ~275, 477, 679, 806, 835, 921, 1033, 1317, 2247 and 2294 cm^{-1}. Perdeuteroazulene shows them at ~272 (ring deformation), 470 (deformation), 612 (deformation), 748 (C—C stretching), 775 (C—D in-plane bending), 798 (combination), 833 (out-of-plane C—D bending), 921 (combination), 960 (in-plane C—D bending), 1115 (in-plane C—D bending), 1134 (in-plane C—D bending), 1390 (combination), 1581 (combination), 2233, 2245, 2256 and 2283 cm^{-1} (all the last four are C—D stretching bands).

4.60 Deuterated and Tritiated Acetylenes

D_2-acetylene molecules which being linear and symmetrical belong to the $D_{\infty h}$ point group can be expected to show only two infrared active fundamentals, ν_3 and ν_5. These bands, together with 37 combination and overtone bands, were indeed observed in the 500–8000 cm^{-1} C_2D_2 absorption spectrum.[302] A number of them (such as the ν_3, ν_5, $\nu_1 + \nu_3$, $\nu_2 + \nu_3$, $\nu_3 + \nu_4$, $\nu_3 - \nu_4$, $\nu_1 - \nu_5$ and the "hot" $\nu_3 + \nu_4 - \nu_4$ and $\nu_3 + 2\nu_5 - 2\nu_5$ bands) were investigated under high resolution, their fine rotational structure being analysed to yield the vibrational and rotational constants of C_2D_2.[302–305]

The photographic infrared spectrum (7500–13,200 cm^{-1}) of D_2-acetylene was observed to show 12 multiple excitation bands[306] (such as $\nu_1 + 2\nu_2 + 3\nu_3$, $5\nu_3$, etc.). Six strong bands in this region were included in a thorough analysis of the rotational structure of the C_2D_2 absorption bands to give precise mean values for its vibrational and rotational constants.[307]

The infrared spectrum of C_2T_2 has also been studied under high resolution to yield values for these constants in the case of this isotopic species.[308] The frequency of the main infrared bands of C_2H_2, C_2D_2 and C_2T_2 is reported in Table 4.44.

TABLE 4.44 The Main Infrared Bands (cm^{-1}) of C_2H_2, C_2D_2 and C_2T_2

Band	Type	C_2H_2 [308a]	C_2D_2	C_2T_2 [308]
ν_3	C—H(D)(T) stretching	3287 (*vs*)^a	2439·22[303]	2072·38
ν_5	C—H(D)(T) bending	729 (*vs*)	538[303]	456
$\nu_1 + \nu_3$	Combination	6500	5097·07[304]	
$\nu_1 - \nu_5$	Difference	2642·5 (*w*)	2167[303]	
$\nu_2 + \nu_3$	Combination		4190·46[304]	
$\nu_2 + \nu_5$	Combination	2701·5 (*m*)		2015·64
$\nu_2 - \nu_5$	Difference			1090
$\nu_3 + \nu_4$	Combination	3897 (*m*)	2944[303]	
$\nu_3 - \nu_4$	Difference	2669 (*w*)	1928[303]	
$\nu_4 + \nu_5$	Combination	1328 (*s*)	1041·48[303]	917
B_0	Rotational constant		0·84770 ± 0·00007[307]	$0{\cdot}6621_5$ (cm^{-1})
D	Rotational constant		7·04 × 10^{-7} (cm^{-1})[307]	

^{a} All values are for vacuum.

An application of the product rule (eqn. 3.17) to the isotopic acetylenic fundamentals gives the expected values of 0·7342 for the $\omega_3(C_2D_2)/\omega_3(C_2H_2)$ and $\omega_5(C_2D_2)/\omega_5(C_2H_2)$ ratios[308a] and 0·565 for the respective $\omega(C_2T_2)/\omega(C_2H_2)$ ratios, which agree with the corresponding 0·741, 0·738 and 0·630, 0·626 values obtained from the various ν_n frequencies reported in Table 4.44.

Since C_2HD does not possess a centre of symmetry all its five fundamentals are infrared active and have indeed been observed as absorption bands[308a] at[309] 3335·62 (in vacuum); (ν_1; C—H stretching; *s*), 1853·78 (ν_2; C≡C stretching; *s*), 2583·60 (ν_3; C—D stretching; *s*), 518 (ν_4; C≡C bending; *s*) and 678 cm^{-1} (ν_5; C—H bending; *s*). Besides these bands an appreciable number of overtone and combination bands in the 1800–9200 cm^{-1} region were also studied under high resolution in order to obtain accurate values for the various vibrational and rotational constants of C_2HD[309–311] and about 8 multiple excitation frequencies were observed in the photographic infrared region[308a] between 9200 and 12,800 cm^{-1}. The more important binary or ternary combinations of this acetylene appear at 1200·30 (*m*; $\nu_4 + \nu_5$),[310] 1338·32 (*m*; $2\nu_5$),[310] 8409 (*m*; $2\nu_1 + \nu_2$)[308a] and 9706 cm^{-1} ($3\nu^1$).

The value of B_0 constant of C_2HD was determined[309] as 0·99156 ± 0·00004 cm^{-1} while its D_0 was found to be equal to 1·17 ± 0·07 × 10^{-6} cm^{-1}.

4.61 LABELLED METHYLACETYLENES

The absorption spectrum of $CD_3C{\equiv}CD$, $CH_3{-}C{\equiv}CD$ and $CD_3C{\equiv}CH$ was measured in the 250–3400 cm^{-1} region, some of the observed bands being examined under high resolution.[312] The fundamentals of $CD_3C{\equiv}CD$

were found at[312b] 2616·45 (ν_1), 2121 (ν_2), 2008·43 (ν_3), 1110·1 (ν_4), 2234·80 (ν_6), 1048·22 (ν_7), 834·43 (ν_8), 497 (ν_9) and 291–5 cm^{-1} (ν_{10}). The fundamentals of $CH_3C{\equiv}CD$ (cm^{-1}) were observed at 2616·3 (ν_1), 2941·7 (ν_2), 2008·3 (ν_3), 1378 (ν_4), 2980·1 (ν_6), 1451·8 (ν_7), 1051·7 (ν_8), 497·9 (ν_9) and 314 (ν_{10}). The analysis of the rotational structure of the ν_1 band yielded the values of 0·2594 and 1×10^{-7} cm^{-1} for the B_0 and D_0 constants of C_3H_3D, respectively, the Coriolis coefficients for the various normal vibrations being determined from the fine structure of the corresponding vibrational–rotational bands.[312c]

The analysis of the high resolution (photographic infrared) spectrum of trideuteromethylacetylene in the 8300–14,000 cm^{-1} region gave the value of 0·24566 cm^{-1} for its B_0 constant.[313] (from the resolution of the $3\nu_1$ and $4\nu_1$ overtone bands). This result confirms the abnormal shortness of the C—C bond (1·467 ± 0·003 Å) in the $CD_3C{\equiv}CH$ molecules.[313] A more recent determination[312b] of this constant, however, has yielded the value of 0·2454 cm^{-1}. Its C—D stretching bands appear at 2121 and 2142 cm^{-1}.

The absorption of gaseous $CF_3C{\equiv}CD$ between 75 and 4000 cm^{-1} shows all the ten fundamentals expected for this compound.[314] Its C—D stretching band appears at 2626 cm^{-1} (for the gas; for the solid at 2617 cm^{-1}), i.e. only about 9 cm^{-1} higher than the respective $CH_3C{\equiv}CD$ band. The C—D bending absorption of gaseous C_3F_3D is shown at 539 cm^{-1} (at 548 cm^{-1} for the solid).

The observed normal frequencies of this compound were used to calculate the potential energy constants of the fluoromethylacetylene molecule force field.[315]

The infrared spectra of gaseous, dissolved and liquid $CD_3C{\equiv}CCl$ in the 280–3300 cm^{-1} range show most of its 10 expected fundamental bands.[316]

These bands appear (for the vapour) at 2142 (ν_1), 2252 (ν_2), 1151 (ν_3), 977 (ν_4) and 547 cm^{-1} (ν_5) in the case of the symmetrical A_1 species and at 2235 (ν_6) and 1071 cm^{-1} (ν_7) in the case of the two E bands observed in the infrared region.

4.62 Higher Deuterated Alkynes

The C—D stretching, C≡C stretching and C—D deformation bands of C_5–C_8 l-deutero-n-alk-l-ynes in carbon tetrachloride solutions appear at about 2593, 1976 and 498 cm^{-1}, respectively.[317] The C—D and C≡C stretching frequencies for the vapour phase of these alkynes are 2610 and 1984 cm^{-1}, correspondingly.[317] The main C—D stretching band is accompanied in each case by one or two weaker ("hot") bands, at an interval of about 10 cm^{-1}, these being due to transitions from low excited deformation levels.[317–319]

When the frequency of such a (C≡)C—D stretching band in solvents which form hydrogen bonds with these groups is compared with the respective free C—D stretching frequency the strength of these bonds can be evaluated. It has thus been observed that such bonds are always weaker than the corresponding C—H····X bonds.[318]

4.63 PERDEUTEROBENZENE

C_6D_6 belongs to point group $\boldsymbol{D_{6h}}$, hence should possess only $1A_{2u} + 3E_{1u}$ infrared active fundamentals. All these frequencies were in fact observed to give rise to intense absorptions in the 300–6000 cm^{-1} vapour spectrum of C_6D_6,[53, 320, 321] besides a number of weaker combination bands. The liquid spectrum of this compound[320] also shows a number of forbidden fundamentals (at *ca.* 579, 664, 869 and 945 cm^{-1}) which are made allowed as a result of the molecular fields existing in the liquid. The frequency of the main infrared bands of normal benzene and C_6D_6 is reported in Table 4.45.

TABLE 4.45 The Main Infrared Absorption Bands (cm^{-1}) of C_6H_6 and C_6D_6[320b, 321]

		C_6H_6		C_6D_6	
Band[321]	Type	Gas	Liquid	Gas	Liquid
ν_{11}	Out-of-plane C—H(D) bending	673	675	496 (*s*)	497
ν_{18}	In-plane C—H(D) bending	1037	1035	814 (*s*)	812
ν_{19}	C—C stretching	1482	1479	1333 (*m*)	1330
ν_{20}	C—H(D) stretching	3064[a]	3053[a]	2288 (*s*)	2276[a]
$\nu_1 + \nu_{11}$	Combination		1669	1450 (*m*)	
$\nu_6 + \nu_{17}$	Combination	1807		1616 (*m*)	

[a] After correction for the effect of Fermi resonance.

An examination of the rotational structure of the 496 cm^{-1} band under a resolution of 0·3 cm^{-1} did not yield satisfactory results because of the low rotational spacing of the C_6D_6 species and the occurrence of many intense "hot" bands in this vicinity.[322]

The calculation of the normal frequencies of D_6-benzene, assuming it to have a Urey–Bradley force field (Section 2.11) modified by including in it the so called "Kekulé constant", resulted in values which agree well with the observed fundamentals.[323] Other calculations of the in-plane[324, 325] and out-of-plane C_6D_6 normal vibrations[326] were also reported.

The good agreement of the observed Teller–Redlich ratios products of the C_6D_6/C_6H_6 active fundamentals with the respective values expected on basis of the D_{6h} symmetry was taken as strong evidence for the correctness of the D_{6h} structure of benzene.[326a]

The 650–5000 cm^{-1} spectrum of the charge-transfer complex $C_6D_6.I_2$[327]

and the absorption of $C_6D_6 . Mo(CO)_3$[328] in the 370–3300 cm^{-1} region have also been measured.

4.64 Other Labelled Benzenes

The infrared spectra of all the 11 possible partially deuterated benzenes were measured both for the vapour and for the liquid state,[329] usually down to 400 cm^{-1}. Figure 4.17 shows the absorption of liquid deuterobenzene in the 400–4000 cm^{-1} region.

All the C—D stretching bands of the liquid partially deuterated benzenes appear at 2276 $\pm$ 13 cm^{-1} as compared with the 2276 cm^{-1} C—D stretching frequency of liquid C_6D_6[321] while its 812 cm^{-1} in-plane C—D deformation band is shown by them between 815 and 877 cm^{-1}, sometimes as a doublet, and its 497 cm^{-1} out-of-plane C—D deformation fundamental is displayed in their spectra, in the 514–608 cm^{-1} range.

Both the planar and the non-planar fundamentals of the various partly deuterated benzenes were calculated satisfactorily from those of C_6H_6 and C_6D_6 by using the so called "complete isotopic rule" (Section 3.6).[321] This rule is based on the relations holding between the secular determinants of the normal frequency symmetry types of the partly labelled compound and those of corresponding types of more symmetrical species and allows an interpolation of the fundamentals of the less symmetrical species in cases where the respective isotopically-homogeneous modification is either linear or planar and highly symmetrical.[330]

The in-plane fundamental frequencies of 1,4-D_2-,[324, 331] 1,3,5-D_3- and 1,2,4,5-D_4-benzene[324] have also been calculated on basis of a normal co-ordinate analysis.

The investigation of the rotational structure of the 531 cm^{-1} absorption band of 1,3,5-D_3-benzene gave the value of 0·1733 $\pm$ 0·001 cm^{-1} for its B_0 constant.[322]

4.65 Deuterated Substituted Benzenes

The C—D stretching vibration of liquid deuteropentafluorobenzene seems to absorb at 2315 cm^{-1} while the 400–3300 cm^{-1} spectrum of gaseous C_6DF_5 appears to show it at[332] 2320 cm^{-1}. In the case of *p*-deuterofluorobenzene[332] and *p*-deuteroiodobenzene[332–334] solutions (in C_2Cl_4), however, this band is displayed nearer to its value in C_6H_5D (2265 cm^{-1}), i.e. at 2273 and 2304 cm^{-1} (a doublet) and at 2283 cm^{-1}, respectively.

The absorption spectra of *o*-, *m*- and *p*-deuterochlorobenzene have also been reported.[333, 334]

The out-of-plane C—H deformation vibration of the *p*-C_6H_4DX compounds (X = F,[332] Br,[335] I[334]) absorbs strongly in the 840–847 cm^{-1} range which is well within the limits of the usual *p*-disubstituted benzene C—H bending region.

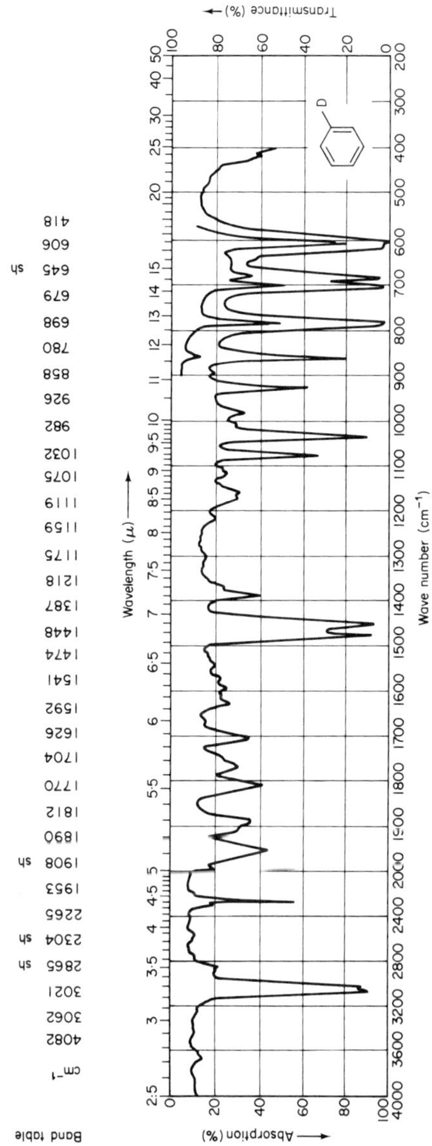

Fig. 4.17 The absorption of deuterobenzene (liquid).

The 100–5000 cm^{-1} spectrum of gaseous 1,2,4,5-tetrachlorodideuterated benzene at various temperatures up to 280°C shows[336, 337] its C—D stretching band at about 2300 (*m*) cm^{-1}, its in-plane C—D bending absorption at 873·5 cm^{-1} and its out-of-plane C—D deformation fundamental at 726·6 cm^{-1}.

The infrared absorption of solid 1,3,5-trichloro-D_3-benzene displays its C—D stretching band[338] at about 2310 cm^{-1}. Its vapour, examined at a number of temperatures, exhibits appreciable deuterium-sensitive fundamentals at[337] 535 (*s*) and 767 (*m*) cm^{-1} (both belonging to an interaction between an out-of-plane C—D bending vibration and a ring deformation fundamental), 841 (*m*; C—D in-plane bending) and 1342 cm^{-1} (*vs*; ring stretching).

The spectrum of *p*-$C_6D_4F_2$ was reported for the whole 130–2400 cm^{-1} region; in the far infrared it differs from that of *p*-$C_6H_4F_2$ only slightly.[339, 340]

The absorption of gaseous *p*-$C_6D_4Cl_2$ was investigated in the 300–2400 cm^{-1} range, that of the solid compound being measured between[341] 100 and 300 cm^{-1}. Its C—D stretching bands appear at 2289 and 2301 cm^{-1} while its in-plane C—D bending modes absorb at 811·7 and 821·8 cm^{-1} and the out-of-plane C—D deformation fundamental is shown at 694·8 cm^{-1}. The observed fundamental frequencies are in agreement with the values calculated for them[342] and obey the product rule.[341]

The spectra of gaseous and liquid C_6D_5X, where X = F,[343], Cl,[344] Br[345] or I,[345] were measured in the whole of the 250–5000 cm^{-1} region and all the observed absorption bands have been assigned to their respective vibration modes assuming these compounds to have the C_{2v} symmetry. Table 4.46 reports their main infrared bands.

4.66 Labelled Toluenes

The infrared absorption of gaseous and liquid perdeuterotoluene and $C_6H_5CD_3$[346] was investigated down to 660 cm^{-1} and resulted in the assignment of almost all of their infrared active fundamentals. The CD_3 symmetrical and non-symmetrical C—D stretching vibrations of $C_6H_5CD_3$ and $C_6D_5CD_3$ are shown at 2130 (*m*) and 2209 (*m*), 2230 (*m*) and 2121 and 2211, 2241 cm^{-1}, respectively, as compared with the corresponding C_7H_8 values of 2921 and 2952, 2979 cm^{-1}. The corresponding frequencies in the case of aliphatic hydrocarbons such as $C_2H_5CD_3$, e.g. are (see Section 4.39) 2075 and 2214, 2225 cm^{-1}, respectively.

As can be expected the aromatic C—D stretching frequencies of C_7D_8 also (see Section 4.64) appear in the near vicinity of 2276 cm^{-1} (2262, 2276, 2293) and this fact makes it easy to distinguish spectroscopically between toluenes deuterated in the aromatic ring and their ω-isomers which usually absorb near 2150 cm^{-1}.[347]

TABLE 4.46 The Main Infrared Absorption Bands (cm^{-1}) of Liquid C_6D_5X Compounds[345]

Species	Approximate Description	X = F	X = Cl	X = Br	X = I
A_1	C—D stretching	2295	2295	2296 (*m*)	2290
A_1	C—D stretching	2270		2270 (*m*)	
A_1	C—C stretching	1578	1563	1551 (*s*)	1550
A_1	C—C stretching	1389	1346	1346 (*vs*)	1330
A_1	C—X stretching	1163	1038	1020 (*vs*)	992
A_1	Ring breathing	959	959	958 (*s*)	952
A_1; B_2	C—D deformations	817	816	817 (*s*)	815
B_2	Ring deformation		667	644 (*s*)	623
B_2	C—D stretching			2280 (*m*)	2273
B_2	C—C stretching	1564	1543	1541 (*vs*)	1535
B_2	C—C stretching	1311	1322	1316 (*s*)	1305
B_2	In-plane C—D deformation	1035	1022	1012 (*vs*)	999
B_2	Out-of-plane C—D deformation	753	747	743 (*m*)	735
B_2	Out-of-plane C—D deformation	627	618	614 (*s*)	607
B_1	Ring deformation	563	546	543 (*s*)	537
B_1	Ring deformation	438	420	403 (*m*)	397

The CD_3 symmetric and asymmetric deformation bands of $C_6H_5CD_3$ appear at 1045 (*m*) and 1050 (*m*) cm^{-1} (C_7D_8: 1041, 1049) while its C—CD_3 rocking vibration is shown at 845 (*m*) cm^{-1} (C_7H_8: 1040).

The symmetric and non-symmetric in-plane C_{ar}—D deformation fundamentals of C_7D_8 are observed at 819, 880 and 869, 972 cm^{-1}, the respective out-of-plane bending absorption being displayed at 785 cm^{-1}.

The exact location of the C—D stretching frequency of monodeuterated toluene depends upon the position of the D atom in the molecule, i.e. whether it is an *o*-, a *m*-, a *p*- or an *ω*-deuterotoluene.[347] A C_2Cl_4 solution of *p*-deuterotoluene shows two such bands, at 2262 and[348] 2288 cm^{-1}. The calculated out-of-plane C—H bending frequencies of *o*-, *m*- and *p*-deuterotoluene (as well as monomeric and polymeric *p*-deuterostyrene) are in good agreement with the respective observed values.[335]

The 400–4000 cm^{-1} spectrum of *p*-deuterobenzyl chloride has also been published.[348]

4.67 OTHER DEUTERATED BENZENE DERIVATIVES

The absorption of gaseous and liquid 1-deutero-2-phenylacetylene and its solutions was measured in the 300–3800 cm^{-1} region[349] in order to facilitate the assignment of the $C_6H_5C{\equiv}CH$ fundamentals. The C—D stretching band of liquid $C_6H_5C{\equiv}CD$ appears at 2585 (*s*) cm^{-1} while its gaseous phase shows it at 2608 cm^{-1} as compared with the value of 2616·3 cm^{-1} observed for gaseous $CH_3C{\equiv}C{-}D$ (Section 4.61).

The 400–3100 cm^{-1} spectra of $(C_6H_5)_2CHD$ and $(C_6H_5)_2CD_2$ (in the 650–1300 cm^{-1} region: in solution in CS_2; above 1300 cm^{-1}: in CCl^4) show their C—D stretching bands at 2156 (*m*) and 2138 (symmetrical stretching; medium) and 2202 cm^{-1} (antisymmetrical stretching; medium), respectively.[350]

The CD_2 compound shows also a third (medium) band in this vicinity, which belongs to an overtone of its CD_2 scissoring vibration absorbing at 1056 cm^{-1}.

Another C—D (out-of-plane) bending fundamental of $(C_6H_5)_2CD_2$ is observed at 920 cm^{-1} while $(C_6H_5)_2$ CHD shows a similar absorption at 957 cm^{-1}.

Deuterotriphenylmethane under the same conditions exhibits its C—D stretching band at 2132 cm^{-1} and its C—D bending fundamental at[350] 989 (*s*) cm^{-1}.

The absorption of perdeuterated triphenylmethyl chloride (and triphenylmethyl fluoroborate) in the 300–4000 cm^{-1} range[351] shows considerable C—D bands at about 616 (*m*), 621 (*s*), 753 (*m*) and 826 (*s*) (out-of-plane bending), 809 (*s*), 839 (*s*), 847 (*m*), 863 (*s*) and 1035 (*m*) (in-plane bending) and 2279 (*m*) cm^{-1} (stretching).

4.68 Labelled Biphenyls

The 250–3500 cm^{-1} spectra of $C_{12}D_{10}$ in the gaseous, solution and vitreous states as well as its absorption in the molten and solid phases,[352, 353] down to 50 cm^{-1}, have all been analysed in order to interpret correctly the vibrational spectra of biphenyl. In solution, $C_{12}D_{10}$ shows considerable bands at: 444 (*m*), 466 (*m*), 551 (*s*), 588 (*s*), 627 (*m*), 656 (*m*), 744 (*s*), 784 (*m*), 816 (*vs*), 826 (*m*), 844 (*s*), 863 (*m*), 952 (*m*), 983 (*m*), 1198 (*m*), 1316 (*m*), 1328 (*m*), 1346 (*vs*), 1376 (*m*), 1441 (*m*), 1568 (*s*), 1603 (*m*), 2278 (*s*), 2283 (*vs*) and 2295 (*m*) cm^{-1}. The absorption frequencies observed agree satisfactorily with the respective calculated values and obey the product rule.[354, 355]

The infrared spectra of 4,4′-D^2, 3,5,3′,5′-D_4- (the parent 3,5-D_2-bromobenzene), 2,6,2′,6′-D_4-, 2,4,6,2′,4′,6′-D_6-, 3,4,5,3′,4′,5′-D_6- and 2,3,5,6,2′,3′,5′,6′-D_8-biphenyl show all their C—D stretching bands in the 2250–2295 cm^{-1} range, i.e. just where the C—D bands of deuterated benzenes (Section 4.64) appear.

The absorption of 2,4,5,2′,4′,5′-D_6-biphenyl in the 700–3000 cm^{-1} region has also been reported.[354]

4.69 Deuterated Naphthalenes

$C_{10}D_8$ which possesses a centre of symmetry and belongs to point group $\boldsymbol{D_{2h}}$ can be expected to show only 20 infrared active fundamentals out of its 48 normal vibrations. Almost all of these bands have indeed been observed in the absorption spectra of $C_{10}D_8$ investigated as a vapour,[356] in solution in CCl_4, CS_2 or Nujol,[357] or in the crystalline state.[356] This absorption was measured down to[358] 60 cm^{-1}.

The naphthalenic C—D stretching frequencies remain the same as in the case of the benzenes, $C_{10}D_8$ itself absorbing at[283] 2232, 2258 (*s*), 2278 (*vs*) and 2299 cm^{-1} (*vs*). Its other infrared active fundamentals were observed[323, 358, 359] at 163, 328, 404, 593, 626, 734, 791, 828, 885, 1050, 1082, 1260, 1290 and 1545 cm^{-1}. These values agree well with the frequencies calculated for these fundamentals by various methods.[323, 360]

The 300–3100 cm^{-1} spectrum of 1,4,5,8-D_4-naphthalene[359] and the 700–3100 cm^{-1} absorption of 2,3,6,7-D_4-naphthalene ("β-$C_{10}H_4D_4$") were also measured and interpreted.[361] The (observed) strongly-shifted fundamentals of "α-$C_{10}H_4D_4$" appear at 407, 929, 1038, 2260 and 2278 cm^{-1}.

4.70 Other Labelled Aromatic Compounds

Perdeuteroanthracene also belongs to point group $\boldsymbol{D_{2h}}$ and can therefore be expected to show only part (28) of its (66) fundamentals as absorption bands in the infrared spectrum. Most of these infrared active frequencies have been observed in the 250–3000 cm^{-1} absorption spectra of $C_{14}D_{10}$ which was investigated as a vapour, in solution (in CCl_4 or CS_2), as an oriented crystal (with polarised radiation), as a polycrystalline film or ground in a potassium bromide pellet.[362, 363] Its main absorption bands for the crystalline state appear at[363] 406, 562, 567, 575, 580, 733, 790, 825 (C—D bending), 894, 920, 1237, 1257, 1380, 1503, 1826, 2275 and 2284 cm^{-1} (the last two: C—D stretchings). The calculated values for the fundamentals of this molecule are in satisfactory agreement with the observed frequencies.[323, 364]

The C—D stretching bands of 1-, 2- and 9-deuteroanthracene and 9,10-D_2-anthracene, in solution in CCl_4 or $CHCl_3$, were all observed in the 2242–2275 cm^{-1} range.[365, 366]

The absorption intensity of a single crystal of perdeuterated phenanthrene in the 200–2500 cm^{-1} range was measured both for light polarized in parallel to one of the crystalline planes and for light polarized perpendicularly to it.[367] The ratio between these intensities for each absorption band was compared with the ratios predicted theoretically for each normal vibration mode using a valence force field derived from the anthracene force constants.

This comparison made possible the assignment of the absorption bands investigated thus.†

The solution spectra of $C_{14}D_{10}$ (in CCl_4 and CS_2) in the 350–4000 cm^{-1} region show their main bands at 559 (*vs*), 609 (*vs*), 649 (*s*), 741 (*m*; all these belonging to out-of-plane C—D deformation modes), 832 (*s*), 839 (*m*; in-plane C—D bending), 845 (*m*), 895 (*s*), 975 (*m*), 2268 (*s*), 2285 (*s*), 2300 (*m*) and 2310 cm^{-1} (*m*; the last four being due to C—D stretching vibrations).

The dichroism of a single crystal of perdeuterated pyrene in the 400–4000 cm^{-1} range was also measured, together with the spectra of an oriented polycrystalline film of $C_{16}D_{10}$, its vapour at 250°C and its solution in various solvents.[368] The comparison of the observed dichroism of crystalline $C_{16}D_{10}$ with the predictions made on basis of its structure made possible the assignment of almost all of its 31 infrared active fundamentals (out of its 72 normal vibrations). The absorption of light polarized parallel to the *ab* plane of a $C_{16}D_{10}$ crystal showed its main bands at 521 (*m*; ring deformation), 571 (*vs*; C—D deformation), 603 (*vs*), 742 (*vs*), 762 (*m*; all three: ring deformations), 803 (*s*), 810 (*vs*), 833 (all three: C—D bendings), 945 (*m*), 1037 (*s*), 1276 (*m*), 1338 (*s*), 1560 (*m*; all five: ring vibrations), 2242 (*m*), 2256 (*m*) and 2278 cm^{-1} (*m*; all three C—D stretchings).

Oxygenated Organic Compounds

4.71 Labelled Methanol

The spectra of CH_3OD, CD_3OH and CD_3OD in the gaseous, liquid vitreous and crystalline states were measured in the 300–4000 cm^{-1} region[369] and all their absorption bands were assigned to their respective vibrational modes. Figure 4.18 reproduces the various CD_3OD spectra observed in this region and Table 4.47 collects the main absorption bands of liquid CH_3OH, CH_3OD, CD_3OH and CD_3OD.

The broadness of the OH(D) stretching bands of liquid methanols is due to the strong hydrogen bonding that exists in the condensed phases of alcohols between neighbouring molecules. In the case of (the monomeric) gaseous methanols the O—H and O—D stretching bands appear near 3680 and 2720 cm^{-1}, respectively.

Dilute CCl_4 solutions of CH_3OD show its monomeric O—D band at about 2698 cm^{-1} while similar benzene solutions show it[370] at 2670 cm^{-1}. Such solutions, if not too dilute, also show the associated CH_3O—D···O(D)CH_3 O—D stretching band at 2596 and 2602 cm^{-1}, respectively. Measuring the

† For the 700–900 cm^{-1} spectrum of 9-deuterophenanthrene (as well as that of α- and β-deuteronaphthalene and 9-deuteroanthracene) see Coppens, G. and Nasielski, J. (1961). *Bull. Soc. chim. belges* **70**, 136. For the spectrum of D_8-biphenylene see Pecile, C. and Lunelli, B. (1968). *J. Chem. Phys.* **48**, 1336.

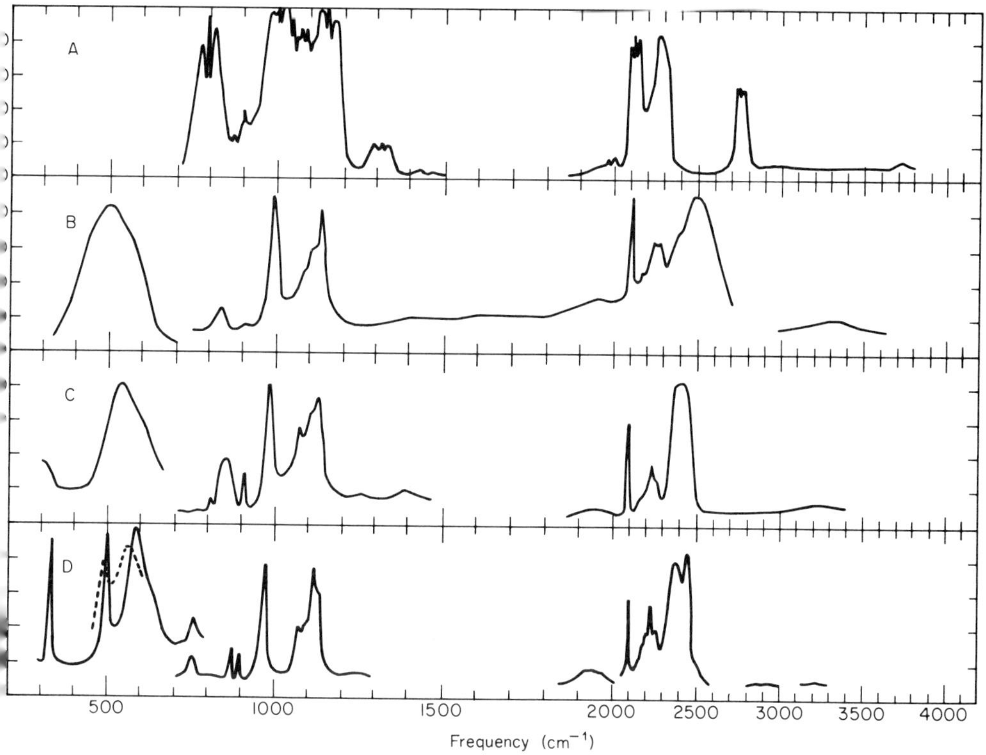

Fig. 4.18 Infrared absorption spectrum of CD_3OD. A—Vapour, 10-cm path-length, 80 mm Hg pressure, room temperature. B—Liquid, room temperature. C—Vitreous solid, $-180°C$. D—Crystalline solid, $-180°C$ (full line) and $-108°C$ (broken line, shown only where the spectrum differs appreciably from that at $-180°C$).[369]

TABLE 4.47 The Main Absorption Bands (cm^{-1}) of CH_3OH, CH_3OD, CD_3OH and CD_3OD[369]

Description	CH_3OH	CH_3OD	CD_3OH	CD_3OD
Out-of-plane O—H(D) bending	655	475 (*s*)	665	483
In-plane O—H(D) bending	~1420	942 (*s*)	1391	818
C—O stretching	1029	1038 (*s*)	982	979
$CH_3(D_3)$ rocking		1235 (*m*)	882	898
$CH_3(D_3)$ bending	1455	1452 (*m*)	1122	1124
$CH_3(D_3)$ asymmetrical bending	1480	1472 (*m*)	1020 (*vw*)	1097
Overtone (2 × 942)	—	1885 (*m*; broad)	—	—
O—H(D) stretching	3337	2485 (*vs*; broad)	3310	2474
$CH_3(D_3)$ symmetrical stretching	2822	2825 (*s*)	2078	2082
$CH_3(D_3)$ asymmetrical stretching	2934	2940 (*vs*)	2232	2225, 2250

solution absorption intensity at all these points as a function of the CH_3OD concentration makes it possible to calculate its association constant in each solvent.[370] The difference between the monomeric and the hydrogen bonded O—D frequencies of CH_3OD, $\Delta\nu$, is used to calculate the energy of this hydrogen bond.[371]

This bonding also affects the associated CH_3OD in-plane O—D bending frequency which in the case of monomeric CH_3OD gives rise to a band at about[372] 865 cm^{-1} as compared with the respective 942 cm^{-1} value for the liquid CH_3OD.

The infrared spectrum of CH_2DOD was also reported.[373]

The observed symmetrical CD_3 deformation frequency of CD_3OH and CD_3OD is in good agreement with the value calculated for it using the expression:

$$\nu_{CD_3} = 317 \times \log \frac{\chi_X}{r^2_{CX}} + 1046 \text{ cm}^{-1} \tag{4.3}$$

where χ_X = the electronegativity of the X atom of the CD_3XH_{m-1} or CD_3XD_{m-1} (m = valence of X) molecule and r_{CX} = the C—X distance.[374]

4.72 Deuterated Ethanols

The absorption of monomeric and polymeric C_2H_5OD, CH_3CHDOH, CH_3CD_2OH, CH_3CHDOD, CD_3CH_2OH, CH_3CD_2OD, CD_3CH_2OD, CD_3CHDOH, C_2D_5OH, CD_3CHDOD and C_2D_5OD in the 350–4000 cm^{-1} region has recently been measured.[375] Liquid C_2H_5OD shows absorption bands at[376] 876, 953 (hydrogen bonded O—D in-plane bending),[372] 1054, 1104, 1386, 1449, 2475 (bonded O—D stretching), 2882, 2928 and 2973 cm^{-1} while monomeric C_2H_5OD shows its in-plane O—D bending absorption at about 875 cm^{-1} and its O—D stretching band at about 2713 cm^{-1}.

Liquid CH_3CD_2OH shows its in-plane O—H deformation band as a broad absorption near[377] 1400 cm^{-1}, which appears for the $C_2H_3D_2OH$ gas at[375] 1280 cm^{-1}, as a result of the dissociation of the $(CH_3CD_2OH)_n$ polymers. In CCl_4 solutions, this band is observed at 1212 cm^{-1}, while in the case of CH_3CD_2OD, it is much weaker.[377]

In the case of gaseous CD_3CH_2OH and other ethanols which lack the —CD_2— group, the in-plane O—H bending vibration is coupled with a non-symmetrical CH_2 or CHD deformation mode and is displaced to the 1220–1260 cm^{-1} region. Since such a coupling is possible only if these two vibrations belong to the same symmetry species, this means that the O—H bending mode is also non-symmetrical, i.e. the hydroxylic hydrogen atom of gaseous ethanols is outside the C—C—O plane.[377] In the liquid phase CD_3CH_2OH shows its main absorption bands at[376] 754, 938, 1036, 1134, 1317, ~1420, 2229, 2885 and 3350 cm^{-1}.

4.73 Labelled Propanols

The 600–4000 cm^{-1} spectrum of gaseous $CH_3CH_2CH_2OD$ was reported[378] to show its (monomeric) O—D deformation band at 841 cm^{-1}. Its other absorption bands, as well as those of liquid C_3H_7OD have also been reported.[379, 380] The monomeric O—D stretching fundamental of gaseous C_3H_7OD was observed at 2726 cm^{-1}.

The spectra of gaseous and liquid $(CH_3)_2CHOD$ and its dilute solutions in non-polar solvents were measured in the 330–3800 cm^{-1} region, the observed absorption bands being assigned to their respective vibration modes.[381] The assignment of their O—D, C—H and CH_3 fundamentals was greatly facilitated by investigating the corresponding spectra of $(CD_3)_2CHOH$ and $(CD_3)_2CHOD$.[381] The monomeric (gaseous) *i*-C_3H_7OD O—D stretching band was observed[382] at about 2702 cm^{-1} while the respective dimeric alcohol band (in an 0·1 molar solution in CS_2) was found at 2595 cm^{-1} and the corresponding absorption due to $(i\text{-}C_3H_7OD)_n$ appeared at 2495 cm^{-1}. The O—D in-plane deformation band of liquid *i*-C_3H_7OD was observed at 983 cm^{-1}, its frequency for the vapour phase being[383] 883 cm^{-1}.

The infrared spectra of gaseous $(CH_3)_2CDOH$ and $(CH_3)_2CDOD$ give evidence for their existence in this phase in an equilibrium between two rotational isomers as compared with only one form in the case of the liquid phase.[384] The main C—D stretching fundamental of $(CH_3)_2CDOH$ was observed at[217b] 2105 cm^{-1}.

Liquid $(CD_3)_2CHOH$ was found to show its absorption bands at[376] 815, 1045, 1075, 1140, 1287, 1339, ~1420, 2230, 2887 and 3350 cm^{-1}, the O—D deformation band of its (liquid) O—D modification appearing at[383] 921 cm^{-1}.

The Nujol (pure liquid paraffin oil) mull of sodium 2-deutero-2-propoxide exhibits its C—D stretching absorption in the vicinity of 1996 cm^{-1} as compared with the corresponding value of 2155 cm^{-1} in the case of $(CH_3)_2CDOH$ itself.[385] This shift is due to the anionic hyperconjugation existing between the α-hydrogen atom and the rest of the alkoxide ion. The β-C—H stretching frequency of *i*-C_3H_6DONa also moves, from 2967 to 2941 cm^{-1}, as a result of this conjugation.

4.74 Deuterated Butanols

Gaseous (and very dilute CCl_4 solutions of) n-C_4H_9OD and $CH_3CHODC_2H_5$ show their monomeric O—D stretching band near the expected, *ca.* 2725 cm^{-1}, frequency [379] while the respective associated O—D band of the liquid phase appears near 2480 cm^{-1}. The in-plane deformation band of this group is observed at about[379] 940 cm^{-1} (885 cm^{-1} for gaseous $CH_3CHODC_2H_5$).[378]

The absorption spectrum of C_3H_7CHDOH in the 670–4000 cm^{-1} region was also reported,[386] as were the C—D stretching bands of $CH_3CDOHC_2H_5$[217b] (*2096*, 2146 cm^{-1}).

The absorption of liquid perdeuterated *t*-butanol[387,388] shows its main bands at 669, 800, 816, 949 (OD bending), 1059, 1084, 1157, 1270, 2062 (*m*), 2080 (*w*), 2131 (*m*), 2140 (*w*), 2230 (*s*) and *ca.* 2500 (OD stretching) cm^{-1}. The values calculated for the normal frequencies of $(CD_3)_3COD$ agree satisfactorily with those observed for its monomer.[389]

The absorption of $(CH_3)_3COD$ in the 200–4000 cm^{-1} region was measured both for the solid and for the liquid phases as well as for solutions in CCl_4 or CS_2.[389,390] Its monomeric O—D stretching band was observed[390] at 2655 cm^{-1}, the respective $((CH_3)_3COD)_n$ band appeared at[389] *ca.* 2500 (*s*) cm^{-1} and the hydrogen bonded O—D in-plane deformation absorption was found at [389] 950 cm^{-1}. The infrared spectrum of a mull of deuterated "potassium *t*-butoxide"[390] (*t*-BuOD + *t*-BuOK) as well as the absorption of a mixture of 25% of $(CH_3)_2COHCH_2D$ and 75% of normal *t*-butanol[391] in the 800–5000 cm^{-1} region have also been recorded.

Liquid $(CD_3)_3COH$ shows its absorption bands at [389] 300, 347, 391, 721, 733, 754, 771, 1052, 1077, 1082, 1102, 1156, 1209, 1367, 2050 (*m*), 2068 (*m*), 2131 (*m*), 2230 (*s*) and *ca.* 3350 cm^{-1}.

4.75 Higher Heavy Alcohols

The absorption of liquid n-$C_5H_{11}OD$, n-$C_6H_{13}OD$, n-$C_8H_{17}OD$ and n-$C_{10}H_{21}OD$ always shows their associated O—D stretching band near 2480 cm^{-1}, the corresponding dilute solution (monomeric) O—D absorption again appearing at about[379,380] 2725 cm^{-1} (a sharp band). Their in-plane (associated) O—D deformation vibration also absorbs as usual,[379,380,392] at about 940 (*w*) cm^{-1}.

The C—D stretching bands of $(C_2H_5)_2CDOH$ and $(CH_3)_2CClCDOHCH_3$ appear at[217b] *2092* and 2146, 2088 and *2133* cm^{-1}, respectively (as compared with *2123* and 2191 cm^{-1} for $CH_3CDOHCH^2Cl$).

The infrared spectra of

$(CH)_3)_2CHCHODCH(CH_3)_2$, $(CH_3)_2CHCDOHCH(CH_3)_2$

and

$(CH_3)_2CDCHOHCD(CH_3)_2$

have been measured in order to assist in the correct assignment of the absorption bands observed for the normal 2,4-dimethyl-3-pentanol.[393]

The absorption of 1-, 2-, 3- and 4-OD-octanols shows that the O—D stretching band of the liquid 2-, 3- and 4-isomers moves to 2490 cm^{-1} while its in-plane bending mode absorbs broadly at 948, 938 and 932 cm^{-1}, respectively. The out-of-plane O—D deformation band of the four deuterated octanols was always found[394] at about 450 (*b*) cm^{-1}.

The O—D stretching frequency of sterically hindered carbinols, such as 2,2,4,4-tetramethyl-pentan-3-ol, 2,2,3,4,4-pentamethylpentan-3-ol and tricyclohexyl carbinol, was observed at about 2680 cm^{-1} for their monomers and at about 2600 cm^{-1} for their polymers.[395]

The main C—D stretching bands of 1-deuterocyclopentanol, 1-deuterocyclohexanol, *cis*- and *trans*-1-deutero-2-chlorocyclopentanol and *cis*- and *trans*-1-deutero-2-chlorocyclohexanol appear at[217b] 2165, 2155, 2174, 2180, 2136 and 2119 cm^{-1}, respectively. A similar effect of the configuration of a compound on its C—D frequency was also observed[217b] for *erythro*- (2167 cm^{-1}) and *threo*-$CH_3CDOHCHCHClCH_3$ (2157 cm^{-1}). The spectra of various deuterated cyclobutanol have also been reported.[364]

4.76 Deuterated Diols

The 300–3500 cm^{-1} spectra of liquid $DO(CH_2)_2OD$[396a,b] and $HO(CD_2)_2OH$[396a] were measured in order to facilitate the interpretation of the ethylene glycol absorption bands. The main infrared bands of these compounds are reported in Table 4.48.

Table 4.48 The Frequency (cm^{-1}) of the Main Ethylene Glycol Bands[396a]

Approximate description	$HO(CH_2)_2OH$	$DO(CH_2)_2OD$	$HO(CD_2)_2OH$
?	523 (*m*)	508 (*w*)	
Out-of-plane O—H(D) bending	~700 (*s*; *b*)	~450 (*m*; *b*)	~650 (*m*; *vb*)
C—C stretching	864 (*m*)	866 (*m*)	740 (*sh*; *s*)
CH_2 rocking	882 (*s*)	840 (*sh*)	
Asymmetric C—O stretching	1041 (*vs*)	1027 (*vs*)	1063 (*s*)[a]
Symmetric C—O stretching	1084 (*vs*)	1077 (*vs*)	1110 (*vs*)
$CH_2(D_2)$ deformation	1203 (*w*)	1214 (*w*)	897 (*m*)
$CH_2(D_2)$ deformation	1332 (*m*)	1333 (*vw*)	967 (*vs*)
$CH_2(D_2)$ deformation		1372 (*m*)	946 (*sh*)
In-plane O—H(D) bending	~1405 (*m*, *vb*)	~971 (*s*; *b*)	~1368 (*m*; *b*)
$CH_2(D_2)$ scissoring	1461 (*m*)	1462 (*m*)	1197 (*s*)
Symmetrical $CH_2(D_2)$ stretching	2877 (*s*)	2882 (*m*)	2108 (*m*)
Asymmetrical $CH_2(D_2)$ stretching	2938 (*s*)		2222 (*m*)
Bonded O—H(D) stretching	3155 (*s*; *b*)	2433 (*s*; *b*)	3150 (*s*)

[a] This assignment seems more plausible than that using the 1076 cm^{-1} shoulder. *w* = weak, *m* = medium, *s* = strong, *v* = very, *sh* = shoulder, *b* = broad.

The 650–3500 cm^{-1} spectrum of 6-deutero-1,6-cyclodecanediol (in a KBr pellet) shows its C—D stretching band[397] at *ca.* 2130 cm^{-1}.

4.77 LABELLED PHENYL CARBINOLS

The absorption of 2-deutero-1,2-diphenylethanol in the 670–3700 cm^{-1} region depends strongly upon its configuration being quite different for the *erythro*-isomer from its form in the case of the *threo*-modification.[398] The C—D stretching band of CCl_4 solutions of the *erythro*-carbinol (and its acetate, benzoate and trimethylbenzoate), e.g. appears at 2160 ± 5 cm^{-1} while the respective *threo*-configuration band is shown at[398] 2175 ± 5 cm^{-1} (as compared with the value of 2230 cm^{-1} for α-deutero-*trans*-stilbene).[398]

The spectrum of a Nujol mull of the perdeuterated-aromatic-ring triphenylcarbinol, $(C_6D_5)_3COH$, shows its fundamentals[351] at 759 (*m*), 766 (*m*), 812 (*w*), 823 (*s*), 833 (*sh*), 838 (*w*) and 863 (*w*; all being due to C—D bending modes), 960 (*m*; ring deformation), 997 (*vs*; C—O stretching), 1046 (*s*; C—D bending), 1111 (*vs*), 1137 (*w*), 1158 (*w*) and 1272 (*m*; all three due to C—C stretching), 1325 (*s*; O—H bending coupled with C—C stretching), 1547 (*vw*) and 1561 (*w*; C—C stretching), 2282 (*m*; C—D stretching) and 3472 cm^{-1} (bonded O—H stretching).

4.78 DEUTERATED PHENOL

The 300–3800 cm^{-1} spectra of C_6H_5OD in the gaseous, liquid and solid phases, as well as the absorption of its solutions in CCl_4, CS_2 or n-hexane, have all been measured and interpreted.[399]

The monomeric O—D stretching fundamental of C_6H_5OD vapour (in a heated cell) gives rise to a triplet band at 2687, 2695 and 2700 cm^{-1} while its solution in carbon tetrachloride shows it at[371] 2668 cm^{-1}. If this solution is not too dilute it also shows the end group O—D stretching band, of its complex $C_6H_5OD \cdots (O(D)C_6H_5)_n$, at *ca.*[370] 2575 cm^{-1} and the respective polymeric O—D absorption at 2490 ± 10 cm^{-1}. Liquid deuterated phenol[399] also shows the last band at 2490 cm^{-1} (*vs*), exhibiting an adjacent end group absorption at 2590 cm^{-1} (*sh*).

The monomeric O—D in-plane deformation absorption of gaseous C_6H_5OD is shown at[399] 910, 915, 918 and 923 cm^{-1} while in the case of its solution it appears at 918 cm^{-1}. Its liquid phase shows the end group O—D deformation band at 932 (*m*) cm^{-1} and the respective polymeric OD absorption at 978 (*s*) cm^{-1}.

The out-of-plane O—D bending fundamental of labelled phenol is observed at *ca.* 480 and *ca.* 520(b) cm^{-1} for the liquid and solid phases, respectively.[399]

The normal vibrations and force field of C_6H_5OD have also been described.[400]

The O—D stretching frequency of deuterated phenol in various "inert" solvents was found to yield almost the same $(\nu_{\text{vapour}}-\nu_{\text{solution}})/\nu_{\text{vapour}}$ ratio, for each solvent, as obtained in the respective case of C_6H_5OH.[401,402]

In the case of electron donating solvents (which form hydrogen bonds with the phenol molecules) this $\Delta\nu/\nu_0$ ratio was observed, however, to be always appreciably larger for C_6H_5OH than for C_6H_5OD,[402] the isotopic effect on the solvent shift varying markedly with its basicity.[402,403]

These changes in the strength of the isotopic effect have been shown to be mainly due to isotope-dependent variations in the anharmonicity of the O—H(D) bond as a result of the hydrogen bonding of this group to the solvent molecules.[402]

The far-infrared spectra of the C_6H_5OD complexes with trimethyl-, triethylamine and pyridine show their D-bond stretching bands at 141, 120 and 130 cm^{-1}, respectively, as compared with the 143, 123 and 134 cm^{-1} values observed for the corresponding normal complexes.[404] These results are consistent with the fact that there is no drastic deuteration effect on the force field of these complexes.†

4.79 Other Aromatic OD Compounds

o-Halophenols usually exist in an equilibrium between the intramolecularly hydrogen bonded *cis* form and the *trans* form (see Fig. 4.19). The O—D

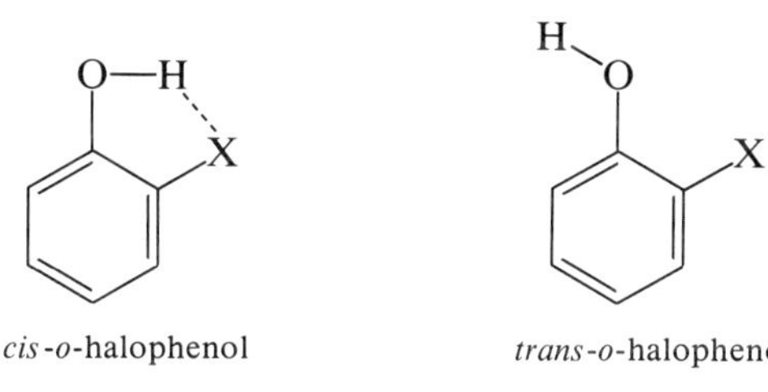

cis-*o*-halophenol   *trans*-*o*-halophenol

Fig. 4.19 The *cis* and *trans* forms of *o*-halophenols. X = F, Cl, Br or I.

stretching frequencies of these forms in the case of gaseous C_6H_4XOD molecules where X = F, Cl, Br and I are given in Table 4.49.

The difference in enthalpy between the *cis* and the *trans* form in each case has been estimated from the measured changes in the relative peak heights of their respective O—D bands following an increase in the temperature of the gaseous samples from their boiling points to 290°C.[405]

The NaCl prism region spectra of (*o*- and) *p*-C_6H_4IOD have also been reported.[406]

† The vapour spectrum of C_6D_5OH has recently been measured down to 225 cm^{-1}.[408]

TABLE 4.49 The O—D Stretching Frequency (cm^{-1}) of Isotopic *o*-X-phenols[405]

Compound	X = H	X = F		X = Cl		X = Br		X = I	
		cis	*trans*	*cis*	*trans*	*cis*	*trans*	*cis*	*trans*
C_6H_4XOH	3653	3635[a]		3583	3645	3562	3646	3538	3647
C_6H_4XOD	2695	2678[a]		2642	2687	2628	2687	2613	2688

[a] Assumed to be for the *cis* form.

The 250–3900 cm^{-1} spectra of liquid (and, in most cases, also solutions of) *p*-$CH_3C_6H_4OD$, *p*-CH_3-2,6-D_2—$C_6H_2D_2OH$ and *p*-CH_3—C_6D_4OH have been studied in order to check the assignment of the *p*-cresol fundamentals.[407] The monomeric and polymeric O—D stretching bands of the C_7H_7OD cresol were observed at 2652 and 2472 cm^{-1}, respectively, while the corresponding in-plane deformation fundamentals appear at 918 (*m*) and 970 (*w*) cm^{-1}, respectively. The C—D stretching vibrations of *p*-CH_3—C_6D_4OH give rise to an absorption band at 2269 (*m*) cm^{-1}.

The non-planar fundamental frequencies of deuterated *o*-, *m*- and *p*-cresol have been calculated.[409]

The —OD torsion bands of 2,6-dimethyl-, *o*-*t*-butyl-, 2-methyl-6-*t*-butyl- and 2,6-di-*t*-butylphenol were observed at 219, 230, 222 and 210 cm^{-1}, respectively, as compared with the values of 295, 303, 302 and 280 cm^{-1} for the corresponding normal phenols.[410] The absorption of crystalline O-deuterated resopcin has also been reported.[411]

The 825–4000 cm^{-1} absorption spectrum of 1-deutero-2-naphthol shows its C—D stretching band at about 2270 cm^{-1} as does also 2,4-D_2-1-naphthol.[412]

4.80 *Gem.*-DIDEUTEROXY-COMPOUNDS

The absorption of mulls of $CCl_3CH(OD)_2$, $CHCl_2CH(OD)_2$, $CCl_3C(OD)_2CCl_3$, $CH_3CHClCCl_2CH(OD)_2$ and ninhydrin deuterate was studied in the 730–1500 and 1900–2500 cm^{-1} regions in order to enable the identification of the in-plane O—H deformation bands of the respective dihydroxy-compounds.[413]

The O—D stretching bands of the chlorinated acetone deuterate appear at 2519 (sharp) and 2577 cm^{-1} while its in-plane O—D deformation fundamentals are found at 930 and 966 cm^{-1}. Ninhydrin deuterate shows these bands at 2410, 2439, 930 and 941 cm^{-1}, respectively, whereas chloral deuterate exhibits them at 2439, 905 and 913 cm^{-1} and α,α',β-trichlorobutyraldehyde deuterate shows the O—D bending absorption at 906 cm^{-1}.

The usual effect of deuteration on the 1000–1400 cm^{-1} region of the *gem.*-dihydroxy-compounds spectra is to increase the number of their absorption bands in this range. This phenomenon is probably due to a decoupling of a number of interacting C—C and C—O vibrations (as a result of the involved changes in the vibrating masses) occurring on deuteration.

4.81 Perdeuterated Ethylene Oxide

Since isotopically homogeneous ethylene oxide molecules belong to point group C_{2v} C_2D_4O can be expected to show $5A_1 + 4B_1 + 3B_2$ infrared active fundamentals. All these bands have been identified in the 375–3800 cm^{-1} absorption spectrum of gaseous perdeuterated ethylene oxide.[414,415] Their frequency is given in Table 4.50.

TABLE 4.50 The Infrared Active Fundamentals (cm^{-1}) of C_2H_4O and C_2D_4O[414]

Species	No.	Description	C_2H_4O	C_2D_4O
A_1	1	C—H(D) stretching	3005	2204
	2	$CH_2(D_2)$ scissoring	1490	1301
	3	Ring breathing	1266	1013
	4	$CH_2(D_2)$ out-of-plane bending	1120	970
	5	Ring deformation	877	755
B_1	9	C—H(D) stretching	3019	2174
	10	$CH_2(D_2)$ scissoring	1470	1145
	11	$CH_2(D_2)$ out-of-plane bending	1153	952
	12	Ring deformation	892	809
B_2	13	C—H(D) stretching	3079	2317
	14	$CH_2(D_2)$ torsion	1143	896
	15	$CH_2(D_2)$ rocking	821	577

4.82 Labelled Dimethyl Ethers

The absorption of perdeuterated dimethyl ether vapour was measured between 280 and 3500 cm^{-1}, the spectrum of its solution in CCl_4 in the 740–3000 cm^{-1} region being also investigated.[416] The main absorption bands of this solution were observed at 824 (*s*; symmetrical C—O—C stretching), 926 (*m*; CD_3 rocking), 1048 (*m*; CD_3 deformation), 1058 (*m*; CD_3 deformation), 1146 (*vs*; asymmetrical C—O—C stretching), 2051, 2182 and 2239 cm^{-1} (all three strong and due to C—D stretching modes).

The symmetrical CD_3 deformation frequency of $(CD_3)_2O$ is given correctly by eqn. 4.3 where χ refers, in this case, to oxygen.[374]

The carbon tetrachloride solution spectrum of CH_3—O—CD_3 in the 830–3020 cm^{-1} range[417,418] shows its fundamentals at 854 (symmetrical C—O—C stretching), 899 (perpendicular CD_3 rocking), 944 (parallel CD_3 rocking), 1057 (asymmetrical CD_3 bending), 1104 (asymmetrical COC stretching; strongest band), 1140 (symmetrical CD_3 deformation), 1209 (parallel CH_3 rocking), 1450 and 1463 (CH_3 deformations), 2054 (symmetrical CD_3 stretching), 2186 (asymmetrical CD_3 stretching), 2244 (another CD_3 stretching), 2814, 2927 and 2987 cm^{-1} (the corresponding three CH_3 stretching bands).

High resolution of the rotational fine structure of the 861 (symmetrical COC stretching), 950 (CD_3 rocking) and 2190 cm^{-1} (asymmetric C—D stretching) absorption bands of gaseous CH_3OCD_3 was also achieved.[417]

The 400–800 and 2000–4000 cm^{-1} spectra of the gaseous, hydrogen bonded, complex formed between $(CH_3)_2O$ and DF were recorded, its absorption bands observed thus being assigned to their respective vibration modes.[419]

4.83 Other Deuterated Saturated Ethers

Fairly complete vibrational assignments for $(C_2D_5)_2O$, $(CH_3CD_2)_2O$ and $(CD_3CH_2)_2O$ have recently been derived.[420]

The 250–600 cm^{-1} spectra of the addition compound of $(CH_3CD_2)_2O$ and aluminium chloride or bromide have also been described, in connection with the interpretation of the respective spectrum of the normal complex.[421]

The absorption of gaseous, liquid and solid perdeuterated, 2,2,5,5-D_4-, 3,3,4,4-D_4- and 2,3,4,5-D_4-tetrahydrofuran in the 350–3200 cm^{-1} region was investigated yielding probable assignments for the tetrahydrofuran absorption bands.[422] Table 4.51 reports the absorption bands observed for liquid C_4D_8O.

TABLE 4.51 The Absorption Bands (cm^{-1}) of C_4H_8O and C_4D_8O

Tetrahydrofuran	D_8-tetrahydrofuran	Probable description
654 (*s*)		CH_2 rocking
908 (*vs*)	842 (*m*)	ring stretching
1030 (*m*; *sh*)	749 (*m*)	$CH_2(D_2)$ rocking
1067 (*vs*)	1051 (*vs*)	asymmetrical ring stretching
1177 (*s*)	1162 (*vs*)	asymmetrical ring stretching
1234 (*m*)		CH_2 wagging
1289 (*m*)		CH_2 wagging
1333 (*w*)		CH_2 wagging
1364 (*m*)		CH_2 wagging
1461 (*s*)	1099 (*vs*)	$CH_2(D_2)$ scissoring
1972 (*m*)	1884 (*w*)	combination
2680 (*m*)	2093 (*vs*)	combination
2861 (*vs*)	2133 (*s*; *sh*)	symmetrical C—D stretching
2977 (*vs*)	2226 (*vs*)	asymmetrical C—D stretching

The 2800–3000 cm^{-1} spectrum of trideuteromethylcyclohexyl ether has also been measured.[423]

4.84 Labelled Aromatic Ethers

The 400–3300 cm^{-1} absorption of 1-deuterofuran vapour shows its fundamentals at[424] 699, 912, 1005, 1089, 1317, 1456, 3128 (A_1) and 805, 1169 and 1230 (B_1) cm^{-1}. In the case of 2-deuterofuran these bands appear at ~720, 886, 1021, 1061, 1295, 1483, 3130 and 829, 1171 and 1242 cm^{-1}. The spectrum of 1,5-D_2-furan was also reported[424] as was the 700–1600 cm^{-1} absorption of D_4-furan and other five of its less deuterated modifications.[425]

The absorption of CD_3-anisole, $C_6H_6OCD_3$, in the 400–4700 cm^{-1} range was investigated in order to enable the isolation of the ring centred fundamentals of anisole.[426] Its deuteration-sensitive normal frequencies for the liquid phase were observed at 955 (parallel CD_3 rocking), 985 (perpendicular CD_3 rocking), 1062 (symmetrical CD_3 deformation), 1081, 1112 (asymmetrical CD_3 deformation), 2069 (symmetrical C—D stretching) and 2250 cm^{-1} (asymmetrical C—D stretching).

The spectrum of a carbon tetrachloride solution of triphenylmethyl D_3-methyl ether[427] shows these bands at 904, 926, 934, 941, 1005, 2069, and 2229/2238 cm^{-1}, respectively.

4.85 Other Deuterated C—O—C Compounds

The 280–3000 cm^{-1} absorption of gaseous and liquid D_2-, D_6- and D_8-dimethoxy-methane, as well as that of their solutions in carbon tetrachloride, was found to be due to one rotational isomer only ("*gauche gauche*"), this possessing a C_2 symmetry.[428] The main absorption bands observed for a CCl_4 solution of $C_3D_8O_2$ below 2000 cm^{-1} appear at 295 (*s*), 370 (*s*), 514 (*m*), 556 (*s*), 829 (*s*), 1068 (*s*), 1093 (*s*) and 1145 (*s*) cm^{-1}. Its gaseous phase C—D stretching fundamentals were found at 2069 and 2076 (*s*; symmetrical) and 2209, 2218, 2241 and 2252 cm^{-1} (asymmetrical).

The 280–3000 cm^{-1} spectra of gaseous and liquid $CD_2(OC_2H_5)_2$ and its solutions in CCl_4 and CS_2 have also been interpreted assuming it to exist at room temperature in a single (rotational isomeric) form only.[429]

The main absorption bands of liquid D_2-diethoxymethane appear at 300 (*m*), 348 (*s*; symmetric COCOC deformation), 411 (*m*), 443 (*s*; symmetric COCOC deformation), 617 (*s*; asymmetric COCOC deformation), 812 (*s*; symmetric COCOC stretching), 969 (*m*; asymmetric C—C stretching), 987 (*s*; asymmetric COCOC stretching), 1072 (*s*; symmetric COCOC stretching), 1136 (*s*; symmetric COCOC stretching), 1391 (*m*), 1443 (*m*), 1481 (*m*; CH_2 scissoring), 2080 (*s*; symmetric C—D stretching), 2171 (*m*), 2214 (*s*; asymmetrical C—D stretching), 2870 (*s*), 2900 (*m*), 2928 (*m*) and 2975 cm^{-1}.

The 400–4000 cm^{-1} spectrum of liquid O-deuterated polyethylene glycols $DO(—CH_2CH_2O—)_nD$, where $n = 2$–7, was measured in connection with the interpretation of the vibrational spectra of high polymers.[396b] The O—D stretching band of these glycols was observed at about 2500 (*s*) cm^{-1} while their in-plane O—D bending modes appeared at about 980 (*m*) cm^{-1}, i.e. their O—D bands are shown almost at the same frequencies as those of O-deuterated ethylene glycol itself.[396]

Gaseous perdeuterated trioxan (which was also investigated in the liquid and solid phases as well as in a CCl_4 solution) was found to show its main absorption bands in the 300–4000 cm^{-1} range[430] at 721 (*m*), 818 (*s*; CD_2 rocking), 887 (*s*), 1065 (*s*), 1089 (*m*), 2083 (*s*; symmetric CD_2 stretching) and 2280 cm^{-1} (*m*; asymmetric CD_2 stretching).

4.86 Heavy Formaldehyde

The isotopically homogeneous formaldehyde molecules belong to point group C_{2v} and can be expected to show all their six normal vibrations as infrared fundamentals. All these bands have indeed been observed in the infrared absorption spectrum of $D_2C{=}O$.[431] The 400–4000 cm^{-1} absorption of gaseous $DHC{=}O$ has also been investigated and has yielded the normal frequencies of this isotopic species.[432] Table 4.52 gives the values found for these frequencies.

TABLE 4.52 The Main Infrared Absorption Bands[a] (cm^{-1}) of Isotopic Formaldehydes

Band[b]	Description	H_2CO[b]	D_2CO[431]	DHCO[432]
ν_1	In-phase $CH_2(D_2)(HD)$ stretching	2780 (*s*)	2055·8	2844·1
ν_2	C=O stretching	1743·6 (*vs*)	1700	1723·4
ν_3	$CH_2(D_2)(HD)$ scissoring	1503 (*s*)	1105·7	1400·0
ν_4	Out-of-phase $CH_2(D_2)(HD)$ stretching	2874 (*vs*)	2159·7	2120·7
ν_5	$CH_2(D_2)(HD)$ rocking	1280 (*s*)	990	1041
ν_6	Out-of-plane C—H(D) bending	1167 (*s*)	938	1074
$2\nu_3$	Overtone	2973 (*s*)	2208	

[a] In vacuum. [b] According to ref. 53.

The DHCO ν_1 band fine structure was resolved, its analysis yielding information on the rotational constants of this molecule.[432]

The isotope effect observed for the D_2CO frequencies has been used to determine all the harmonic force constants of the formaldehyde molecule[433] and the form of its normal vibrations, both for the normal and for the deuterated species.[434]

4.87 Other Deuterated Acyclic Aldehydes

1-Deuteroacetaldehyde,[435] 1-deuteropropionaldehyde[436] CF_3CDO[437] and 1-deutero-n-butyraldehyde[437, 438] all show their aldehyde C—D stretching band at about 2060 (*s*) cm^{-1}. Their in-plane C—D bending vibration absorbs at about 1092 (*s*) and 1095 cm^{-1}, in the case of propionaldehyde vapour,[436] and a CCl_4 solution of n-butyraldehyde,[437, 438] respectively. Other labelled propionaldehyde bands, which were observed to shift as a result of deuteration, appear at[436] 794 (*s*; shifted by 54 cm^{-1}; CH_3 rocking), 856 (*s*; shifted by 39 cm^{-1}; also CH_3 rocking), and 1738 cm^{-1} (*vs*; shifted by 16 cm^{-1}; C=O stretching).

The 400–3300 cm^{-1} absorption of gaseous D_2-glyoxal was observed to show only three fundamentals,[439] at 1008 (C—D bending), 1712 (C=O stretching) and 2141 cm^{-1} (C—D stretching). This fact proves the presence of a centre of symmetry, which brings about a mutual exclusion of the Raman and infrared active fundamentals, in the glyoxal molecules.

The 400–5000 cm^{-1} spectrum of 2-deutero-malonaldehyde anion (in a mull of $NaDC(CHO)_2$) and deuterobromomalonaldehyde, which exists as DO—CH=CBrCHO, was investigated in connection with the assignment of the vibration modes involving the labile protons of the respective normal compounds.[440] The O—D stretching vibration of a mull of this bromoaldehyde was observed at 2220 cm^{-1} (very broad) while its O—D deformation fundamental was found at 1208 cm^{-1}.

The calculated fundamental frequencies of deuterated acetaldehyde were also reported.[441]

4.88 Labelled Benzaldehydes

The 670–3200 cm^{-1} spectrum of liquid C_6H_5CDO and its CCl_4 and CS_2 solutions[438, 442] was investigated in order to elucidate the origin of the characteristic aldehydic C—H bands. This spectrum shows its C—D stretching band at about 2082 (*s*) cm^{-1} near another band at about 2035 cm^{-1}, which is probably due to some combination.[443] Other deuterium-sensitive bands were observed at 733 (*s*), 790 (*s*), 939 (*m*), 1215 (*s*) and 1232 (*s*) cm^{-1}, as compared with the corresponding values of 753, 830, 1008, 1170 and 1289 cm^{-1}, found for the normal benzaldehyde.[438]

The infrared absorption of CCl_4 solutions of *p*-Cl-benzaldehyde and piperonal, both deuterated in the formylic group, was also reported.[438] *p*-Cl—C_6H_4CDO shows its C—D stretching band at 2095 (*s*) cm^{-1} near another band at 2045 cm^{-1} while labelled piperonal absorbs at 2070 and 2025 cm^{-1}.

4.89 Deuterated Aliphatic Ketones

The 250–4000 cm^{-1} spectra of gaseous and liquid perdeuterated and $\alpha\alpha\alpha$-D_3-acetone, as well as their absorption in CCl_4, CS_2 and *trans*-$C_2H_2Cl_2$ solutions and an argon matrix (at liquid helium temperature), have all been measured in order to effect the correct assignment of the vibrational spectra of acetone.[444, 445] Figure 4.20 reproduces the spectrum of gaseous C_3D_6O.

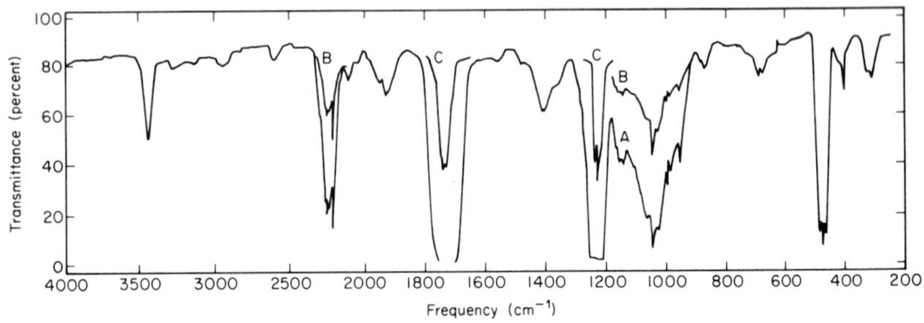

Fig. 4.20 Gas-phase infrared spectrum of D_6-acetone recorded on a Perkin–Elmer model 521 spectrometer. Path length 9 cm; pressures A: 150, B: 30; C: 5 torr.

The frequency of the main infrared bands observed for these compounds is given in Table 4.53.

The C—D stretching bands of CH_2DCOCH_3 appear at [217b] *2183* and 2217 cm^{-1}.

The infrared absorption of CCl_4 and CS_2 (below 1300 cm^{-1}) solutions of perdeuterated diethyl ketone, $(CH_3CD_2)_2CO$, $(CD_3CH_2)_2CO$, $CD_3CH_2COC_2H_5$, $C_2D_5COC_2H_5$ and $CH_3CD_2COC_2H_5$ has been investigated in order to isolate the various CH_3 and CH_2 bands of normal diethyl ketone.[446] It has thus been found that the CD_3 group gives rise to bands at 2078, 2129 and 2233 (*s*; asymmetric C—D stretching) while the CD_2 groups bring about an absorption at 2173 cm^{-1}. The C=O stretching absorption of $(C_2D_5)_2CO$ was observed at 1714 cm^{-1} as compared with the 1720 cm^{-1} value in the case of $(C_2H_5)_2CO$.

The C—D stretching of 2,4-D_2-2,4-dimethyl-3-pentanone showed its C—D stretching bands at 2140 and 2175 cm^{-1} and its C—D deformation frequency at[447] 1052 cm^{-1}.

$(CH_3)_2CDCOCH(CH_3)_2$ exhibits its C—D stretching bands at [217b] 2141 and 2178 cm^{-1}.

TABLE 4.53 The Main Infrared Bands (cm^{-1}) Observed for Liquid $(CH_3)_2CO$, $(CD_3)_2CO$ and CH_3COCD_3[444]

Band[a]	Approximate Description	$(CH_3)_2CO$	$(CD_3)_2CO$	CH_3COCD_3
ν_1	C—H stretching	—	—	3004 (*s*)
ν_2	C—H(D) stretching	2924 (*s*)	2111 (*w*)	2925 (*s*)
ν_3	C=O stretching	~1710 (*vs*)	~1700 (*vs*)	~1705 (*vs*)
ν_4	Asymmetric $CH_3(D_3)$ bending	~1430 (*s*)	1035 (*s*)	~1420 (*s*)
ν_5	Symmetric $CH_3(D_3)$ bending	—	1085 (*m*)	~1350 (*s*)
ν_6	$CH_3(D_3)$ rocking	—	890 (*m*)	—
ν_{13}	C—H(D) stretching	3004 (*s*)	2255 (*s*)	2255 (*m*)
ν_{15}	Asymmetric $CH_3(D_3)$ deformation	~1410 (*s*)	—	1006 (*s*)
ν_{16}	Symmetric $CH_3(D_3)$ deformation	1361 (*vs*)	1005 (*m*)[445]	1029 (*s*)
ν_{17}	C—C stretching	1220 (*vs*)	1244 (*vs*)	~1225 (*vs*)
ν_{18}	$CH_3(D_3)$ rocking	902 (*m*)	—	779 (*w*)
ν_{20}	C—H(D) stretching	2964 (*s*)	2222 (*s*)	2220 (*m*)
ν_{22}	$CH_3(D_3)$ rocking	1092 (*m*)	960 (*m*)[445]	—
$2\nu_3$	Overtone	3414 (*m*)	3418 (*m*)	3406

[a] The notation given refers to the case of $(CH_3)_2CO$.

4.90 HEAVY CYCLIC KETONES

The 250–3000 cm^{-1} spectra of liquid and gaseous $\alpha,\alpha,\alpha',\alpha'$-$D_4$-cyclobutanone were used in the determination of the force constants of the cyclobutanone molecules.[448a] The main bands observed for gaseous $C_4H_2D_4O$ appeared at 627, 673 (CH_2 rocking), 752, 770, 852, 862, 1024 (CD_2 torsion), 1094 (CD_2 scissoring), 1237 (CD_2 out-of-phase scissoring), 1252 (out-of-plane C—D deformation), 1464, 1808 (C=O stretching), 2200, 2247 (both, C—D stretching), 2975 and 3000 cm^{-1}.

The 30–120 cm^{-1} absorption of this ketone exhibits a series of sharp (Q branches of the) $0 \rightarrow 1, 1 \rightarrow 2, \ldots, 7 \rightarrow 8$ ring puckering bands, the first of which appearing at *ca.* 30 cm^{-1} and the last at[449] 78·9 cm^{-1}. The values of these frequencies suggest that the cyclobutanone molecules are planar.[449]

The infrared spectra of 2-chloro- and 2-bromo-2,4,4-D_3-cyclobutone have been reported.[448b,c]

2-Deuterocyclopentanone was observed to show its C—D stretching band at 2197 cm^{-1} while 2-deuterocyclohexanone absorbs at[217b] 2159 and *2202* cm^{-1}.

2,6-D_2-cyclohexanone (in a CCl_4 solution) shows a strong C=O stretching band at 1760 cm^{-1} besides a very weak band at 1727 cm^{-1}. The former is due to its diequatorial conformation while the weaker band belongs to its diaxial modification which usually exists to a much lower extent.[450] The

C=O stretching bands of CCl_4 solutions of α,α'-D_4-cyclohexanone and its 4,4'-dimethyl and 4-*t*-butyl derivatives are exhibited at 1714, 1715 and 1719 cm^{-1}, respectively, while α-chloro-α,α',α'-D_3-cyclohexanone in CCl_4 solutions absorbs mainly at 1727 cm^{-1}, its bromo-analogue at 1718 cm^{-1} and α,α-dichloro-α',α'-D_2-cyclohexanone displays a band at 1746 cm^{-1}. A further number of deuterated halogenocyclohexanones have also been investigated in this region in various solvents.[451]

The 650–3500 cm^{-1} spectrum of 6-deutero-6-methylcyclodecanone[397] shows the C—D stretching band of its KBr pellet at 2130 cm^{-1}.

The absorption of 2,3-D_2-1,5-dimethylbicyclo(3,3,1)nonan-9-one was investigated in connection with the conformation of bicyclo(3,3,1)nonane.[452]

The C—D stretching bands of 3-deutero-camphor, 2-deutero-indanone and 2-deutero-tetralone were observed at[217b] 2193, 2173 and 2202 cm^{-1}, respectively.

4.91 Deuterated Olefinic Ketones and Similar Compounds

The 380–3200 cm^{-1} absorption of liquid perdeuterated methyl-β-chlorovinyl ketone, $CH_3COCD{=}CDCl$ and $CD_3COCH{=}CHCl$, as well as the 1500–1750 cm^{-1} spectra of their solutions in acetonitrile and $CCl_2{=}CCl_2$, were all investigated in connection with the interpretation of the vibrational spectra of methyl β-chlorovinyl ketone.[453,454]

These compounds exist in an equilibrium between the "*s-cis*" and the "*quasi s-trans*" rotational isomers, a fact which is responsible for the splitting observed in some of their absorption bands, e.g. the C=O stretching band of C_4D_5OCl which appears at 1668 (*vs*) and 1690 cm^{-1} (*vs*; *s-cis*). Its C—D stretching frequencies are shown at 2270 (*w*) and 2310 (*w*) cm^{-1} while its symmetrical and asymmetrical CD_3 bending modes absorb at 1036 (*m*) and 1070 (*w*) cm^{-1}, respectively. The in-plane =C—D bending vibration of this ketone is displayed at 956 (*m*) cm^{-1} and the respective out-of-plane fundamental is observed at 705 (*s*) cm^{-1}.

D_2-cyclopropenone (in a CCl_4 solution) shows its two double bond region bands at 1628 and 1842 cm^{-1} as compared with the 1633 and 1861 cm^{-1} values found for the corresponding bands of C_3H_2O.[455,456] The low isotopic shifts observed in this case are the result of the high rigidity of this ring which makes the higher vibration a deformation of the whole ring and the lower one its "breathing" vibration, rather than the localized C=O and C=C stretching modes, respectively.

The absorption of 4-deutero-2,2-dichloro-3-phenylcyclobutenone and 2-deutero-2,4-dichloro-3-phenylcyclobutenone in the 2940–3120 cm^{-1} region has also been reported.[457]

The fundamental frequencies of deuterated propiolaldehyde were calculated on basis of a normal co-ordinate analysis.[441]

4.92 Labelled Diketones

The infrared spectra of liquid and gaseous perdeuterated biacetyl as well as the absorption of its solutions in CCl_4, CS_2 and hexane have been investigated[458] down to 300 cm^{-1}. Absorption bands (for solutions) were observed at 314 (*w*), 480 (*m*), 770 (*w*), 975 (*s*; CD_3 rocking), 1025 (*m*; symmetrical CD_3 bending), 1133 (*s*; asymmetrical CD_3 bending), 1715 (*vs*), 2110, 2194, 2225, 2264 (all four: weak C—D stretching bands).

The absorption of a Nujol mull of D_8-cyclohexane-1,4-dione in the 660–5000 cm^{-1} region was examined in connection with the problem of its conformation. It was found to show bands at 730, 755, 773, 843, 855, 872, 890, 903, 952, 985, 1043, 1061, 1121, 1166, 1203, 1261, 1294, 1719 and 2179 cm^{-1} (C—D stretching), all being due to its single, "twist", conformer of C_2 symmetry[459] (Fig. 4.21).

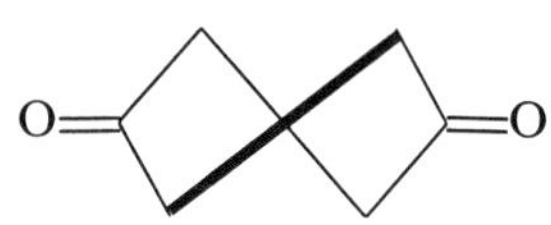

Fig. 4.21 The "twist" conformation of cyclohexane-1,4-dione.

4.93 Heavy Ketenes

The 250–4000 cm^{-1} spectra of gaseous and solid $D_2C{=}C{=}O$ and (mixtures of $H_2C{=}C{=}O$, $D_2C{=}C{=}O$) $DHC{=}C{=}O$, as well as their absorption in a matrix, made possible the interpretation of the infrared spectra of ketene and the evaluation of its force constants.[460] The analysis of the ν_5, ν_6 and ν_9 vibration–rotation bands of $CD_2{=}C{=}O$ and the ν_9 band of $DHC{=}C{=}O$ yielded their rotational constants. Table 4.54 gives the frequency of the isotopic ketene fundamentals.

Table 4.54 Fundamentals (cm^{-1}) of Gaseous Isotopic Ketenes[460]

Band	Species[a]	CH_2CO	CD_2CO	CHDCO
ν_1	A_1	3070	2267	3115
ν_2	A_1	2152	2120	2142
ν_3	A_1	1388	1228	1292
ν_4	A_1	1118	927	~1085
ν_5	A_1	3166	2383	2320
ν_6	B_1	978	855	867
ν_7	B_1	438[b]	371[b]	398[b]
ν_8	B_1	588 (591)[b]	548[b]	555[b]
ν_9	B_2	528	418	496

[a] In the case of CH_2CO and CD_2CO.
[b] Observed in a noble gas matrix.

The 250–4000 cm^{-1} spectrum of the perdeuterated dimer of ketene, D_4-diketene—which exists as a β-lactone (see Fig. 4.22)—shows the following

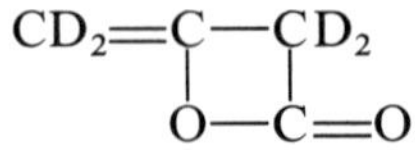

Fig. 4.22 The structure of D_4-diketene.

bands for its vapour phase:[461] 442 (*w*), 664 (*s*; γ-CD_2 wagging), 817 (*s*), 895 (*w*; α-CD_2 wagging), 908 (*w*; α-CD_2 twisting), 944 (*vs*), 1004 (*s*), 1032 (*w*; CD_2 deformation), 1048 (γ-CD_2 twisting), 1220 (*s*), 1678 (*vs*), 1910 (*vs*), 2210 (*vw*; α-CD_2 symmetric stretching) and 2257 cm^{-1} (*vw*; γ-CD_2 symmetric stretching).

4.94 Deuterated Quinones

The 650–4000 cm^{-1} spectrum of gaseous perdeuterated *p*-benzoquinone as well as the absorption of its (CCl_4, CS_2, CH_3CN and cyclohexane) solutions in the 75–4000 cm^{-1} region have both been investigated in order to arrive at the correct interpretation of the vibrational spectra of *p*-benzoquinone.[462]

These solutions show fundamental bands at about 106 (*m*), 402 (*s*), 418 (*w*), 635 (*vw*), 793 (*vs*), 873 (*m*), 1160 (*m*), 1226 (*m*), 1258 (*s*; in Fermi resonance with the former frequency) 1560 (*m*), 1663 (*vs*: in resonance with the following), 1690 (*vs*) and 2277 cm^{-1} (*w*; C—D stretching).

Monodeuterated *p*-benzoquinone, its 2,5-D_2- and its 2,6-D_2-derivative have also been examined under the same conditions.[463] The C—D stretching absorption of $C_6H_3DO_2$ appears at 2290 (*vw*) cm^{-1} while in the case of the 2,5- and 2,6-D_2-modifications it is shown at 2286 and 2275 cm^{-1}, respectively.

The observed fundamentals of these quinones are in satisfactory agreement with the values calculated for them.[464]

The 33–4000 cm^{-1} spectra of solutions and oriented crystals of 99·4 atom % perdeuterated 9,10-anthraquinone were reported.[465, 466] The symmetry species of its observed absorption bands was determined (on the basis of the dichroic relations obtained for them from the polarized light spectra of its oriented crystals) and their vibrational assignment was proposed.[465]

The spectra of $DSbCl_6$ salts of various perimethoxyanthraquinones have also been measured.[467]

4.95 2-Deuteroxytropone

Deuterated tropolone, $C_7H_5DO_2$, in solution in CCl_4 shows its intramolecularly hydrogen bonded O—D stretching bands at[468] 2342 and 2400

cm^{-1} (as compared with the corresponding O—H stretching frequency of 3150 cm^{-1}) and the 1267 and 1490 cm^{-1} bands of normal tropolone, which probably belong to its O—H deformation vibrations, shifted to 932 and 1095 cm^{-1}, respectively.

4.96 Labelled Carbohydrates

The infrared absorption of C_1-deuterated α- and β-D-glucose, α-and β-D-mannose, α- and β-D-galactose, β-D-arabinose, α-D-xylose, the penta-acetates of these glucoses, their methyl glucosides and methyl β-D-glucoside tetraacetate was investigated.[469] It was found that while the observed equatorial C_1—D stretching bands appeared between 2180 (for α-D-glucose and β-D-arabinose; *m*) and 2245 cm^{-1} (for the α-D-glucose acetate; *m*) the axial C_1—D frequencies were always 40–140 cm^{-1} lower than the corresponding equatorial C—D values.

The deformation bands of such equatorial C_1—D groups were observed between 998 (*m*; in methyl α-glucoside) and 1132 cm^{-1} (*s*; in α-galactose) and in the 965 (α-glucose) 1038 cm^{-1} (*s*; α-mannose) range. The respective axial C_1—D regions are 1000 (*s*; β-mannose) 1115 (*m*; β-galactose) and 912 (*m*; β-mannose) 972 cm^{-1} (*s*; β-glucose).

4.97 Heavy Acid Halides

The 500–5000 cm^{-1} spectrum of gaseous DCOF was observed to show absorption bands at[470] 657·5 (ν_5), 857·4 (ν_6), 967·9 (ν_3; C—D bending), 1073·2 (ν_4), 1624·5 ($\nu_3 + \nu_5$), 1705·8 ($\nu_4 + \nu_5$), 1796·8 (ν_2), 1928·4 ($\nu_4 + \nu_6$), 1930·6 ($2\nu_3$), 2028·6 ($\nu_3 + \nu_4$), 2137·8 ($2\nu_4$), 2261·7 (ν_1; C—D stretching) and 3579·4 ($2\nu_2$) cm^{-1}. The analysis of the fine structure of the $\nu_1 - \nu_5$ bands yielded the value of its B'' (0·3649 cm^{-1}) as well as its other rotational constants.

The normal co-ordinate analysis of its normal vibrations has also been carried out.[441,471]

The 250–4000 cm^{-1} absorption frequencies of gaseous D_3-acetyl chloride have been observed and assigned to the respective CD_3COCl vibrations.[472] Its CD_3 stretching bands occur at 2087 (symmetric) and 2276 cm^{-1} (asymmetric) while its symmetric C—D bending mode absorbs at about 1040 cm^{-1} and the asymmetric analogue is exhibited at 1135 (*m*) cm^{-1}. The in-plane and out-of-plane rocking vibrations of this CD_3 group are shown at 818 and 877 cm^{-1}, respectively, and its CD_3—C stretching band appears at 958 cm^{-1}.

The main CD_3COBr absorption frequencies have been reported[473] at 2275 (*w*; C—D stretching), 1825 (*vs*; C=O stretching), 1122 (*vs*; C—D asymmetric bending), 1099 (*s*), 1082 (*m*), 1043 (*w*; symmetric C—D bending), 1010 (*m*), 973 (*m*), 951 (*vs*), 855 (*w*), 808 (*w*), 568 (*m*), 525 (*s*), 440 (*m*) and 345 (*s*) cm^{-1}.

4.98 Deuterated Formic Acid

As formic acid exists predominantly in the monomeric form only at relatively high temperatures the absorption of monomeric formic acids[474] had to be measured on gaseous samples at about 150°C.

The 625–3800 cm^{-1} spectra of gaseous $(DCOOD)_2$, $(HCOOD)_2$ and $(DCOOH)_2$ as well as the 475–4000 cm^{-1} absorption of crystalline DCOOD and the 70–4000 cm^{-1} spectra of crystalline HCOOD and DCOOH were also investigated.[475,476] Table 4.55 reports the main bands observed for gaseous isotopic formic acids. The 33–400 cm^{-1} infrared active "hydrogen bond" fundamentals[477] (which are due to vibrations of the two units constituting the dimer, against each other) are also included in this Table.

The O—D stretching frequency of very dilute solutions of (monomeric) HCOOD (0·01 M or less) in various solvents varies from 2299 cm^{-1}, in the case of diethyl ether, to 2599 cm^{-1}, in the case of carbon tetrachloride.[478]

Since the solvent-induced shifts of such a frequency reflect the strength of the bonding existing in each case between the proton-donating acid and the solvent molecules and since this bonding is a function of the pK_a of the solvated acid the comparison of the O—D frequency shifts with the corresponding O—H values yielded the relative acidity of HCOOD as compared with that of HCOOH[478] (1:1·11).

The ν_{OH}/ν_{OD} ratio of monomeric formic acid in solution usually differs from its 1·36 value for the vapour state and, in the case of acetone solutions, becomes as low as 1·33. This "spectroscopic isotope" effect is probably due to the increase in the anharmonicity of the hydroxylic bond as a result of the association effects operating in solution.[478]

The normal co-ordinate analysis of the normal vibrations of DCOOD, HCOOD and DCOOH, which among others yielded the values of their force constants, has been carried out.[441,479]

4.99 Labelled Acetic Acids

The 400–4000 cm^{-1} spectra of monomeric, gaseous CD_3COOD, CH_3COOD and CD_3COOH (at 150°C) exhibited[480a] the fundamentals given in Table 4.56.

Liquid acetic acid is mainly composed of cyclic dimers and linear polymers of acetic acid units. The absorption of liquid CD_3COOD[480b] showed its main bands[481] at 425 (*m*), 581 (*m*), 692 (*m*; out-of-plane O—D bending), 831 (*m*; symmetrical CD_3 rocking), 926 (*m*; asym. CD_3 rocking), 1047 (*m*; symmetrical CD_3 deformation), 1087 (*m*; O—D bending), 1355 (*s*), 1708 (*vs*), 2096 (*m*) and 2278 cm^{-1} (*s*; O—D stretching).

Liquid CH_3COOD absorbs[481] mainly at 472 (*m*), 605 (*m*), 697 (*m*; out-of-

TABLE 4.55 Fundamental Bands (cm^{-1}) of Isotopic Formic Acids

Assignment	HCOOH[475]	$(HCOOH)_2$	DCOOD	$(DCOOD)_2$	HCOOD	$(HCOOD)_2$	DCOOH	$(DCOOH)_2$
Out-of-plane O—H(D)···O bending		68 (*m*)				68 (*m*)		
Ring torsion		164 (*s*)				158 (*s*)		
H(D)— bond stretching		248 (*vs*)				240 (*vs*)		
		307 (*w*)				299 (*w*)		
		329 (*w*)						
O—H(D) out-of-plane		395 (*m*)				390 (*m*)		
bending	636	917		678		693		930
OCO bending	636	697		642	667	651	678	695
CH(D) out-of-plane bending	1033		872	890	1007			890
C—O stretching	1105	1218	1171	1246	1178	1259	1143	1239
COH(D) bending	1229	1450	945		972	1037	1206	
C—H(D) bending	1387	1365	1040	976, 987		1387	970	996
C=O stretching	1770	1754		1720	1770	1745		1726
C—H(D) stretching	2943	2957	2232	2226	2948	2960	2220	2224, 2251
O—H(D) stretching	3570	3110	2632	2323	2632	2314	3567	3098

TABLE 4.56 Fundamental Bands (cm^{-1}) of Isotopic Acetic Acids

Approximate description	CH_3COOH	CD_3COOD	CH_3COOD	CD_3COOH
Out-of-plane O—H(D) bending	534	408 (*w*)	415	560
C—C—O bending	581		543	478
C—C—O out-of-plane deformation		528 (*m*)		560
C(=O)—O bending	642	540 (*m*)	603	615
C—C stretching	847		840	
$CH_3(D_3)$ symmetrical rocking	989	813 (*w*)	990	819
$CH_3(D_3)$ asymmetrical rocking	1048	917 (*w*)	1052	918
C—O stretching	1182[a]	1300 (*s*)	1270[a]	1167[a]
O—H bending	1264[a]	1004 (*m*)	955	1338[a]
$CH_3(D_3)$ symmetrical bending	1382[a]	1080 (*m*)	1383[a]	1078
$CH_3(D_3)$ asymmetrical bending	1430	1056 (*m*)	1440	1055
C=O stretching	1788	1775 (*vs*)	1775	1783
$CH_3(D_3)$ symmetrical stretching	2944		2952	
$CH_3(D_3)$ asymmetrical stretching	2996 3051	2240 (*vw*) 2275 (*vw*)	2997 3039	2230 2285
O—H(D) stretching	3583	2642 (*m*)	2642	3582

[a] A coupled vibration.

plane OD deformation), 1000 (*m*), 1055 (*m*), 1322 (*s*) 1388 (*s*), 1715 (*vs*), 2051 (*m*), 2127 (*m*) and 2268 cm^{-1} (*s*; O—D stretching). The main absorption bands of CD_3COOH were observed[481] at 434 (*m*), 598 (*m*), 823 (*m*), 924 (*m*), 1297 (*s*), 1408 (*m*), 1710 (*vs*), 2565 (*m*), 2670 (*m*), 2927 (*s*), 3028 (*s*) and 3088 (*m*) cm^{-1}.

The 30–240 cm^{-1} spectra of liquid and solid CD_3COOD and CH_3COOD have also been reported and interpreted.[480b, 482]

The absorption of CD_3COOD, CH_3COOD and CD_3COOH was investigated in the case of a number of solvents[481] and the O—D stretching frequency of CH_3COOD solutions was found[478] to vary between 2341 (Et_2O) and 2614 cm^{-1} (CCl_4), its relative acidity (as compared with that of CH_3COOH) being again 1:1·11.

The room temperature spectral measurements of gaseous (mostly) $(CD_3COOD)_2$, $(CH_3COOD)_2$ and $(CD_3COOH)_2$[481] were extended in the latter case also to the 33–400 cm^{-1} region.[477] Bands were thus observed at 50 (*m*; out-of-plane O—H····O bending), 157 (*s*; ring torsion) and 180 cm^{-1} (*s*; H-bond stretching).

The 2100–3400 cm^{-1} spectra of the mixed cyclic dimers

$CH_3COOH—CH_3COOD$ and $CD_3COOH—CD_3COOD$

show differences only in their O—H(D) bands.[483]

The normal vibrations of CD_3COOD, CH_3COOD, CD_3COOH, CF_3COOD[479] and deuterated $CH_2ClCOOH$[441] have been given a normal co-ordinate treatment.

The O—D stretching frequency of various solutions of monomeric CCl_3COOD, $CHCl_2COOD$ and $CH_2ClCOOD$ was observed in the 2196–2602 cm^{-1} region.[478]

Crystalline CCl_3COOD absorbs in the 1500–3700 cm^{-1} region[484] at 1750 (*s*), 1870, 1983, 2090 (*s*), 2150, 2180, 2258 (*s*), 2625 and 2860 cm^{-1}. The bands appearing near 2100 cm^{-1} are due to summation vibration modes involving the COOD group and are intensified by Fermi resonance.[484]

The infrared absorption of the α, β and γ polymorphic forms of crystalline 2-D_2-chloroacetic acid as well as the spectra of its solutions in CS_2 and acetonitrile were also described.[485]

4.100 Other Deuterated Aliphatic Acids

The study of the absorption of many liquid or crystalline RCOOD acids in the 500–1500 cm^{-1} span shows that their C—OD stretching band appears at 1350 ± 50 cm^{-1} while their in-plane O—D bending mode gives rise to an absorption at 1050 ± 10 cm^{-1} and the corresponding out-of-plane vibration is exhibited at[486,487] 675 ± 25 cm^{-1}. Table 4.57 reports the exact values of these bands in a number of cases.

TABLE 4.57 The C—O Stretching and O—D Bending Frequencies (cm^{-1}) of Various RCOOD Acids[486]

Acid	C—O stretching	In-plane bending	Out-of-plane bending
Acetic	1320	1058	670
Propionic		1043	
n-Valeric	1390	1048	650
Lauric	1367	1055	665
Stearic	1366	1058	666
Oxalic	1228	1046	
Malonic		1046	672
Succinic	1383	1046	680
Glutaric	1332	1049	675
Adipic	1300	1050	675
Trimethylacetic	1344	1049	
Trichloroacetic	1341	1050	

$C_2H_5CHDCOOH$ shows its C—D stretching band at 2165 cm^{-1} (*vw*) while $(CH_3)_2CDCOOH$ exhibits two such bands,[488] at 2092 (*vw*) and 2169 cm^{-1} (*vw*), and $(CH_3)_2CDCOOD$ absorbs at 2101 (*m*) and 2283 cm^{-1} (*m*; O—D stretching).

Perdeuterated trimethylacetic acid shows bands at[387] 750, 818, 1030, 1060 (last three due to CD_3 rocking modes), 1090, 1220, 1280, 1340 (last three belong to CD_3 bendings), 2222 and 2256 cm^{-1} (C—D stretchings).

$(CH_3)_3CCOOD$ solutions display their O—D stretching band in the 2334–2610 cm^{-1} span, depending upon the solvent.[478]

Completely deuterated succinic acid[489] exhibits its main bands in the 400–4000 cm^{-1} region at 617 (*m*), 685 (*m*; out-of-plane OD bending), 947 (*m*), 1010 (*m*), 1043 (*m*; OD bending), 1058 (*m*; CD_2 bending), 1365 (*vs*) and 1700 cm^{-1} (*vs*). The spectra of $(CH_2COOD)_2$ and $(CD_2COOH)_2$ have also been measured.[486, 489]

The deuterium-induced shifts observed for the fundamentals of crystalline $(CD_2CD_2COOD)_2$, $(CH_2CH_2COOD)_2$ and $(CD_2CD_2COOH)_2$ were used to assign the infrared bands of normal adipic acid in the 400–4000 cm^{-1} span.[490] $C_6D_{10}O_2$ absorbs at 481, 557 (CD_2 rocking), 655, 680 (out-of-plane OD bending), 718 (CD_2 rocking), 813, 946 (CD_2 twisting), 958 (CD_2 wagging), 1045 (CD_2 bending), 1069 (OD bending), 1086 (CD_2 bending), 1185, 1360, 1690, 2116, 2220, 2238 (all the last three are due to C—D stretchings) and 2270 cm^{-1} (O—D stretching).

4.101 Various Heavy Acids

Deuterated cyclohexadecarboxylic acid, $C_6H_{11}COOD$, shows its O—D bending fundamental at 1054 cm^{-1}, i.e. in the range of the aliphatic COOD acids.[486]

α-Deuterated cyclohexanecarboxylic acid shows its C—D stretching band at 2165 cm^{-1} (*vw*) while its O—D derivative absorbs at[488] 2101 (*m*; C—D stretching) and 2268 cm^{-1} (*m*; O—D stretching).

The infrared absorption of gaseous deutero-oxalic acid[491] and the 1180–5000 cm^{-1} spectrum of a Nujol mull of O-deuterated D_2-maleic acid[492a] were reported as was also the absorption of the adduct between maleic anhydride and deuterated sorbic acid.[492b]

A D_2-propynoic acid solution in CCl_4 shows absorption bands at[493] 861 (*w*), 881 (*w*), 1050 (*m*; OD bending), 1181 (*w*), 1327 (*s*), 1700 (*s*), 1870 (*w, b*), 1987 (*s*), 2096 (*m*), 2218, 2278, 2331, 2356 (all four *m, vb* and due to bonded O—D stretchings), 2600 (*m*; C—D stretching) and 2631 cm^{-1} (*w, b*).

The D_2-acetylenedicarboxylic acid[486] exhibits its C—O stretching absorption at 1360 cm^{-1} and its O—D bending fundamental at 1054 cm^{-1}. Crystalline C_6H_5COOD shows these bands at 1367 and 1042 cm^{-1}, respectively, and its out-of-plane O—D deformation band at[486] 648 cm^{-1}. Above 1500 cm^{-1} it shows bands at[484] 1680, 1853, 1910, 2070 (*s*), 2173, 2240 (*s*) and 3067 cm^{-1}. Its CCl_4 solution spectrum was also reported.[484]

The O—D stretching band of a CCl_4 solution of $C_6H_5CH_2COOD$[478] appears at 2612 cm^{-1}.

4.102 Salts of Deuterated Acids

The absorption of a KBr pellet containing about 0·2% deuterated sodium formate displayed fundamentals at[494] 768 (*m*; OCO scissoring), 915 (*w*; C—D out-of-plane bending), 1014 (*m*; C—D in-plane bending), 1335 (*s*; C—O stretching), 1595 (*vs*; C=O stretching) and 2132 (*s*) cm^{-1} while its aqueous solution showed the C—D in-plane bending fundamental at *ca.* 1023 cm^{-1} and the C—D stretching band at 2031 cm^{-1}.

The calculated $DCOO^-$ ion fundamentals have also been reported.[441]

Solid sodium D_3-acetate shows its fundamentals at[495,496] 419, 523, 613, 828 (CD_3 in-plane rocking), 880, 929 (out-of-plane rocking), 1034 (symmetric CD_3 deformation), 1081 (asymmetric CD_3 deformation), 1427, 1562, 2232 (symmetric C—D stretching) and 2257 cm^{-1} (asymmetric C—D stretching). These values agree with the frequencies calculated for these fundamentals on basis of a C_{2v}, Urey–Bradley force field model.[495,496]

The spectra of solid CHDClCOONa and $CD_2ClCOONa$ were also recorded and interpreted.[497]

The 830–4000 cm^{-1} absorption of mulls of deuteroxy-aluminium dilaurate and $(C_4H_9CH(C_2H_5)COO)_2AlOD$ shows their O—D stretching band[498] at about 2720 cm^{-1}.

The hydrogen bonded, symmetric, O-deuterated acid maleate ion (see Fig. 4.23) absorbs at[499] 309, 416, 546, 695, 860, 980, 1060, 1195, 1218 (O····D····O

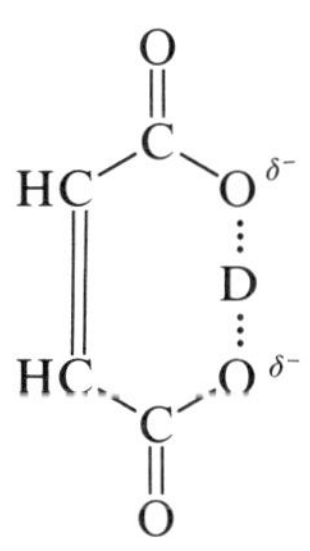

Fig. 4.23 The O-deuterated acid maleate ion.

bending), 1363, 1385, 1485, 1630 and 2856 cm^{-1} while its D_2—C-deuterated analogue[499] exhibits bands at 309, 416, 543, 665, 805, 900 (C—D bending), 980, 1050, 1195, 1375, 1475, 1650, 1700 and 2020 cm^{-1} (C—D stretching).

The C—D stretching frequency of deuterated sodium propynoate in a KBr pellet appears at 2576 (*m*) cm^{-1}, its in-plane and out-of-plane C—D deformation bands being shown at 557 (*m*) and 508 (*m*) cm^{-1}, respectively.[493]

The 5300–7500 cm^{-1} spectrum of deuterated sodium potassium tartrate (Rochelle salt) was also reported.[500]

The deuterate of lithium pyruvate absorbs characteristically at 1023, 1273 and 2250 cm^{-1} (O—D stretching).[413]

4.103 Labelled Esters

The observed 250–3600 cm^{-1} spectra of gaseous $DCOOCH_3$ and $HCOOCD_3$ as well as their absorption in solution were all interpreted on basis of a planar *cis* configuration[501a] (with respect to the carbonyl oxygen and methyl carbon atoms). Table 4.58 gives the frequency of their measured fundamentals.

Table 4.58 Fundamental Frequencies (cm^{-1}) of Gaseous $HCOOCH_3$, $DCOOCH_3$ and $HCOOCD_3$[501a]

Band	Approximate description	HCO_2CH_3	DCO_2CH_3	HCO_2CD_3
ν_1	$CH_3(D_3)$ asymmetrical stretching	3045	3041 (*s*)	2284 (*s*)
ν_2	$CH_3(D_3)$ symmetrical stretching	2969	2967 (*s*)	2087 (*s*)
ν_3	CH(D) stretching	2943	2216 (*s*)	2931 (*s*)
ν_4	C=O stretching	1754	1712 (*vs*), 1751 (*vs*)[a]	1754 (*vs*)
ν_5	$CH_3(D_3)$ asymmetrical bending	1465	1468 (*m*)	1063 (*m*)
ν_6	$CH_3(D_3)$ symmetrical bending	1445	~1441 (*m*)	1102 (*s*)
ν_7	CH(D) bending	1371	1048 (*m*)	1368 (*m*)
ν_8	C—O stretching	1207	1213 (*vs*)	1210 (*vs*)
ν_9	$CH_3(D_3)$ rocking	1168	1157 (*vs*)	985 (*m*)
ν_{10}	C—O stretching	925	878 (*s*)	877 (*s*)
ν_{11}	OCO bending	767	762 (*m*)	714 (*m*)
ν_{12}	COC bending	325	304 (*s*)	305 (*s*)
ν_{13}	$CH_3(D_3)$ asymmetrical stretching	3012	3007 (*s*)	2258 (*s*)
ν_{14}	$CH_3(D_3)$ asymmetrical bending	1454	1447 (*m*)	1060 (*m*)
ν_{15}	{$CH_3(D_3)$ rocking	—	1033 (*m*)	—
ν_{16}	CH(D) bending}	1032	—	906 (*w*)

[a] Fermi resonance splitting.

The infrared absorption of $DCO_2CD_2CD_3$ has also been reported.[501b]

Perdeuterated methyl acetate[502] absorbs in the 300–4000 cm^{-1} region mainly at 412 (*s*), 526 (*s*), 590 (*s*), 744 (*s*), 828 (*m*), 919 (*s*), 1091 (*s*; CD_3 bending), 1175 (*s*; CD_3 bending), 1283 (*vs*), 1415 (*m*), 1443 (*m*), 1484 (*m*), 1735 (*vs*), 1830 (*m*), 2018 (*s*), 2064 (*s*), 2078 (*s*), 2125 (*s*), 2195 (*s*; last four bands due to C—D stretchings) and 2257 (*s*) cm^{-1}.

Comparing the 650–3200 cm^{-1} spectra of (CCl_4 and CS_2) solutions of

CD_3COOCD_3, CD_3COOCH_3 and CH_3COOCD_3[503] shows that the OCD_3 group is characterized by absorption bands at 2016, 2082, 2128 (last two: symmetrical C—D stretchings), 2196, 2254 and 2272 cm^{-1} (last three: asymmetrical C—D stretchings).

The absorption of $CD_3CO_2C_2D_5$, $CH_3CO_2CD_2CH_3$, $CH_3CO_2CH_2CD_3$, $CD_3CO_2C_2H_5$, $CH_3CO_2C_2D_5$, $CD_3CO_2CD_2CH_3$ and $CD_3CO_2CH_2CD_3$ solutions in the 700–4000 cm^{-1} region[503] was investigated. $C_4D_8O_2$ absorbs at 718, 780, 815, 917 (CD_3CO), 958, 988, 1047 (C_2D_5), 1059 (C_2D_5), 1082 (CD_3CO), 1097 (C_2D_5), 1178 (C_2D_5), 1194 (C_2D_5), 1737 (CO), 2074 and 2120 (both probably C—D stretchings of the ethoxy CD_3 group), 2151 (CD_2), 2186 (CD_2), 2230 (C_2D_5) and 2242 cm^{-1}.

The 650–3200 cm^{-1} solution spectra of seven deuterated modifications of methyl laurate (containing terminal CD_3, α-CD_2 or OCD_3 groups or their combinations) together with those of

$CD_3—(CH_2)—CD_3$, $CD_3(CH_2)_{10}—CD_2OH$, $C_{11}H_{23}CD_2OH$, $C_{11}H_{23}CD_2Br$ and $CD_3(CH_2)_{10}CD_2Br$

assisted in the assignment of all the observed absorption bands of methyl laurate.[504] In this case also the presence of a OCD_3 group gives rise to bands at about 2080 (molecular extinction coefficient, ε = *ca.* 24 1 $mole^{-1}$ cm^{-1}), 2125 ($\sim$15), 2195 (18–26), 2250 (23–35) and 2272 cm^{-1} (16). The terminal CD_3 group was found to absorb at 2077 (8–26), 2120 (7–18) and 2210 cm^{-1} (17–67) while the α-CD_2 characteristic band was observed at 2220 (3) cm^{-1}.

The 650–1350 and 1600–1800 cm^{-1} spectra of CS_2 solutions of *cis*- and *trans*-$CH_3CD{=}CDCO_2CH_3$ show their prominent bands at 791 (70), 1089 (90), 1130 (350), 1256 (*ca.* 550), 1277 (100), 1620 (80), 1718 (*ca.* 800) and 718 (120), 1080 (240), 1244 (*ca.* 450), 1260 (*ca.* 500), 1623 (110) and 1720 cm^{-1} (*ca.* 800), respectively.[505]

The 700–5000 cm^{-1} absorption of the ethyl ester of ethoxy-deuteroacetic acid has also been reported.[506]

4.104 Deuterated Acid Anhydrides

Liquid D_4-succinic anhydride absorbs at 770 (*m*), 782 (*vw*), 872 (*s*), 932 (*m*), 970 (*vs*), 1035 (*m*), 1087 (*m*), 1088 (*m*), 1149 (*vs*), 1172 (*w*), 1232 (*vs*), 1276 (*w*), 1370 (*w*), 1410 (*w*), 1790 (*vs*), 1852 (*vs*), 1873 (*w*), 2020 (*vw*), 2180 (*w*), 2260 (*w*) and 2298 (*vw*) cm^{-1}. The 400–4000 cm^{-1} polarized light spectra of its crystals have also been measured, in order to assist in the assignment of its absorption bands.[507, 508]

Deuterated Organic Nitrogen Compounds

4.105 Deuterated Non-cyclic Amines and Their Salts

The infrared absorption of gaseous perdeuterated methylamine,[509] CH_3ND_2[509–511] and CD_3NH_2[509] (of the C_s point group) was measured in the 300–3500 cm^{-1} region, the vibration–rotation overtone band (of the asymmetrical ND_2 stretching) of CH_3ND_2, centred at 5070·4 cm^{-1}, being also investigated, under good resolution.[510] The fundamental bands observed thus for these amines are reported in Table 4.59.

TABLE 4.59 Fundamental Bands (cm^{-1}) Observed for Isotopic Methylamines[a]

Band	Species	Approximate description	CH_3NH_2	CD_3ND_2	CH_3ND_2	CD_3NH_2
ν_1	A'	$NH_2(D_2)$ symmetrical stretching	3361	2477	2479	3361
ν_2	A'	$CH_3(D_3)$ asymmetrical stretching	2961	2202	2961	2203
ν_3	A'	$CH_3(D_3)$ symmetrical stretching	2820	2073	2817	2077
ν_4	A'	N—H(D) bending	1623	1227	1234	1624
ν_5	A'	C—H(D) out-of-phase bending	1473	—	1468	—
ν_6	A'	C—H(D) in-phase bending	1430	1123	1430	1142
ν_7	A'	$CH_3(D_3)$ rocking	1130	880	1117	913
ν_8	A'	C—N stretching	1044	942	997	973
ν_9	A'	$NH_2(D_2)$ wagging	780	601	625[512]	740
ν_{10}	A''	$NH_2(D_2)$ asymmetrical stretching	3427	2556	2556	3427
ν_{11}	A''	$CH_3(D_3)$ asymmetrical stretching	2985	2238	2985	2236
ν_{12}	A''	C—H(D) out-of-phase bending	—	1077	—	1077
ν_{13}	A''	N—H(D) deformation	—	—	—	1416
ν_{14}	A''	$CH_3(D_3)$ rocking	—	910	1187	—

[a] Unless stated otherwise, values are from ref. 509.

The 220–3400 cm^{-1} absorption spectrum of gaseous N-deuterated dimethylamine shows fundamental bands at[513, 514] 225, 252, 383, 587, 912, 1064, 1117, 1234, 1245, 1412, 1460, 2520 (N—D stretching), 2792, 2831, 2955 and 2982 cm^{-1}. It should be noted that the N—D stretching frequency of $(CH_3)_2ND$ (2520 cm^{-1}) is almost exactly equal to the mean of the two N—D stretching frequencies of CH_3ND_2 (= (2479 + 2556)/2 = 2518 cm^{-1}). These frequencies can be seen to have an ν^D/ν^H ratio of *ca.* 0·74, as compared with the respective ratio of 0·73, calculated for the diatomic N—H(D) frequencies (Table 4.1).

The 70–3200 cm^{-1} absorption of the deuterated trimethylamines was also reported.[515, 516]

The absorption of gaseous, liquid and solid (in this case in polarized light) D_4—(N-deuterated)ethylenediamine shows that this compound exists in all these phases in the same, probably the *cis*, rotational structure.[517] Its N—D stretching bands appear at 2342, 2411 and 2479 cm^{-1} while its ND_2 bending absorption is shown at 1191 cm^{-1} and other ND_2 fundamentals are observed at 745, 765 (wagging) and 923 cm^{-1} (twisting; the respective three $H_2N(CH_2)_2NH_2$ values are 920, 951 and 1249 cm^{-1}).

The 222–667 cm^{-1} spectra of *cis*- and *trans*-N-deuterated D_4-chromium-bis (ethylenediamine) $(NCS)_2$, SCN were very helpful in assigning the bands of the normal complexes.[518]

Liquid perdeuterated aniline[519] shows its main absorption bands at 711 (*m*), 738 (*s*), 752 (*m* to *s*), 814 (*m–s*), 951 (*m*), 1025 (*m*), 1130 (*m*), 1245 (*m–s*), 1253 (*m–s*), 1287 (*s*), 1400 (*vs*), 1460 (*m*), 1570 (*vs*), 1580 (*vs*), 2238 (*m*), 2262 (*m–s*), 2276 (*m*), 2377 (*m*), 2438 (*s*), 2498 (*m*) and 2573 cm^{-1} (*m–s*).

In a dilute CCl_4 solution C_6H_5NHD shows its free N—D stretching band at 2545 cm^{-1} while $C_6H_5ND_2$ exhibits its symmetrical free ND_2 stretching band at 2493 cm^{-1} and its asymmetrical one at[520] 2600 cm^{-1}. The free N—H stretching fundamental frequency of C_6H_5NHD (in cyclohexane solution) appears[521] at 3440 cm^{-1} while its first overtone is shown at 6720 cm^{-1}. Various ring substituents[521] (such as 2—CO_2CH_3) can shift this fundamental up to 3484 cm^{-1} and its overtone up to *ca.* 6820 cm^{-1}.

The appearance of such a (D)N—H band half-way between the already present (symmetric and asymmetric) NH stretching bands, following partial deuteration of amines, is a good criterion for the presence of the primary NH_2 group in them.[520b]

The 4800–5200 cm^{-1} spectra of ND_2-anilines usually show three strong bands which probably belong to the overtone and combination vibrations of the fundamental N—D stretching frequencies.[521]

The infrared spectrum of 2,4,6-D_3-aniline[352] as well as the 2800–3000 cm^{-1} absorption of N-trideuteromethylaniline[423] have also been reported.

The spectra of N-deuterated N-methyl- and N-ethylaniline, *p*-$ClC_6H_4NDCH_3$ and *o*-, *m*- and *p*-$CH_3C_6H_4NDC_2H_5$ were investigated to make possible the identification of the absorption bands associated with the NH group of their normal analogues[522] while the absorption frequencies of $C_6H_5N(CD_3)_2$ and its salt with methyl *p*-toluenesulphonate were measured in order to be able to calculate the equilibrium isotope effect of this salt formation.[523]

Ortho-phenylenediamine in dilute CCl_4 solutions is doubly internally hydrogen bonded, each amino group acting both as a proton donor and as a proton acceptor. The partly N-deuterated diamine solutions therefore show both "free" (*trans*) and "bonded" (*cis*) (D)N—H as well as (H)N—D stretching bands, at 3415 and 3371 and 2521 and 2492 cm^{-1}, respectively.[524] Its symmetric and asymmetric ND_2 stretching absorptions are observed at 2431, 2450 (in Fermi resonance) and 2565 cm^{-1}, respectively.[524] Various ring substituted (as, e.g. 4-chloro- or 4,5-dimethyl-) NHD-*o*-phenylenediamines show these bands at 3400–3438 ("free" N—H), 3355–3389 ("bonded" N—H), 2514–2529 ("free" N—D) and 2482–2498 cm^{-1} ("bonded" N—D) while their ND_2-analogues absorb at 2446–2458 (symmetric ND_2 stretching) and 2554–2573 cm^{-1} (asymmetric ND_2 stretching).[524]

As can be expected *p*-NHD-phenylenediamine solutions show only one N—H (3412 cm^{-1}) and one N—D stretching (2522 cm^{-1}) frequencies, the corresponding ND_2 bands appearing at 2475 and[524] 2522 cm^{-1}.

The infrared spectrum of $(C_6H_5)_2ND$ has also been measured.[522]

The 163–360 and 650–3200 cm^{-1} absorption of various crystalline phases of $CD_3ND_3^+X^-$, $CH_3ND_3^+X^-$ and $CD_3NH_3^+X^-$ (X = Cl, Br, I) was investigated at a number of temperatures down to −190°C.[525] The observed fundamentals of the room temperature modification of a number of isotopic methylammonium chlorides are reported in Table 4.60.

TABLE 4.60 Fundamental Bands (cm^{-1}) of α-CH_3NH_3Cl, α-CD_3ND_3Cl, α-CH_3ND_3Cl and CD_3NH_3Cl

Band	Approximate description	CH_3NH_3Cl	CD_3ND_3Cl	CH_3ND_3Cl	CD_3NH_3Cl
ν_1	Symmetric $NH_3(D_3)^+$ stretching	2993	2180 (*vs*)	2197	2990
ν_2	Symmetric $CH_3(D_3)$ stretching	2917	—	—	2100 (*vw*)
ν_3	Symmetric $NH_3(D_3)^+$ bending	1534	1184 (*s*)	1178	1534
ν_4	Symmetric $CH_3(D_3)$ bending	1427	—	1427 (*w*)	1115
ν_5	C—N stretching	1004	900 (*vw*)	954	945
ν_7	Asymmetric $NH_3(D_3)^+$ stretching	3080	2285 (*vs*)	2275	3080
ν_8	Asymmetric $CH_3(D_3)$ stretching	2963	—	2970 (*vw*)	—
ν_9	Asymmetric $NH_3(D_3)$ bending	1578	1150 (*s*)	1150	1578
ν_{10}	Asymmetric $CH_3(D_3)$ bending	1463	1058 (*w*)	1462	1062
ν_{11}	Rocking vibration	1265	1015 (*vw*)	1188	1177
ν_{12}	Rocking vibration	954	692 (*m*)	775	790

The frequency of corresponding fundamentals belonging to the respective bromides and iodides is usually quite similar to that of the chlorides.

A comparison of the methylammonium ND_3^+ stretching and deformation bands with their ND_2 analogues (of methylamine; see Table 4.59) shows immediately that in the deuterated ammonium case also the free positive charge on the nitrogen atom brings about a considerable decrease of its N—D force constants that is reflected strongly in its stretching and bending frequencies.

The 33–3800 cm^{-1} spectra of Nujol mulls of $R_4N^+(ClDCl)^-$ (R = CH_3, C_2H_5, C_3H_7, C_4H_9, C_5H_{11}) and $C_{16}H_{33}(CH_3)_3N^+(ClDCl)^-$ as well as of their solutions in acetonitrile or benzene have been reported.[526] Similar $(ClDBr)^-$[527] and $(BrDBr)^-$[528, 529] salts, etc. have also been investigated.

The 660–3800 cm^{-1} absorption of a Nujol mull of $NH_2(CD_2)_2NH_2.2HBr$ does not show any significant C—D stretching band although the parent D_4-ethylenediamine does show[530] such bands at about 2100 cm^{-1}.

4.106 Labelled Cyclic Amines

Both piperidine and its N-deuterated modification exist, in the liquid as well as in the solid phase (although in this case mainly in the equatorial form), in an equilibrium between axial and equatorial conformations, as is apparent from their 400–3400 cm^{-1} spectra.[531,532] This explains why $C_5H_{10}ND$ shows *two* N—D stretching bands for the vapour as well as for the liquid phase[533] (at 2422 and 2439 cm^{-1}).

The N-deutero-piperidine deuterochloride shows a number of ND_2^+ stretching bands in the 1900–2400 cm^{-1} region[534] and its ND_2^+ deformation bands, probably, at 1238 and 1297 cm^{-1}.

The 625–5000 cm^{-1} spectrum of liquid 11-deutero-11-azabicyclo(4,4,1)-1-undecene[535] (Fig. 4.24) shows its N—D stretching band at 2400 cm^{-1}.

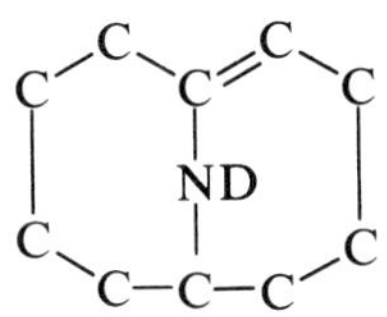

Fig. 4.24 11-deutero-11-azabicyclo(4,4,1)-1-undecene.

4.107 Deuterated Benzanilines

Deuterated benzanilines seem to be the only labelled Schiff's bases whose infrared absorption has been reported in the literature.† The C—D stretching band of their characteristic —CD=N— group was found to absorb[536] at about 2151 cm^{-1}.

The solution, crystalline powder and single crystal (in this case in polarized light) spectra of $C_6H_5CH{=}N{-}C_6D_5$ were measured in the 400–3200 cm^{-1} region.[519] The main bands of this (crystalline) compound are shown at 475 (*s*), 514 (*m–s*), 523 (*m*), 555 (*s*), 625 (*m*), 687 (*s*), 690 (*s*), 747 (*s*), 759 (*s*), 772 (*m–s*), 810 (*m–s*), 860 (*m*), 878 (*m–s*), 920 (*m*), 973 (*m–s*), 976 (*m*), 985 (*m*), 1070 (*m*), 1150 (*s*), 1156 (*m–s*), 1220 (*m*), 1310 (*m*), 1346 (*m–s*), 1378 (*s*), 1450 (*s*), 1557 (*m–s*), 1575 (*m*), 1625 (*s*), 2260 (*m*), 2267 (*m–s*), 2274 (*m–s*), 2848 (*m*), 2882 (*m–s*), 2977 (*m*), 3030 (*m–s*), 3060 (*m–s*) and 3080 (*m*) cm^{-1}.

4.108 DCN and TCN

The (linear) hydrogen cyanide molecule possesses three normal modes of vibration: ν_1 which can be described as a C≡N stretching mode, the

† See, however, Perrier-Datin, A. and Lebas, J. M. (1969). *Spectrochim. Acta* **25A**, 169, for the 100–3300 cm^{-1} spectrum of $(C_6H_5)_2C{=}ND$.

degenerate ν_2 bending vibration and ν_3 which is practically a stretching of the carbon–hydrogen bond. The absorption of gaseous DCN and its solutions in various solvents has been measured extensively, under high resolution, in the whole 500–6000 cm^{-1} region.[537–541]

The 400–4000 cm^{-1} spectrum of gaseous TCN has also been reported.[542] Table 4.61 gives the wave number of the fundamental bands observed for HCN, DCN and TCN and the corresponding zero-point frequencies of DCN.

TABLE 4.61 The Fundamental Frequencies (cm^{-1}) of Gaseous HCN, DCN and TCN

Band	Type	HCN[541]	DCN[541]	TCN[542]
ν_1	parallel	2096·7 (*vw*)	1925·3	1724
ν_2	perpendicular	711·9 (*vs*)	569·1	513
ν_3	parallel	3311·4 (*s*)	2630·3	2460
ω_1^0			1932·4[538]	
ω_2^0			569·2[538]	
ω_3^0			2650[538]	

The analysis of 16 of the vibration–rotation bands of DCN yielded all its rotational constants[538, 540] ($B_0 = 1{\cdot}2077\ cm^{-1}$, $D_0 = 1{\cdot}9 \times 10^{-6}\ cm^{-1}$, etc.) as well as many of its quadratic and cubic anharmonicity constants.[540]

The agreement of the observed DCN fundamental frequencies with the values calculated for them is very good[538] (the largest deviation between them being 0·2 cm^{-1}). The observed splitting of the higher overtone levels of the degenerate ν_2 mode ("l splitting"—see Section 2.9) also agrees with theory[540] and the l-type doubling constants calculated for DCN.[543]

The DCN ν_1, ν_2 and ν_3 fundamentals of its solutions in cyclohexane appear at 1920·5, 577·5 and 2612·2 cm^{-1}, respectively, as compared with the corresponding 1898·5, 613·0 and 2568·0 cm^{-1} values in the case of acetonitrile solutions.[541]

4.109 DEUTERATED NITRILES AND ISOCYANIDES

Isotopically homogeneous acetonitrile molecules, belonging to point group $\boldsymbol{C_{3v}}$, possess four totally symmetric (A_1) and four degenerate (E) fundamentals, all of them being active in infrared absorption. Their frequency in the case of gaseous CD_3CN[544] is given in Table 4.62.

The somewhat anomalous fact that the ν_2 C≡N stretching frequency of the *heavier* isotopomer CD_3CN is *higher* (2278 cm^{-1}) than the corresponding CH_3CN fundamental (2267 cm^{-1}) is the result of a mechanical interaction

TABLE 4.62 The Fundamental Bands (cm^{-1}) of CH_3CN and CD_3CN [544a]

Band	Species	Approximate description	CH_3CN	CD_3CN
ν_1	A_1	$CH_3(D_3)$ symmetrical stretching	2954	2126
ν_2	A_1	C≡N stretching	2267	2278
ν_3	A_1	$CH_3(D_3)$ symmetrical bending	1389	*ca.* 1110
ν_4	A_1	C—C stretching	920	831·3
ν_5	E	$CH_3(D_3)$ asymmetrical stretching	3009	2256·6
ν_6	E	$CH_3(D_3)$ asymmetrical bending	1454 (*vs*)	1046·4 (*vs*)
ν_7	E	$CH_3(D_3)$ rocking	1041	846·6
ν_8	E	C—C≡N bending	361[a]	331[a]

[a] Value calculated from overtone and difference bands.

between ν_2 and ν_1 in CD_3CN.[544a] This interaction has been brought about by the deuteration-induced decrease in the ν_1 frequency (which has now become very near to that of ν_2) and has "kicked apart" the ν_1 and ν_2 CD_3CN frequencies.[544a] Its ν_2 fundamental has therefore become higher than expected while its ν_1 mode has suffered a decrease in its frequency (as can be deduced from the fact that the $\nu_1^{CD_3CN}/\nu_1^{CH_3CN}$ ratio is equal to 2126/2954 = 0·72 as compared with the ν^{CD}/ν^{CH} = 0·73 ratio calculated (see Table 4.1) for the diatomic hydride).

The analysis of four perpendicular vibration–rotation bands of CD_3CN yielded many of its rotational and vibrational–rotational constants.[544a]

The complex formed by CD_3CN and zinc chloride in aqueous solutions showed its C≡N stretching infrared absorption band at 2310 cm^{-1} (*vs*) as compared with the 2265 cm^{-1} (*vs*) value for the uncomplexed nitrile under similar conditions. This increase in the CN frequency of the complex reflects an increase in the C≡N stretching force constant as a result of the attachment of the zinc ion to the nitrogen lone-pair electrons of the nitrile.[545]

In the case of the solid complexes of $AgNO_3.CD_3CN$ (at 20°C) and $AgNO_3.2CD_3CN$ (at 77°K) also the C≡N stretching band was observed to appear at higher frequencies (2274 and 2278 cm^{-1}, respectively) relative to the corresponding uncomplexed nitrile values (2263 and 2260 cm^{-1}, respectively). Their NO_3^- characteristic bands as well as their aqueous solution spectra are also affected by the complex formation.[546]

The 300–2700 cm^{-1} spectrum of liquid D_2-malononitrile[547] shows its absorption bands at 356 (*m–s*), 577 (*m–s*), 714 (*w*), 795 (*w*; CD_2 rocking), 829 (*m–s*), 858 (*m*), 892 (*vw*), 905 (*vw*), 1037 (*s*; CD_2 scissoring), 1142, 1165 (both *m* and due to a Fermi resonance between the CD_2 wagging and a combination frequency), 2146 (*m–s*; symmetric C—D stretching), 2230 (asymmetric C—D stretching) and 2272 (*m*) cm^{-1}. Figure 4.25 reproduces the spectra of liquid $CD_2(CN)_2$ and $CHD(CN)_2$ in the 700–4000 cm^{-1} region.[547]

The agreement between the frequencies observed for the $CD_2(CN)_2$ and

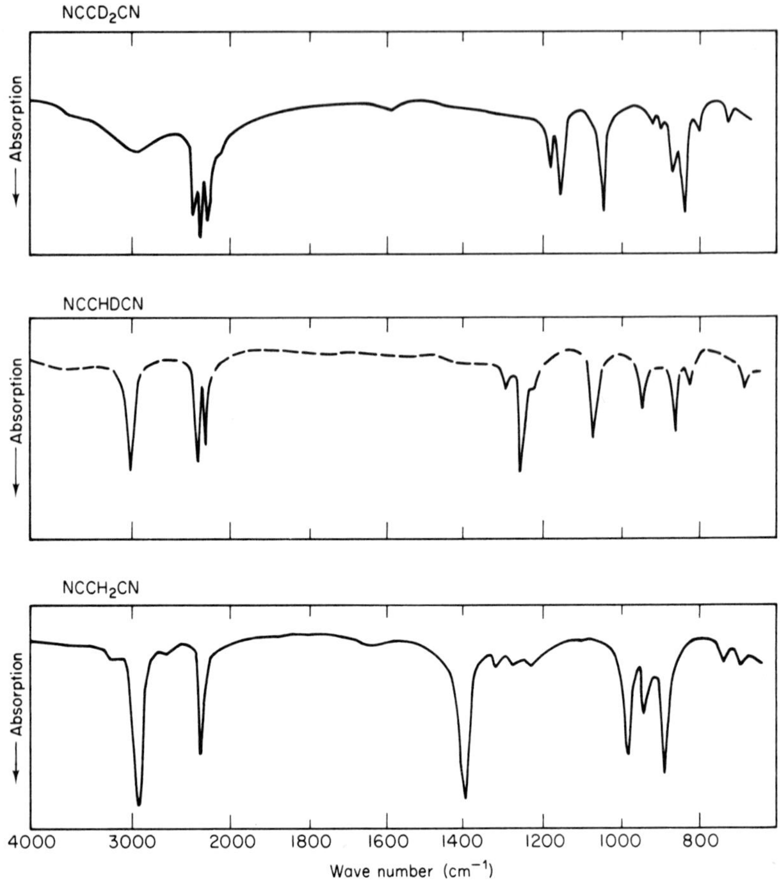

Fig. 4.25 Infrared spectra of malononitrile and its deuterated molecules in the NaCl region.

$CHD(CN)_2$ fundamentals and those calculated for them on basis of a Urey–Bradley force field is usually within 5% of their values.[547]

The 50–4000 cm^{-1} absorption of liquid D_5-benzonitrile was investigated in connection with the vibrational assignment of benzonitrile.[548] Its C—D stretching band appears at 2283 (*m*) cm^{-1} (as compared with 2270 cm^{-1} for C_6D_5F) while its in-plane C—D bending modes result in bands at 823 (*m–s*), 838 (*m–s*) and 842 cm^{-1} (*sh*) and its out-of-plane C—D deformation vibrations are shown at 572 (*m*), 638 (*m*), 695 (*w–m*), 769 (*w–m*) and 780 cm^{-1} (*vw*).

The C—D stretching band of *o*-, *m*- and *p*-deutero-benzonitrile appears[549] at 2280, 2270 and 2265 cm^{-1}, respectively.

The absorption of aromatic ND_2-enaminonitriles was also reported.[550]

The only isocyanide infrared absorption spectrum reported so far in the literature seems to be that of CD_3NC (of point group C_{3v}) which was measured in the vapour phase in the 670–5000 cm^{-1} region.[551] All its fundamentals were thus found and are reported in Table 4.63.

TABLE 4.63 The Fundamental Vibrations (cm^{-1}) of CH_3NC and CD_3NC

Band	Species	Approximate description	CH_3NC	CD_3NC
ν_1	A_1	Symmetric C—H(D) stretching	2965·8	2250·6
ν_2	A_1	N≡C stretching	2166·0	2165·0
ν_3	A_1	$CH_3(D_3)$ symmetric deformation	1429·0	1117·4
ν_4	A_1	C—N stretching	944·6	876·7
ν_5	E	Asymmetric C—H(D) stretching	3014·3	2262·9
ν_6	E	Asymmetric $CH_3(D_3)$ deformation	1466·9	1058·2
ν_7	E	$CH_3(D_3)$ rocking	1129·3	900·1
ν_8	E	C—N≡C bending	262·7[a]	248·9[a]

[a] Calculated from bands involving ν_8.

The analysis of the ν_5, ν_6 and ν_7 vibration–rotation bands of CD_3NC yielded its corresponding Coriolis constants.[551]

4.110 LABELLED HETEROCYCLIC COMPOUNDS CONTAINING ONE NITROGEN ATOM

The 400–4000 cm^{-1} spectrum of perdeuterated pyrrole, N-deuterated pyrrole and symmetrical D_4-pyrrole was investigated both for the liquid and for the vapour phases, the observed absorption bands being assigned to their respective modes of vibration.[552] The main bands of liquid C_4D_5N in the 750–1900 cm^{-1} region appear at[553, 555] 789 (*vs*), 813 (*vs*), 852 (*s*), 893 (*s*), 910 (*s*), 961 (*m*), 1025 (*s*), 1085 (*s*), 1147 (*s*), 1210 (*m*), 1341 (*s*), 1389 (*vs*) and 1415 (*s*) cm^{-1} while N-deuterated pyrrole is characterized by strong bands at[553] 766, 835, 915, 1015 (*vs*), 1048 (*vs*), 1074 (*vs*), 1384, 1418, 1428 and 1464 cm^{-1}. The infrared spectrum of the perchlorate of C_4D_5N has also been reported.[555]

The main absorption bands of liquid perdeuterated pyridine in the 250–3500 cm^{-1} region[554] are shown at 371 (*m*), 530 (*vs*), 567 (*m*), 582 (*s*), 762 (*m*), 823 (*s*), 833 (*m*), 886 (*m*), 963 (*s*), 1011 (*m*), 1228 (*m*), 1302 (*s*), 1322 (*m*), 1364 (*m*), 1530 (*vs*), 1542 (*s*), 1550 (*m*), 1645 (*m*), 2245 (*vs*),[556] 2261 (*vs*),[556] 2268 (*s*)[556] and 2290 (*s*) cm^{-1}.[556] The last four bands belong to the C—D stretching vibrations of C_5D_5N and correspond to the 2256 and 2289 cm^{-1} bands of gaseous D_5-pyridine.[556] Their first and second overtone frequencies as well as their values in solution (and for the solid state) have also been measured.[556]

The frequencies calculated for the 19 in-plane and 8 out-of-plane normal vibrations of C_5D_5N, on basis of force constants transferred from the force field of benzene, agree with the values observed for them within an average error of[557] 4%. The 400–3200 cm^{-1} spectra of liquid 2-, 3-[558] and 4-deuteropyridine[558, 559] were investigated to yield almost all their 27 fundamental frequencies. Their C—D stretching bands appear at 2258, 2289 and 2285 cm^{-1}, respectively, i.e. practically where the deuterated benzene C—D stretching fundamentals absorb (2263–2289 cm^{-1}, see Section 4.64). Their in-plane C—D bending frequencies, shown at about 834, 848 and 862 cm^{-1}, respectively, are also in the corresponding benzene C—D region (812–877 cm^{-1}; Section 4.64). The deuteropyridine out-of-plane C—D deformation fundamentals appear, however, in the 632–638 cm^{-1} vicinity as compared with the parallel 497–608 cm^{-1} span in the case of deuterated benzenes (*ibid*).

The vibrational spectra of 2-, 3- and 4-deuteropyridine have been calculated on basis of a normal co-ordinate analysis.[560–562]

The 670–1620 cm^{-1} absorption of a Nujol mull of the $ZnCl_2$ complex of 2-deuteropyridine was also reported.[563]

The C—D stretching bands of gaseous 2,6- and 3,5-D_2-pyridine appear at 2252, 2260 and 2281, 2297 cm^{-1}, respectively,[559] the difference in frequency between these isomers probably reflecting a lower C—D bond strength in the case of the 2 and 6 positions as compared with that of the 3 and 5 C—D groups. The normal frequencies of these D_2-pyridines have been calculated.[560–562]

The 375–3650 cm^{-1} spectrum of liquid D_7-γ-picoline (and its solutions in cyclohexane or CCl_4) yielded most of its fundamentals which were found to agree with the values calculated for them.[564] Its CD_3 stretching absorption appears mainly at 2216 (*m*) cm^{-1} while its ring C—D stretching modes of vibration absorb at 2243 (*sh*), 2255 (*s*) and 2277 cm^{-1} (*sh*).

The C—D stretching bands of liquid 2- and 6-deutero-3-picoline are shown at 2244 and 2259 cm^{-1}, respectively.[563] The 670–1620 cm^{-1} absorption of Nujol mulls of their zinc chloride complexes was also measured.[563]

The 300–4000 cm^{-1} spectrum of a potassium bromide pellet of N-methyl-D_5-pyridinium iodide as well as the 1000–4000 cm^{-1} absorption of an aqueous solution of its chloride show no unexpected major differences between their D_5-pyridinium centred fundamentals and the corresponding C_6D_5F frequencies.[565] The C—D stretching bands of the solid iodide appear at 2270 and 2291 cm^{-1} while its CH_3—N^+ stretching band is observed at 1144 cm^{-1} as a result of an interaction between this vibration and another fundamental.

The symmetric ND_2 stretching absorption of a dilute CCl_4 solution of 2-ND_2-pyridine appears at 2472 and 2510 cm^{-1} while its asymmetric ND_2 band is observed at 2626 cm^{-1}. The N—D stretching fundamental of 2-NHD-pyridine gives rise to one band only (at 2565 cm^{-1}) indicating thus that an internal hydrogen bonding in this case is not probable.[520b]

While N-deuteropyrrole solutions in CCl_4 show its free N—D stretching fundamental as a single band (at 2586 cm^{-1}) similar solutions of its fused benzene ring derivative, indole, exhibit this vibration as a doublet band at[566] 2587 and 2607 cm^{-1}. A study of the frequency and relative intensity of these branches in various solvents as well as in the solid phase (at -145°C) demonstrated that this splitting is the result of a Fermi resonance between the N—D stretching vibration and some combination frequency.

Deuteration of indole in the 3-position brings about the disappearance of its characteristic bands at 732 and 767 cm^{-1} and the appearance of new bands at about 808 (*s*), 830 (*m*) and[567] 1660 cm^{-1}.

4.111 Deuterated Diazomethane†

Isotopically homogeneous diazomethane molecules belong to point group C_{2v} and possess 9 normal vibrations, all of which are infrared active. The 230–4000 cm^{-1} absorption spectra of gaseous and solid CD_2N_2, as well as CH_2N_2 which contained appreciable amounts of $CHDN_2$, yielded 7 of the normal frequencies of CD_2N_2 and $CHDN_2$.[568] Their values are given in Table 4.64.

Table 4.64 Fundamental Frequencies (cm^{-1}) of Gaseous CH_2N_2, CD_2N_2 and $CHDN_2$[568]

Band	Species	Approximate description	CH_2N_2	CD_2N_2	$CHDN_2$
ν_1	A_1	In-phase C—H(D) stretching	3077·1	2244 (*s*)	3135
ν_2	A_1	N=N stretching	2102·2	2096 (*vs*)	2097
ν_3	A_1	CH(D) in-phase deformation	1414·4	1213 (*s*)	1309
ν_4	A_1	C=N stretching	1170	970 (*m*)	1157
ν_5	B_1	C—H(D) out-of-phase stretching	3184·5	~2414 (*w*)	2330
ν_6	B_1	CH(D) out-of-phase deformation	1109	902·6	—
ν_9	B_2	C=N=N out-of-plane bending	406	318·3	368

The values given in Table 4.64 agree satisfactorily with the respective calculated values.[568]

Force field refinement calculations for CD_2N_2 were also reported.[569]

4.112 Deuterated Cyanamides

The 290–4000 cm^{-1} absorption of liquid D_2-cyanamide yields a product rule ratio (see Section 3.4), for the symmetrical amino group stretching and

† For the infrared spectra of partially deuterated mono- and dimethylhydrazines see Anthoni, U., Larsen, C. and Nielsen, P. H. (1968). *Acta Chem. Scand.* **22**, 1025.

deformation and C—N and C≡N stretching frequencies, that fits a non-planar structure for the cyanamide molecules, with a barrier of about 660 cm^{-1} to an inversion at the amino nitrogen atom.[570, 571]

The main absorption bands of D_2-cyanamide appear at 540, 720 (ND_2 rocking), 815, 1020 (*vs*; C—N stretching), 1222 (*vs*; ND_2 deformation), 2220 (*vs*; C≡N stretching), 2368 (N—D in-phase stretching), 2458 (*s*) and 2528 cm^{-1} (N—D out-of-phase stretching).

The strongest characteristic bands of liquid HDNCN appear at[570] 772 (NHD rocking), 1442 (NHD deformation), 2240 (C≡N stretching), 2881 (2 × 1442) and 3300 cm^{-1} (N—H stretching).

The values of the normal frequencies of D_2NCN and HDNCN can be reproduced to within an average error of 0·5% by using a ten constants valence forces field.[572]

4.113 Labelled Cyclic Compounds Containing Two Nitrogen Atoms

The 600–4000 cm^{-1} spectra of gaseous and liquid N-deuteropyrazole† as well as the absorption of its CCl_4 solutions were investigated in connection with the elucidation of the pyrazole vibrational spectra.[573] The N—D stretching band of its CCl_4 solution appears at about 2622 cm^{-1} while the N—D deformation fundamental is shown at 864 cm^{-1}.

The N—D stretching frequency of deuteropyrazole solutions in various solvents depends strongly upon the strength of the ND····X hydrogen bond formed between it and the solvent molecules, becoming as low as 2380 cm^{-1} in the case of dimethylsulphoxide.[574]

The 30–4000 cm^{-1} absorption of D_4-imidazole,[575, 576] N-deuteroimidazole,[575] its tribromo-[575, 576] and triido-derivatives[576] and 2,4,5-D_3-imidazole[575] was investigated in connection with the assignment of the imidazole bands. The main bands of solid $C_3N_2D_4$ appear at 656 (*s*), 672 (*s*; out-of-plane C—D bending), 792 (*s*), 809 (*vs*; C—D bending), 913 (*s*; C—D bending), 961 (*vs*; C—D bending), 1118 (*s*; broad), 1422 (*vs*), 1860 (*s*), 1930 (*s*), 2015 (*vs*), 2100 (*vs*), 2140 (*vs*), 2160 (*vs*), 2215 (*vs*) and 2260 cm^{-1} (*s*; all the last 8 bands due to N—D stretching modes). The C—D stretching vibrations of deuterated imidazoles appear only as shoulders or very weak bands at about 2360 cm^{-1}.

The 250–1700 cm^{-1} spectrum of a solution of 1-deuteroimidazole in D_2O has also been reported.[575]

The frequency of its monomeric N—D stretching band varies in different solvents between 2390 (for dimethylsulphoxide) and 2574 cm^{-1} (CCl_4). The value of the $(\nu_{gas} - \nu_{solution})/\nu_{solution}$ quotient in the case of the X—D stretch-

† For the infrared spectrum of D_6-3,3-dimethyldiazinine see ref. 577.

ing vibration has always been found to be lower than the corresponding value in the case of the respective X—H vibration, for imidazoles as well as for pyrroles, phenols and hydrochloric acids. The $(\Delta\nu/\nu)_{XH}/(\Delta\nu/\nu)_{XD}$ ratio for chloroform solutions of imidazoles has been observed to be as high as 1·79, as compared with the value of 1·22 for their ethyl ether solutions and still lower ratios in the case of the other compounds.[574]

The reason for this phenomenon is not yet clear.

Liquid N—CD_3-imidazole (which has also been investigated as a gas, as a solid and in solution) shows its CD_3 deformation bands at 1011 (*m*; symmetrical) and 1094 (*m*) cm^{-1} and absorbs in the C—D stretching region at[578] 2073 (*m*), 2119 (*w*), 2131 (*w*) and 2230 (*w*) cm^{-1}.

Liquid N—CH_3—D_3-imidazole[578] shows its out-of-plane C—D deformation bands at 590 (*m*) and 715 (*w*) cm^{-1}, its in-plane C—D bending fundamentals at 785 (*s*), 846 (*vs*) and 884 (*m*) cm^{-1} and its C—D stretching frequencies at 2314 (*m*; 2-position) and 2350 cm^{-1} (*w*; 4- and 5-positions).

N-deuterated benzimidazole (in CCl_4) exhibits its free N—D stretching band[566] as a (Fermi resonance) doublet at 2583 and 2598 cm^{-1} while 4-methylimidazole shows only one such band, at 2577 cm^{-1}.

The 270–4000 cm^{-1} spectrum of liquid perdeuterated pyridazine[579] shows its main bands at 326 (*s*), 563 (*vs*, *b*), 639 (*s*, *b*), 833 (*vs*, *b*), 853 (*vs*), 914 (*m*), 951 (*vs*), 1038 (*vs*), 1272 (*vs*), 1319 (*m*), 1347 (*m*), 1528 (*vs*), 2259 (*sh*), 2277 (*s*) and 3323 (*m*) cm^{-1}. Its 3,6-D_2-derivative displays its C—D stretching band[579] at 2293 (*m*) cm^{-1} while 4,5-D_2-pyridazine absorbs in this region at[579] 2291 (*s*) and 2299 (*s*) cm^{-1}.

The 300–4000 cm^{-1} absorption of gaseous and liquid D_4-pyrimidine as well as the polarized light spectra of its crystals yielded the values for all but three of its 24 fundamentals.[580]

Liquid $C_4N_2D_4$ shows absorption bands at 304 (*w*), 656 (*s*), 812 (*m*), 826 (*sh*), 845 (*m*), 873 (*w*), 906 (*m*), 930 (*w*), 973 (*s*), 1030 (*w*), 1040 (*m*), 1055 (*w*), 1165 (*w*), 1273 (*s*), 1320 (*w*), 1378 (*w*), 1415 (*vw*), 1505 (*sh*), 1526 (*vs*), 1573 (*w*), 2242 (*m*), 2260 (*s*), 2275 (*sh*) and 2295 (*w*) cm^{-1}.

The spectra of gaseous[581, 582] and liquid[582] D_4-pyrazine as well as its oriented polycrystalline films[581] (both in light polarized perpendicularly and in light polarized in parallel to their growth direction) and its solution in CCl_4 or CS_2[581, 582] were measured down to 400 cm^{-1}. The main $C_4N_2D_4$ solution bands were observed at[581] 599 (*s*; out-of-plane C—D deformation), 611 (*m*), 875 (*vs*; C—D bending), 1025 (*vs*), 1106 (*m*), 1170 (*s*), 1264 (*vs*), 2258 (*m*) and 2276 cm^{-1} (*vs*; C—D stretching).

The absorption of a Nujol mull of a 1:1 complex of D_4-pyrazine and $HgCl_2$ has also been reported.[583]

While the C—D stretching frequency of *cis*-D_2-pyrazine is identical with that of the perdeuterated compound its C—D deformation fundamentals are appreciably higher, appearing at 640 (*vs*) and 918 (*vs*) cm^{-1}, respectively.[581]

The D_3-pyrazine isotopomer[582] is characterized by strong bands at 965 and 1315 cm^{-1}.

The planar fundamentals of deuterated pyrazine molecules were calculated.[584]

The 400–4000 cm^{-1} polarized light spectra of oriented, polycrystalline, normal and perdeuterated phenazine films (with the help of a modified Urey–Bradley field calculation of their infrared active normal frequencies) yielded the following values for the observed $C_{12}N_2D_8$ fundamentals:[585] 438, 574 (*vs*), 603 (*vs*; out-of-plane C—D deformation), 626, 628 (*s*; out-of-plane C—D deformation), 744 (*vs*), 757 (out-of-plane C—D deformation), 820, 830 (C—D bending), 895 (C—D bending), 922, 1047 (C—D bending), 1166 (*vs*), 1255 (*vs*), 1385 (*vs*), 1418, 1491 (*vs*), 1595, 2265, 2286 and 2303 cm^{-1} (the last three due to C—D stretching modes).

4.114 Deuterated Acetamidinium and Guanidinium Salts

The fundamental bands observed for the $[CH_3—C(ND_2)_2]^+$ ion on a mull of its chloroplatinate[586] are reported in Table 4.65. The assumed point group of these ions is C_{2v}.

TABLE 4.65 Fundamental Bands (cm^{-1}) of Isotopic Acetamidinium Ions

Band	Species	Approximate description	$CH_3C(NH_2)_2^+$	$CH_3C(ND_2)_2^+$
ν_1	A_1	$NH_2(D_2)$ out-of-phase stretching	3417	2560 (*s*)
ν_2	A_1	$NH_2(D_2)$ in-phase stretching	3368	2425 (*m*)
ν_4	A_1	$NH_2(D_2)$ scissoring	1667	1176 (*m*)[a]
ν_5	A_1	skeletal band	1520	1499 (*s*)
ν_6	A_1	CH_3 in-phase bending	1379	1381 (*m*)
ν_7	A_1	$NH_2(D_2)$ rocking	1200	943 (*w*; *b*)[a]
ν_8	A_1	skeletal band	876	829 (*w*)
ν_9	A_1	in-plane skeletal deformation	540	480 (*m*; *b*)
ν_{13}	B_1	$NH_2(D_2)$ out-of-phase asymmetrical stretching	3417	2531 (*vs*)
ν_{14}	B_1	$NH_2(D_2)$ in-phase asymmetrical stretching	3368	2397 (*vs*)
ν_{15}	B_1	CH_3 stretching	2950	2950 (*vw*)[a]
ν_{16}	B_1	skeletal band	1690	1642 (*vs*)
ν_{18}	B_1	CH_3 asymmetrical deformation	1411	1412 (*m*)[a]
ν_{20}	B_1	CH_3 rocking	1020	1020 (*vw*)
ν_{24}	B_2	C—CH_3 deformation	1053	1050 (*w*)
ν_{25}	B_2	out-of-plane skeletal deformation	720?	621 (*m*)
ν_{26}	B_2	$NH_2(D_2)$ deformation	610	450 (*s*; *b*)

[a] Shared also by another normal frequency.

The observed 400–4000 cm^{-1} absorption spectrum and the calculated A_1 and E fundamentals of perdeuterated guanidinium ion have also been reported.[587]

4.115 Labelled Azides

The absorption of gaseous D_3-methyl azide in the 80–2400 cm^{-1} region[588] shows its C—D stretching bands at 2118 (*vs*; symmetric) and 2226 cm^{-1} (*m*; asymmetric) and its C—D bending fundamentals at 1094 (*m*; symmetric) and 1285 cm^{-1} (*s*; asymmetric). The torsion of the CD_3 group gives rise to a weak band at 90 cm^{-1} (in the case of CH_3N_3 at 126 cm^{-1}).

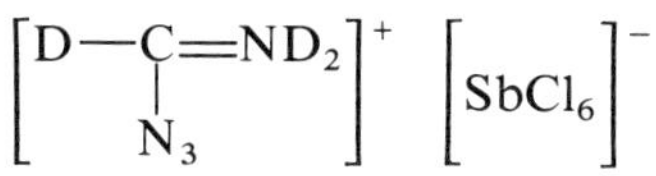

Fig. 4.26 The structure of D_3-imidiumazide hexachloroantimonate.

A mull of D_3-imidiumazide hexachloroantimonate (Fig. 4.26) shows its ND_2 stretching bands at[589] 2405 (*s*; symmetric) and 2545 cm^{-1} (*m–s*; asymmetric), its in-plane ND_2 deformation band at 1140 (*s*) cm^{-1} and an ND_2 rocking fundamental at 967 (*s*) cm^{-1}. The out-of-plane ND_2 deformation absorption appears in this case at 583 (*m*) cm^{-1}.†

4.116 Deuterated Cyclic Compounds Containing More Than Two Nitrogen Atoms

The free N—D stretching band of N-deuterated 1,2,4-triazole in CCl_4 solution[566] appears at 2571 cm^{-1}, showing a ν_H/ν_D isotopic frequency quotient of 1·349.

The infrared spectra of gaseous and solid perdeuterated 1,3,5-triazine was reported.[590] The main absorption bands for its gas phase appear at 577 (*s*), 662 (*s*), 861 (*m*), 1284 (*s*), 1530 (*vs*) and 2280 (*m*) cm^{-1} and agree well with the values calculated for its fundamentals.

The 400–3000 cm^{-1} absorption of D_{15}-2,4,6-triphenyl-1,3,5-triazine has also been measured[591] as well as the infrared spectrum of D_2-s-tetrazine.[592] The N—H stretching band of mono-N-deuterated 5,6-dihydro-2-aminoimidazo[4,5,1-ij]quinoline in a $CHCl_3$ solution is shown at 3458 cm^{-1} while its N—D stretching band appears at 2542 cm^{-1} and the corresponding ND_2 isotopomer absorbs at 2500 (symmetric ND_2 stretching) and[593] 2618 cm^{-1} (asymmetric ND_2 stretching).

† For the absorption of various N-deuterated formazans see Ottig, W. and Neugebauer, F. A. (1968). *Z. Naturforsch*, **23b**, 1064.

The 2180–3400 cm^{-1} spectrum of a mull of partially N-deuterated purine has been reported.[594, 595]

A mull of N-deuterated 9-methyladenine showed its (bonded) ND_2 stretching bands at 2350 (symmetric) and 2500 cm^{-1} (unsymmetric) while its ND_2 wagging absorption appeared at[596] 480 cm^{-1}. In the case of a chloroform solution of N-deuterated 9-ethyladenine the free N—D stretching bands were also observed.[597] They appear at 2483, 2495 (both symmetric) and 2640 cm^{-1} (asymmetric). If one of these deuterium atoms is hydrogen bonded the N—D stretching frequency involving mostly the free atom is 2560 cm^{-1}.

The 650–4000 cm^{-1} absorption of a layer of sublimed D_2—N-deuterated phthalocyanine shows N—D bands at[598, 599] 964, 977, 1076, 1143 and 2480 cm^{-1}.

The infrared spectra of some deuterated tin[600] and germanium[601] phthalocyanines and hemiporphyrazines have also been measured and interpreted.

4.117 Labelled Amides

The infrared absorption frequencies of liquid $DCOND_2$, $DCONH_2$ and $HCOND_2$ were measured in order to derive Urey–Bradley force field constants for formamide, assuming it to possess a C_s symmetry.[602] The C—D stretching band of $DCONH_2$ appears at 2170 cm^{-1} while its in-plane C—D bending vibration gives rise to an absorption at 984 cm^{-1} and the respective out-of-plane fundamental is shown at 881 cm^{-1}. The ND_2 scissoring band of $HCOND_2$ is exhibited at 1118 cm^{-1} as compared with the corresponding NH_2 value of 1608 cm^{-1}.

The 260–3600 cm^{-1} absorption of liquid $DCONHCH_3$, $HCONDCH_3$ and $DCONDCH_3$ is generally in harmony with the values calculated for their fundamental vibrations, assuming them to be in a planar *trans* (C_s) configuration.[603] $DCONDCH_3$ shows its absorption bands at[603] 312, 530 (N—D out-of-plane deformation), 606, 751, 864, 891, 951 (in-plane N—D bending), 1095, 1158, 1405, 1425, 1645, 2154, 2196, 2452, 2505 and 2942 cm^{-1}. Gaseous $HCONDCH_3$ shows its main bands in the 800–1800 cm^{-1} region at[604] 1727 (C═O stretching) and 1372 cm^{-1} (C—N stretching).

Mulls of CD_3COND_2 show strong bands at[605] 1030 (CD_3 asymmetric bending), 1177 (ND_2 bending), 1425 and 1645 cm^{-1} while solid CH_3COND_2 absorbs characteristically at 1189 (ND_2 scissoring) as well as 2315, 2395, 2505 and 2920 cm^{-1}. The CD_3 deformation bands of solid CD_3CONH_2 appear at 1058 (asymmetric) and 1086 cm^{-1} (symmetric), in good agreement with the values calculated for these fundamentals.[605]

The 400–3500 cm^{-1} spectra of liquid and crystalline perdeuterated methylacetamide,[606, 607]

$CH_3CONDCH_3$[604, 606, 607] $CD_3CONHCH_3$, $CH_3CONHCD_3$, $CD_3CONDCH_3$, $CH_3CONDCD_3$ and $CD_3CONHCD_3$[606, 607] made possible the refinement of the interpretation of the vibrational spectra of methylacetamide.[606]

The main absorption bands of liquid $CD_3CONDCD_3$ appear at 506 (*m*), 950 (*m*), 1010 (*m*), 1046 (*m*), 1110 (*m*), 1473 (*s*), 1635 (*vs*), 2406 (*s*), 2444 (*s*) and 2508 cm^{-1} (*m–s*).

The absorption intensity of N-methyl C—D stretching bands was observed to be much higher than that of the corresponding bands of the acetyl groups of deuterated N-methylacetamides.[607]

$CH_3CONDCH_3$, shows its principal bands at[606, 607] 439 (*m*), 510 (*m*; out-of-plane N—D deformation: V′ amide band), 628 (*m*; IV′ amide band: O—C—N deformation), 1000 (*m–s*; III′ amide band: ND bending + C—N stretching + skeletal vibration), 1184 (*m*), 1404 (*s*), 1446 (*m*), 1503 (*s*; II′ amide band: ND bending + C—N stretching), 1647 (*vs*; I′ amide band: C=O stretching), 2420, 2480 (both *s* and due to a Fermi resonance between the N—D stretching and the 1647 + 1000 cm^{-1} vibrations), 2889 (*m*) and 2945 (*m*) cm^{-1}.

The frequencies observed for the planar fundamental vibrations of the various deuterated N-methylacetamides investigated were used to calculate the planar force field of this molecule.[608]

The infrared spectrum of N-deuterated-N-ethylacetamide has also been reported.[604, 609]

The C—N stretching vibration of liquid D_9—N,N-dimethylacetamide gives rise to a strong band at 1430 cm^{-1} as compared with the values of 1388 cm^{-1} for the corresponding band of D_7—N,N-dimethylformamide, 1421 cm^{-1} for that of $CHCl_2CON(CD_3)_2$ and 1398 cm^{-1} in the case of the respective trichloro-amide.[610] These high values for such a fundamental are the result of strong contributions to the actual state of these amides from polarized resonance structures such as:

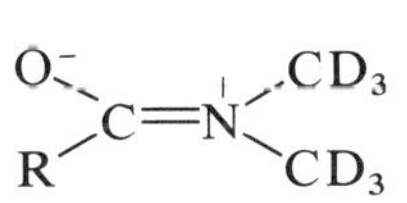

which give these C—N bonds a partial double bond character.

The C=O stretching band of liquid $CD_3CON(CD_3)_2$ appears at 1620 cm^{-1} while that of $CCl_3CON(CD_3)_2$ is shown at 1673 cm^{-1}.

In the vapour phase, contributions from polar structures (which are increased by the dipole dipole interaction operating in the liquid phase) are reduced, hence the C—N stretching frequencies are decreased and those due to the C=O stretching are increased.

The CD_3—N stretching vibration of tertiary amides results in a strong absorption band at[610] *ca.* 1240 cm^{-1}.

The vibrational spectra of D_2-oxamide[611] and D_2—N,N′-dimethyloxamide[604] have been investigated, the latter compound (as a mull) being observed to show its I′, II′, III′, IV′ and V′ amide bands at 1650, 1445, 935, 763 and 567 cm^{-1}, respectively. The infrared absorption of $C_6H_5CONDC_2H_5$,[609] of the boron trifluoride complex of $C_6H_5COND_2$[612] and of $CH_3CONDC_6H_5$[613] has also been reported.

In the case of amides of the general structure R_1CONDR_2, the frequency of the N—D stretching, C═O stretching and II′ amide bands depends appreciably upon the exact identity of R_1 and R_2. If, e.g. R_1 = 2-pyridyl and R_2 = 4-$C_2H_5OC_6H_4$, these bands appear, for a chloroform solution, at 2490, 1682 and 1534 cm^{-1}, respectively, while if R_1 is changed for 3-pyridyl these values become 2552, 1677 and 1518 cm^{-1}, correspondingly.[614]

The 430–4000 cm^{-1} spectrum of melted D_3-diformamide shows its absorption bands[615] at 501 (*vw*), 589 (*m*), 754 (*s*), 873 (*w*; out-of-plane C—D bending), 967 (*m–s*; out-of-plane N—D deformation), 1038 (*w*; C—D bending), 1070 (*vw*), 1228 (*w–m*; N—D bending), 1322 (*s–vs*), 1650 (*vs*), 1714 (*sh*), 1742 (*w–m*), 2070 (*sh*; C—D stretching), 2202 (*w*; C—D stretching), 2325 (*sh*; N—D stretching) and 2415 cm^{-1} (*w–m*; C—D stretching). The main bands of a mull of the mono-sodium derivative of this amide appear at[615] 727 (*s*), 1001 (*m–s*; C—D bending), 1252 (*s*), 1331 (*s*), 1545 (*vs*; *b*), 1646 (*s*) and 2110 and 2153 cm^{-1} (both *m–s* and due to C—D stretching).

4.118 Deuterated Lactams

The absorption of N-deuterated 2-pyrrolidone (γ-butyrolactam) was investigated both for the liquid and for the solid states (at 2° and −180°C) in the 250–3500 cm^{-1} region.[616] The fact that this lactam does not show the 1375 cm^{-1} (II amide) band exhibited by the normal 2-pyrrolidone indicates that it belongs to the N—H bending mode of the latter. The main absorption bands of the liquid N-deuterated pyrrolidone are displayed at 470 (*s*), 488 (*s*), 575 (*s*; out-of-plane N—D deformation), 636 (*s*), 685 (*m*; C═O bending), 798 (*m*), 944 (*m*), 1165 (*m*), 1209 (*m*), 1289 (*s*; C—N stretching), 1413 (*s*), 1460 (*m*), 1492 (*m*) and 1686 cm^{-1} (*s*; C═O stretching).

The isotopic shifts observed for the absorption bands of N-deuterated 2-pyridone assisted in the correct assignment of these bands.[617, 618] For CCl_4 solutions the normal pyridone bands at 723, 991, 1544, 1620, 1658 and 1682 cm^{-1} (*sh*) are shifted in the case of the labelled lactam to 684, 980, 1539, 1610, *ca.* 1618 and 1667 cm^{-1}, respectively, as a result of an appreciable coupling of the C═O stretching, C—N stretching and N—H bending vibration modes in 2-pyridone.[617]

The 2000–2400 cm^{-1} spectrum of partially N-deuterated 2-methyl-4-carbethoxy-5-cyano-6-pyridone was also reported.[619]

4.119 Labelled Oximes

Acetaldoxime exists in the solid state in the *anti* form and when O-deuterated shows its in-plane O—D deformation band at 1078 cm^{-1} and its C—H bending absorption at 1318 cm^{-1}, as compared with the corresponding (coupled) $CH_3CH{=}NOH$ bands at 1435 and 1313 cm^{-1} (see Fig. 4.27). The

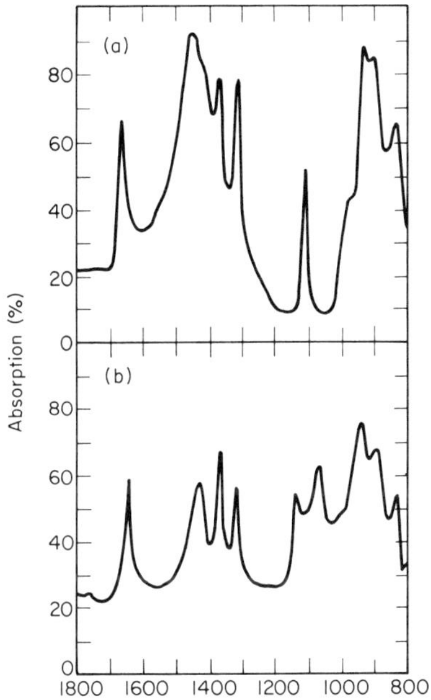

Fig. 4.27 Infrared spectra of solid acetaldoxime (*anti* isomer): (*a*) hydrogen form; (*b*) O-deuterated.[620]

solid C-deuterated acetaldoxime exhibits parallel bands at 1439 and 1062 cm^{-1}, respectively.[620] In the case of liquid phases of acetaldoximes both *syn* and *anti* isomers are present. For 0·1 M solutions in CCl_4 the O-deuterated *syn* isomer shows bands at 955 and 1318 cm^{-1}. The corresponding values for the C-deuterated isotopomers are 1350 and 1058 cm^{-1} for the *syn* isomer and 1327 and 1058 cm^{-1} for the *anti* modification.[620]

O-deuterated benzaldoxime, in solution in CCl_4, displays *syn* bands at 996 and 1360 cm^{-1} and *anti* isomer bands at 1025 and 1361 cm^{-1}. The corresponding values for the *syn* C-deuterated species are 1362 and 1072 cm^{-1}.

Similar bands were observed for solid, O-deuterated *syn*- and *anti*-benzaldoxime, *p*-chlorobenzaldoxime, furfuraldoxime, and *syn*-pyridine-2- and 4-aldoximes as well as for CCl_4 solutions of *syn*- and *anti*-furfuraldoximes and *syn*-*p*-chlorobenzaldoxime.[620]

The in-plane O—D deformation frequency of bis(salicylaldoximato)Me^{II} (Me = Ni, Cu, Co, Pd, Mn) solutions in CCl_4 was studied.[621]

The C=N stretching vibration of 2,4-dimethylpentan-3-one oxime moves as a result of O-deuteration[393] from 1656 to 1642 cm^{-1}.

The 500–4000 cm^{-1} spectrum of solid O-deuterated cyclohexanone oxime shows bands at[622] 580 (associated OD out-of-plane deformation), 955 (N—O stretching), 1080 (associated OD bending), 1652 (C=N stretching) and 2400 cm^{-1} (associated OD stretching). CS_2 or CCl_4 solutions of this oxime absorb at 535 (OD deformation), 930 and 944 (free N—O stretching), 955 (associated N—O stretching), 971 and 1015 (free O—D bending), 1080 (hydrogen bonded OD bending), 1652 (C=N stretching) and 2670 cm^{-1} (free O—D stretching).

The infrared absorption of O-deuterated isonitrosoacetone and diacetyl monoxime has also been reported.[623]

Solid O-deuterated monoxime of *p*-benzoquinone exhibits its C=N stretching band at 1508 cm^{-1} as compared with the value of 1555 cm^{-1} found for the respective band of the normal quinone.[624] Its O—D stretching vibrations absorb at 2140 and 2300 cm^{-1} while the bending of this group is shown at 1092 cm^{-1} and the N—OD stretching band appears at 1030 cm^{-1}. Similar absorptions have been found for crystalline O-deuterated *p*-naphthaquinone monoxime and 1,2-naphthaquinone 2-oxime. 1,2-naphthaquinone 1-oxime, however, exists in the chelated form and displays its O—D stretching band at[624] *ca.* 2000 cm^{-1}.

The absorption of $(CH_3C{=}NOD)_2$[625] and its nickel (nickel dimethylglyoxime) and copper (II) derivatives[626] as well as the spectrum of O-dideuterated 1,2-cyclohexanedionedioxime[625] and 1,2-cycloheptanedionedioxime[625] have all been measured.

4.120 Other Oxygen and Nitrogen Containing Deuterated Compounds

The 400–3000 cm^{-1} absorption of deuterated isocyanic acid vapour shows its six fundamental vibrations at[627] 2634·9 (ν_1: N—D stretching), 2235 (ν_2: asymmetric N=C=O stretching), 1310 (ν_3: symmetric N=C=O stretching), 460 (ν_4: N—D deformation), 766·8 (ν_5: NCO in-plane bending) and 602·9 cm^{-1} (ν_6: out-of-plane NCO deformation). Part of its rotation constants have also been evaluated.[627]

A comparison of the 400–4000 cm^{-1} spectra of gaseous and liquid D_3-isoxazole with the corresponding spectra of normal isoxazole led to a complete vibrational assignment which agrees with the product rule (Section 3.4) within[628] 2%. The main absorption bands of liquid C_3D_3NO appear at 500 (*vs*; ring deformation), 572 (*s*; ring deformation), 612 (*vs*; out-of-plane C—D deformation), 763 (*vs*; in-plane ring deformation), 883 (*s*; in-plane

C—D bending), 886 (*m*), 927 (*m*; ring stretching), 955 (*m*; in-plane C—D bending), 1109 (*vs*; ring breathing), 1270 (*m*; ring stretching), 1401 (*vs*; ring stretching), 1503 (*m*; ring stretching), 2325 and 2381 cm^{-1} (both *w*; C—D stretching).

4.121 Labelled Amino Acids

The fully deuterated ammonium salt of the nonexisting simplest amino acid, carbamic acid, formed by CO_2 and ND_3 at $-25°C$, was investigated[629] as a thin film at $-190°C$. It shows its ND_2COO^- fundamental bands in the 600–4000 cm^{-1} region at 656, 691 (asymmetric CO_2^- wagging), 748 (ND_2 rocking), 822, 968 (ND_2 wagging), 1113 (symmetric CO_2^- stretching), 1199 (ND_2 bending), 1455 (N—C stretching), 1520 (asymmetric CO_2^- stretching), 2424 (symmetric N—D stretching) and 2583 cm^{-1} (asymmetric N—D stretching). The ND_4^+ ion absorbs at 913, 1073, 1154, 1172, 1232, 2071, 2133, 2242, 2310 and 2330 cm^{-1}.

The absorption of crystalline α-D_5-glycine,[630] mono-,[631] di-[631] and tri-N-deuterated glycine,[630,631] C-dideuterated α-glycine[630] and N-deuterated γ-D_3-glycine[632,633] has been investigated down to quite low frequencies. Figure 4.28 reproduces a number of such spectra and Table 4.66 collects the frequencies of their main bands.

The vibration frequencies of the various glycines were used to calculate their force field.[630]

The absorption spectra of C_2-deuterated as well as of N-deuterated bis(glycino)copper(II) monohydrates were also reported,[636,640] as were the spectra of the sodium salts of these glycine isotopomers and the silver salt of ND_2CH_2COOH.[639]

The $[D_3N^+CH_2COOD]X^-$ (X = Cl, Br) salts show[631] C—H stretching bands at 3008 cm^{-1} (asymmetric), bonded N—D deformation absorptions at 1150, 1162 and 1178 cm^{-1}, ND_3 rocking fundamentals at 752 and 768 cm^{-1} and an out-of-plane C—OD deformation vibration at 648 cm^{-1}. The strong bands observed for the deuterochloride of the N-deuterated glycine at[634] 1970, 2067, 2106, 2164, 2192, 2208, 2250 and 2388 cm^{-1}, are probably due, at least partly, to its hydrogen bonded N—D stretching modes.

The infrared spectra of N-deuterated diglycine deuterochloride and deuterobromide have also been reported.[631]

The absorption of solid N-deuterated D_2-sarcosine was measured in the 400–1800 cm^{-1} region.[641] Its ND_2^+ rocking vibrations give rise to bands at 738 (*s*), 788 (*s*), 848 (*s*), 889 (*s*) and 1112 (*s*) cm^{-1} while its ND_2^+ bending fundamental appears at 1268 cm^{-1}. The 250–4000 cm^{-1} spectrum of N-deuterated N,N-dimethylglycine[642] shows its ND^+ deformation band at 1040 (*m*) cm^{-1}. In the N—D^+ stretching region it shows strong bands at 1880, 1970 and 2040 cm^{-1}. The 250–700 cm^{-1} absorption of D_2—N-acetylglycine was also measured.[643]

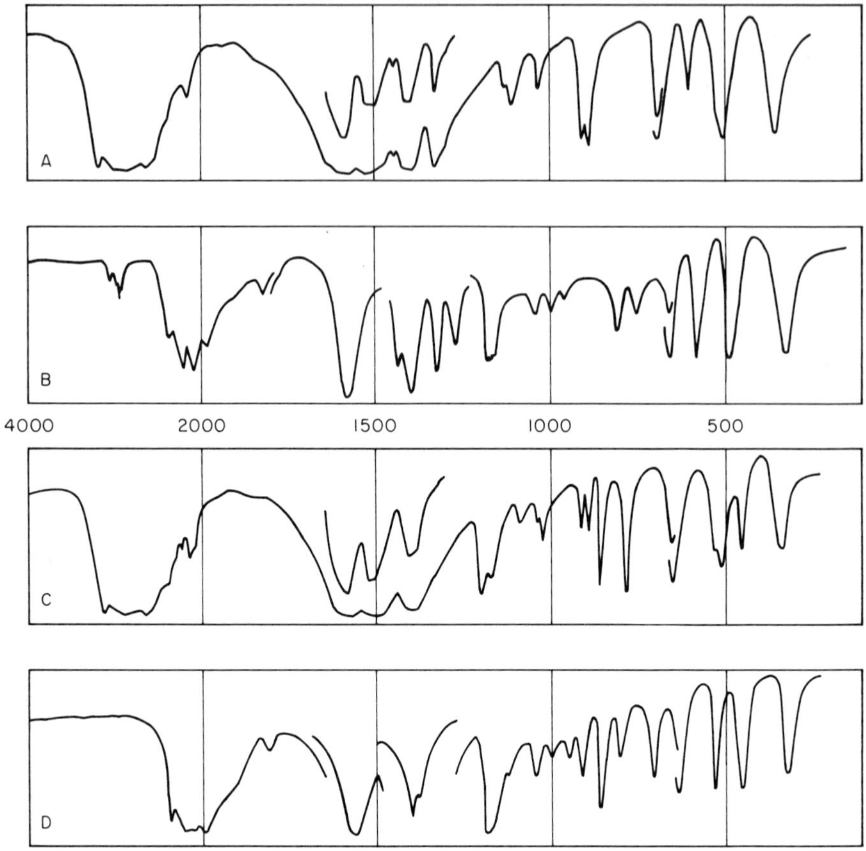

Fig. 4.28 Infrared absorption spectra of the α-form crystals of (A) glycine, (B) glycine-D_3, (C) glycine-D_2 and (D) glycine-D_5.[630]

Perdeutero-L-alanine[644] absorbs at 732 (*m*), 792 (*s*), 806 (*s*), 868 (*s*), 901 (*m*), 937 (*m*; C—D bending), 1034 (*m*), 1048 (*m*), 1058 (*m*), 1135 (*m*), 1180 (*s*), 1395 (*s*) and 1595 (*s*) cm^{-1} while $CH_3CD(NH_3^+)CO_2^-$ shows its main bands at[644] 823, 880 (*m*; C—D bending), 1165, 1219, 1277 (*m*), 1369 (*s*), 1398 (*s*), 1450 (*s*), 1594 (*vs*, *b*) and 1628 cm^{-1}. N-deuterated D_3—DL-α-alanine[634] exhibits its ND_3^+ torsion band at 394 (*m*) cm^{-1} and its ND_3^+ bending vibration at 1180 cm^{-1}. In the ND_3^+ stretching region it shows six strong bands between 1995 and 2300 cm^{-1}.

The infrared spectra of $CH_3CD(ND_3^+)CO_2^-$ and $CD_3CD(NH_3^+)CO_2^-$ were also measured[644] as was the absorption of the copper salt of deuterated DL-α-alanine[640] plus sodium and silver salts of ND_2-alanine.[639]

The deuterochloride of N-deuterated alanine[634] shows its ND_3^+ torsion band at 368 (*m*) cm^{-1}, its out-of-plane C—OD deformation vibration at

TABLE 4.66 Main Absorption Bands (cm^{-1}) of Various Isotopic Glycines

α-Glycine[631]	D_5-Glycine[630]	$ND_3^+CH_2CO_2^-$ [634]	$NH_3^+CD_2CO_2^-$ [630]	Assignment
356	—	330 (*s*)	—	CCN bending
505, 516	—	493 (*s*)	545	NH_3/ND_3 torsion[635]
608	529	596 (*s*)	520	CO_2 wagging
697	646	667 (*s*)	668	CO_2 rocking[635]
895	872	823 (*m*)[a]	872	CO_2 bending[635]
911	804	965[630] (*w*)[a]	799	CH_2/CD_2 rocking
1033	954[a]	1043 (*m*)	1040[a]	C—N stretching
1114	702	763 (*m*)	1185	NH_3/ND_3 rocking
1132	1015	—	1210	NH_3/ND_3 rocking
1334	1074[a]	1324 (*s*)	1056[a]	CH_2/CD_2 wagging
1414	1392, 1411	1403 (*s*)	1396, 1417	symmetrical CO_2 stretching
1445	1134[a]	1441 (*m*)	1110[a]	CH_2/CD_2 scissoring
1520	1168, 1178, 1197	1166, 1176, 1188 (*s*)	1499, 1507, 1521	NH_3/ND_3 deformation
1591	1576	1592 (*s*)	1603	antisymmetrical CO_2 stretching[635]
2830,[635] 2970	—	2972 (*m*)	—	CH_2/CD_2 symmetrical stretching[635]
2915,[635] 3007	—	2940,[635] 3007 (*m*)	2163	CH_2/CD_2 antisymmetrical stretching
3180[635]	—	1914–2341 (*s*)[b]	—	NH_3/ND_3 stretching

[a] In resonance with another mode.
[b] Nine bands.

592 (*s*) cm^{-1}, its O—D bending fundamental at 1024 (*s*) cm^{-1} and its ND_3^+ bending absorption at 1173 (*s*) cm^{-1}. The ND_3^+ stretching vibrations are observed at 1960 (*m*), 2030 (*m*), 2170 (*s*) and 2250 (*s*) cm^{-1}.

The ND_3^+ deformation bands of D_3-α-amino-n-butyric acid,[645, 646] D_3-DL-norvaline,[645, 646] D_3-DL-valine,[641] D_3-DL-glutamic acid,[645, 646] D_3-β-phenylalanine[634] and the deuterochlorides of norvaline and O-deuterated D_3-glutamic acid appear at 1175, 1180, 1177 (*s*), 1173, 1166, 1173 and 1168 cm^{-1}, respectively. Their ND_3^+ rocking fundamentals are exhibited at 840, 829 (and 809), 756 (*m*), 843 (and 821), 800, 806 (and 785) and 779 (and 751) cm^{-1}, respectively.

The 680–5000 cm^{-1} spectra of all the possible deuterated L-aspartic acids and various deuterated DL-aspartic acids were reported,[647] these being used for arriving at the correct assignment of the 1100–1700 cm^{-1} absorption peaks of aspartic acids. A KBr pellet of perdeutero-L-aspartic acid shows its ND_3^+ rocking band at 760 (*m*) cm^{-1}, its ND_3^+ deformation absorption at 1185 (*s*) cm^{-1} and its O—D bending fundamental at 1090 (*s*) cm^{-1}. Its carbon–oxygen bands appear at 1355 (*s*; C—O stretching), 1415 (*s*; CO_2^- symmetrical stretching), 1610 (*s*; antisymmetric stretching) and 1680 cm^{-1} (*s*; C=O stretching).

The infrared spectrum of a KBr pellet of acetyl-α,β,β-D_3-DL-aspartic acid was also measured.[647]

The out-of-plane C—OD deformation band of the deuterochloride of N—D_3-phenylalanine[634] is shown at 616 (*s*) cm^{-1} while its in-plane bending vibration mode absorbs at 1028 (*s*) cm^{-1} and the ND_3^+ deformation absorption is centred at 1160 (*s*) cm^{-1}. The absorption spectrum of the Na^+ and Ag^+ salts of N—D_2—phenylalanine has also been reported.[639]

The ND_3^+ rocking band of N-deuterated *m*-aminobenzoic acid[648] appears at 938 (*w*) cm^{-1} and its ND_3^+ deformation fundamentals are observed at 1076 (*m*; symmetric) and 1241 cm^{-1} (*m*; antisym.) while its ND_3^+ stretching vibrations absorb at 1920 (*m*; *b*), 2040 (*m*; *b*), 2180 (*m*; *b*) and 2270 cm^{-1} (*m*; *b*). Its deuterochloride shows these bands at 934 (*m*), 1085 (*m*), 1233 (*m*) and 2070 and 2190 cm^{-1} (*s*, *b*), respectively.[648] The potassium salt of this acid[648] shows its N—D deformation band at 1086 (*m*) cm^{-1} and its N—D stretching bands at 2330 (*s*), 2410 (*s*) and 2550 (*s*) cm^{-1}.

A dilute CCl_4 solution of N-deuterated ethyl *p*-aminobenzoate[520b] exhibits bands at 2462, 2497, 2507 and 2619 cm^{-1} (free ND_2 antisymmetric stretching) besides a NH—D stretching band at 2551 cm^{-1}.

The infrared spectrum of deuterated copper-histidine[637] and various N-deuterated metal complexes of l-cysteine[638] was also studied.

4.122 Deuterated Nitro Compounds

The absorption spectrum of perdeuterated nitromethane vapour[649] shows its fundamental bands at 424 (*vw*), 563 (*vw*), 632 (*w*), 879 (*m*), 944 (*w*; CD_3 rocking), 1051 (*w*; CD_3 bending), 1093 (*vw*; CD_3 bending), 1384 (*s*), 1571 (*vs*), 2195 (*vw*; CD_3 symmetrical stretching) and 2290 cm^{-1} (*vw*; asymmetrical stretching). The products of the observed CD_3NO_2 fundamental frequencies belonging to the A_1, B_1 and B_2 species (assuming CD_3NO_2 to belong to the $\boldsymbol{C_{2v}}$ point group) are in agreement with the values calculated for them according to the Product Rule.

The 400–3500 cm^{-1} spectra of liquid $ClCD_2NO_2$ and $BrCD_2NO_2$[650] show their CD_2 rocking bands at 507/520 (*w*) cm^{-1}, their CD_2 twisting absorption peaks at 863/805 (*w*) cm^{-1}, the CD_2 wagging fundamentals at 1024/994 (*s*) cm^{-1}, the CD_2 scissoring vibrations at 1043/1036 cm^{-1} and their C—D stretching frequencies at 2203/2202 (*w*; symmetric) and 2303/2309 (*w*; asym.) cm^{-1}, respectively.

The CD_2 wagging, rocking, bending and stretching bands of a mull of the sodium salt of fully deuterated nitromethane ($CD_2{=}NO_2Na$) appear at[651] 536 and 547 (*m*), 876 and 884 (*s*), 1205 (*w*), and 2124 (*w*; symmetric) and 2308 cm^{-1} (*w*; asymmetric), respectively.

The C—D stretching absorption of deuterated dinitromethane was observed[652] at 2380 cm^{-1}. The NO_2 stretching frequencies of deuterotrinitromethane were also reported.[653]

The fundamental bands of gaseous 1-deutero-1-nitroethylene are shown at[654] 95 (*vw*), 315 (*w*), 546 (*w*), 589 (*w*), 643 (*m*), 763 (*s*), 873 (*s*), 897 (*s*), 958, 962 (*vs*), 1135 (*m*), 1350 (*s*), 1376 (*vs*), 1553 (*vs*), 1622 (*m*), 2250 (C—D stretching), 3048 (*w*) and 3120 (*w*) cm^{-1}.

The 620–5000 cm^{-1} absorption spectrum of D_5-nitrobenzene shows only two slight shifts in the location of the NO_2-centred fundamentals[655] (for those at *ca.* 850 and at 1300 cm^{-1}). In the case of D_3-trinitrobenzene also the vibrations involving the NO_2 groups remain practically unchanged upon deuteration.[656] Its out-of-plane C—D deformation band appears at 518 cm^{-1}, its in-plane C—D bending absorption is shown at 806 cm^{-1} and its C—D stretching fundamental is exhibited at 2305 cm^{-1}.

The hydrogen bonded O—D stretching frequencies of dimeric *o*-, *m*- and *p*-nitrophenol in CCl_4 solutions[657] are shown at 2438, 2518 and 2446 cm^{-1}, respectively while the free O—D stretching vibrations of *m*- and *p*-nitrophenol appear at 2662 and 2655 cm^{-1}, respectively.

The free ND_2 stretching frequencies of 2- and 4-nitroaniline in CCl_4 solutions appear at 2471, 2491 (symmetric), 2636 (asymmetric) and 2471 (symmetric) and 2626 cm^{-1} (asymmetric), respectively, while the N—D frequencies of their NHD species are observed at 2566, 2551 and 2560 cm^{-1}, respectively.[520b]

The 430–3260 cm^{-1} spectra of crystalline N-deuterated ethylenedinitramine and tetramethylenedinitramine have also been reported.[658]

4.123 Other Labelled Compounds Containing One Nitrogen Atom

O-deuterated N-acetylmorpholine salts,

$$X^{-}\cdots D—O^{+}{=}\underset{\displaystyle CH_3}{\underset{|}{C}}—N(CH_2CH_2)_2O$$

(X = Cl, Br, I, $SbCl_6$) show their O—D stretching bands between 1700 (for X = Cl) and 2100 cm^{-1} (X = I), the ν_{OH}/ν_{OD} ratio of these salts varying between [659] 1.21 and 1.32. The respective O—D bending frequencies have been observed in the 930–992 cm^{-1} range, the corresponding ν_{OH}/ν_{OD} ratios being 1·27–1·32.

The 1000–3500 cm^{-1} spectra of deuterated acid iminodiacetates of a number of metals have also been studied.[660]

4.124 Deuterated Oxygen Compounds Containing Two Nitrogen Atoms

The 400–4000 cm^{-1} spectrum of a mull of N-deuterated urea shows bands at[661,662] 520 (*w*; N—C—N deformation), 550 (*w*; N—C—N in-plane rocking), 778 (*m*; out-of-plane N—CO deformation), 890 (*m*; ND_2 rocking), 1005 (*m*; N—C—N symmetrical stretching), 1055 (*m*; NHD rocking), 1212 (*w*; in-plane NHD bending), 1247 (*w*; in-plane ND_2 bending), 1460 and 1500 (*vs*, *b*; asymmetrical N—C—N stretching), 1675 (*vs*; C=O stretching), 2375 (*sh*; combination), 2430 (*vs*; ND_2 symmetrical stretching), 2500 (*vs*; NHD stretching) and 2585 cm^{-1} (*vs*; asymmetrical ND_2 stretching).

The 2000–3200 cm^{-1} absorption of 1% solutions of O-deuterated 2-azophenylphenol and 1-azophenyl-2-naphthol in CCl_4 shows their intermolecularly hydrogen bonded O—D stretching bands at *ca.* 2200 and 2155 cm^{-1}, respectively.[663]

N-deuterated D_2-diformylhydrazine which is assumed to be in the planar *trans* form (point group: C_{2h}) can be expected to exhibit 8 planar (B_u) and 4 other (A_u) infrared active fundamentals. The first were observed at[604] 540 (N—D bending), 751 (O—C—N bending), 985 ("III amide band"), 1338 ("II amide band"), 1389 (C—H bending), 1605 (C=O stretching), 2250 (N—D stretching) and 2920 cm^{-1} (C—H stretching).

The main absorption bands of crystalline N-deuterated D_2-diacetylhydrazine appear at[604] 430 ("V"), 588 ("VI′"), 629 ("IV"), 1413 ("II") and 1595 cm^{-1} ("I").

The normal vibrations of N-deuterated D_2-diketopiperazine have also been studied,[664] their out-of-plane N—D deformation mode being found to appear at 652 cm^{-1} as compared with the 510–570 cm^{-1} range observed for the corresponding vibration in the case of secondary amides.[604]

The absorption of various D_2-3-cyano-2,5-pyrrolidiones was measured in the 1150–3500 cm^{-1} region.[665]

The 400–4000 cm^{-1} spectrum of a mull of N-deuterated 1-methylthymine shows its N—D out-of-plane deformation band at[666] 650 cm^{-1} while a $CHCl_3$ solution of N-deuterated 1-cyclohexyluracil[667] exhibits its free N—D stretching absorption at 2501, 2526 and 2548 cm^{-1} (as compared with 2560 cm^{-1} for the free N—D stretching frequency of N-deuterated 9-ethyladenine in chloroform solution).[667] The 400–3600 cm^{-1} spectrum of 9-methyladenine has also been reported.[666]

The fact that partly deuterated "1,2-dihydro-2-imino-4H-3,1-benzoxazin-4-one" shows a *new* (N—H(D) stretching) band (at 3480 cm^{-1}) between the (symmetric and antisymmetric) NH stretching bands, at 3422 and 35352 cm^{-1}, already exhibited by the original compound, proves that these substances really exist as the amino-tautomers[668] (see Fig. 4.29).

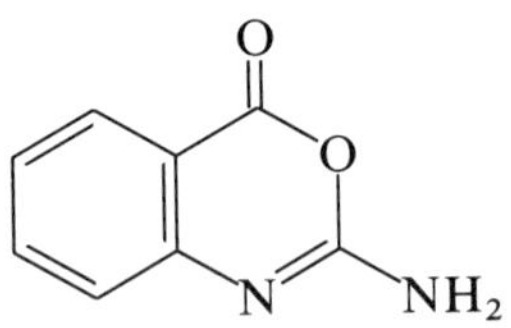

Fig. 4.29 The structure of "1,2-dihydro-2-imino-4H-3,1-benzoxazin-4-one".

The 130–4000 cm^{-1} spectrum of D_2-parabanic acid[669] shows bands at 176 (*m*), 207 (*m*), 366 (*w*), 426 (*s*), 482 (*m*), 531 (*w*; out-of-plane N—D bending), 560 (*m*; out-of-plane N—D bending), 593 (*s*), 660 (*m*), 758 (*m*), 762 (*m*), 782 (*w*), 805 (*m*), 860 (*m*; N—D bending), 902 (*m*; N—D bending), 992 (*m*), 1126 (*m*), 1216 (*m*), 1350 (*s*), 1368 (*s*), 1733 (*s*), *ca.* 1760 (*vs*), *ca.* 1780 (*vs*), 1860 (*m*), 2352 (*s*; N—D stretching), 2427 (*s*; N—D stretching) and *ca.* 3160 (*m*) cm^{-1}.

The N—D stretching bands of N-deuterated anhydrous barbituric acid appear at 2250 (*w*), 2300 (*m*) and 2350 (*m*) cm^{-1} while its hydrate with (2 molecules of) D_2O shows them at 2200 (*m*) and 2290 (*m*) cm^{-1}, and N- deuterated diethyl barbituric acid exhibits them at[670] 2278 and 2309 cm^{-1}. The N—D stretching fundamental of anhydrous deuterated orotic acid[670] was observed at 2300 cm^{-1} (2273 cm^{-1} for the heavy hydrate).

The 1250–5000 cm^{-1} spectrum of N-deuterated 5-(methoxycarbonylmethylene) hydantoin was also studied.[670]

4.125 Labelled Compounds having Three Nitrogen Atoms

The 400–3800 cm^{-1} spectrum of D_5-biuret mulls shows its N—D stretching bands at[671] 2280, 2330, 2405, 2510 and 2590 cm^{-1}. Its N—D bending fundamentals appear at 915 and 1370 cm^{-1} while its N—D out-of-plane deformation vibration is exhibited at 560 cm^{-1}.

The absorption of mulled D_3-1,5-dipropionylbiuret[671] shows its N—D stretching fundamentals at 2295 (symmetric) and 2395 cm^{-1} (asymmetric), its N—D bending vibrations at 987 and 1300 cm^{-1} and its N—D out-of-plane deformation mode at 583 cm^{-1}.

In the case of D_4-1-propionylbiuret[671] the N—D stretching bands appear at 2270, 2292, 2328, 2390 and 2567 cm^{-1} while the N—D bending fundamentals are shown at 890, 975, 1380 and 1405 cm^{-1}, and the N—D out-of-plane vibration is displayed at 545 cm^{-1}.

The 500–3600 cm^{-1} spectrum of a sublimed film of 50–60% deuterated cyanuric acid[672] exhibits its N—D stretching bands at 2310 (*s*) and 2410 (*s*) cm^{-1} and its N—D out-of-plane deformation absorption at 570 cm^{-1} (*w*; sharp).

Various Other Deuterated Organic Compounds

4.126 Beryllium-containing Molecules

The Nujol mull of the complex formed between dimethylberyllium and sodium deuteride, $Na_2(CH_3)_4D_2Be_2$ shows its absorption bands below 1350 cm^{-1} at[673,674] 595 (*vw*), 719 (*w*), 775 (*s*), 870 (*vs*), 920 (*vs*), 1006 (*vs*), 1075 (*sh*), 1141 (*m–s*) and 1198 (*m–s*) cm^{-1}. The 870 and 920 cm^{-1} bands (corresponding to the 1165 and 1325 cm^{-1} bands of $Na_2(CH_3)_4H_2Be_2$[679]) are due to the stretching vibrations of the bridges formed between two $(R_2DBe)^-$

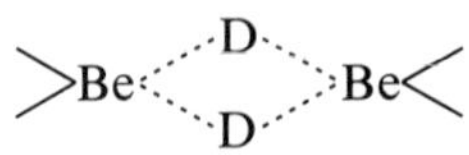

moieties.[673] In the case of the $Na_2(C_2H_5)_4D_2Be_2$ complex these bands have been observed at 835 (*vs*) and 951 (*s*) cm^{-1}, respectively.[673]

The similar $Na_2(CD_3)_4H_2Be_2$ compound[674] absorbs at 600 (*vw*), 678 (*s*), 713 (*m*), 899 (*s*), 934 (*s*), 1043 (*s*), 1164 and 1334 cm^{-1} (the last two due to the BeH_2Be bridge stretching vibrations).

The BeD_2Be bridge stretching frequency of the $(CH_3DBe.N(CH_3)_3)_2$ co-ordination complex (in cyclohexane solution) appears[673] at 1020 cm^{-1}.

The 400–1350 cm^{-1} spectrum of a cyclohexane solution of the complex formed between $(CD_3)_2Be$ and trimethylamine shows a very strong band at 1022 cm^{-1}, corresponding to the 1198 cm^{-1} band of the respective normal complex.[674]

Cyclohexane solutions of $[CD_3BeN(CH_3)_2]_3$ and $[CH_3BeN(CD_3)_2]_3$ show their N—CH_3/CD_3 deformation bands at 1445, 1466 and 1475, and 1064, 1086 and 1101 cm^{-1}, respectively while their N—CH_3/CD_3 rocking modes appear at 1175 and 1233, and 966 and 1064 cm^{-1}, correspondingly.[674]

The spectrum of the monomeric chelate complex of $(CD_3)_2Be$ with tetramethylenediamine exhibits its symmetrical CD_3 deformation band[674] at 969 cm^{-1} ($\nu_H/\nu_D = 1{\cdot}22$).

Fully deuterated di-*t*-butylberyllium[675] displays its CD_3 bands at 680 (*w*, *b*; CD_3 rocking), 806 (*m*; CD_3 rocking), 866 (*s*; CD_3 rocking), 1018 (*m*; symmetric CD_3 deformation), 1064 and 1069 (*s*; asymmetric CD_3 deformation), 2025 (*m*; *sh*), 2057 (*vs*), 2102 (*m*), 2176 (*vs*), 2202 (*vs*), 2213 (*s*; *sh*) and 2225 cm^{-1} (*m*, *sh*; all seven bands due to C—D stretching).

The C—D stretching bands of the deuterated compound have an intensity pattern which is quite different from that of the respective C—H stretching bands and extend over a wider frequency range, this fact being characteristic of $(CD_3)_3$—C— compounds in general.[675]

4.127 Boron-containing Compounds

Gaseous perdeuterated trimethylborane shows its fundamentals at[676] 870 (CD_3 rocking), 1018 and 1070 (CD_3 deformation), 1205 (B—C stretching), 2185 (symmetric C—D stretching) and 2230 cm^{-1} (asymmetric C—D stretching). These values agree within 6% with the frequencies calculated for these fundamentals assuming $(CD_3)_3B$ to have a C_{3h} symmetry.[676]

Gaseous $(C_2D_5)_3B$ shows its main bands at[676] 725 (*m*; CD_2 rocking), 881 (*m*; C—C stretching), 945 (*m*; CD_3 rocking), 1028 (*m*; CD_3 deformation),

1067 (*m–s*; CD_2 scissoring), 1110 (*m–s*; B—C stretching), 1238 (*vs*; out-of-plane CD_2 deformation), 2080 (*s*; symmetric C—D stretching) and 2220 cm^{-1} (*vs*; asymmetric CD_3 stretching).

The vibrational spectra of $B_2D_5.CD_3$, $B_2H_5.CD_3$ and $B_2D_5.CH_3$ have also been calculated.[677]

Gaseous B-deuterated dimethoxyborane[678, 679] shows its out-of-plane B—D deformation band at 770 cm^{-1} (*m–s*) and its B—D stretching absorption at 1905 (*s*) cm^{-1}. In the case of $DB(OC_2H_5)_2$ these bands appear at 762 (*m–s*) and 1890 cm^{-1}, respectively.[680]

Liquid D_9-trimethoxyborane[681] shows its CD_3 rocking band at 905 cm^{-1} (*vw*), its CD_3 bending fundamentals at 1060 (*vw*) and 1145 (*vs*) cm^{-1}, and a CD_3 stretching absorption at 2078 (*vs*), 2135 (*m*), 2228 (*s*) and 2270 cm^{-1} (*sh*).

D_4-1,3-dioxa-2-chloroboracyclopentane[682] exhibits a weak CD_2 torsion band at 915 cm^{-1}, out-of-plane CD_2 deformation absorptions at 940 (*w*) and 975 (*m*) cm^{-1} and C—D stretching vibrations at 2120 (*m*), 2140 (*w*), 2180 (*vw*) and 2245 (*m*) cm^{-1}.

The 630–3500 cm^{-1} spectrum of D_9-trimethoxyboroxine[681] displays its CD_3 bands at 900 (*vw*; rocking), 1060 (*vw*), 1112 (*w*; bending), 1124 (*m*) and 2081, 2138 (*m*) and 2240 cm^{-1} (*m, b*; all three due to stretching modes).

Gaseous N-deuterated D_2-B-dimethylaminoborane[683] absorbs at 682 (*s*), 692 (*s*), 744 (*m*), 882 (*s*), 932 (*m*), 1075 (*m–s*), 1145 (*vs*; ND_2 bending), 1160 (*s*), 1172 (*s*), 1327 (*vs*), 1470 (*s*), 2500 (*m*; symmetrical ND_2 stretching), 2620 (*m–s*; asymmetrical ND_2 stretching), 2860 (*m–s*), 2940 (*s*) and 3460 (*s*) cm^{-1}.

Trimethylamino-D_3-borane[684] shows its (symmetric and asymmetric) BD_3 stretching bands at 1656 and 1780 cm^{-1} while its B—D bending fundamentals appear at 840 and 860 cm^{-1}, respectively, and the corresponding BD_3 rocking vibration is observed at 720 cm^{-1}.

The 650–6000 cm^{-1} absorption spectrum of gaseous $B_2D_5.N(CH_3)_2$ exhibits its main bands at[685, 686] 791 (*m*), 835 (*m*), 891 (*s*), 895 (*s*), 1031 (*vs*), 1037 (*vs*), 1042 (*vs*), 1163 (*s*), 1205 (*s*), 1236 (*m*), 1244 (*s*), 1261 (*m*), 1275 (*m*), 1290 (*m*), 1310 (*m*), 1459 (*m*), 1471 (*m*), 1477 (*m*), 1486 (*m*), 1775 (*s*), 1829 (*s*), 1926 (*vs*), 2818 (*m*), 2869 (*s*), 2881 (*s*), 2946 (*vs*), 2961 (*vs*), 2987 (*vs*), 3028 (*vs*) and 4421 (*m*) cm^{-1}.

D_{10}-1,3-dioxa-2-dimethylaminoboracyclopentane shows its C—D centred modes of vibration at[682] 840 (*m*; CD_2 rocking), 920 (*w*; CD_2 torsion), 982 (*m*; CD_2 out-of-plane deformation), 1065 (*w*; CD_3 deformation), 1100 (*sh*; CD_3 deformation), 2060 (*s*), 2100 (*w*), 2136 (*w*), 2177 (*m*) and 2225 cm^{-1} (*m*; all five due to C—D stretching modes).

The infrared spectrum of the D_4-isotopomer of this compound has also been reported.[682]

The 630–3500 cm^{-1} absorption of (CCl_4 or CS_2) solutions of various

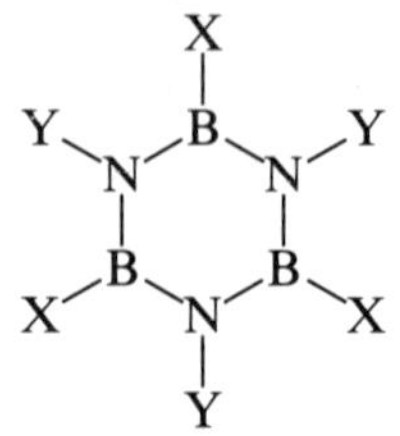

Fig. 4.30 The structure of 1,3,5-borazines.

1,3,5-borazines which contain N—CD_3 groups (Fig. 4.30, Y = CD_3; X = H, D, Cl, F, CD_3, N_3, NCO, NH_2) usually shows five C—D stretching bands between[687] 2070 and 2300 cm^{-1}. Their CD_3 rocking band appears at about 895 (*m*) cm^{-1} while their CD_3 bending vibrations give rise to absorptions near 1070 (*w*) and 1125 (*s*) cm^{-1}. Such borazines which possess B—D groups,[687] (Fig. 4.30, X = D; Y = CH_3, CD_3) exhibit 3–5 B—D deformation bands in the 740–785 cm^{-1} region and a B—D stretching absorption at 1865–1870 (*vs*) and 1910–1940 (*w*) cm^{-1}.

The spectra of other similar borazines have also been measured.[688, 689]

Borazines having B—OCD_3 groups[681] (Fig. 4.30, X = OCD_3; Y = H, CH_3, C_6H_5) display their CD_3 rocking band at 930–950 (*w*) cm^{-1}, their asymmetric and symmetric CD_3 deformation fundamentals at *ca.* 1060 (*w*) and 1120 cm^{-1} (*m–s*), respectively, and the C—D stretching vibrations at 2065–2070 (*s*), 2135 (*w*) and 2225–2240 (*m–s*) cm^{-1}.

The 650–3500 cm^{-1} spectra of solutions of borazines containing B—$N(CD_3)_2$ groups (Fig. 4.30, X = $N(CD_3)_2$; Y = H, CH_3, C_6H_5) show their CD_3 centred vibrations at[690] 830–918 (*w*; rocking), 1050–1060 (*m–s*; asymmetrical deformation) and 2049–2240 cm^{-1} (one strong (at 2049–2057 cm^{-1}) + 3 other C—D stretching bands).

4.128 Silicon Compounds

The 400–5000 cm^{-1} spectrum of gaseous $CH_3.SiHD_2$ exhibits its SiD_2 stretching bands at 1558 (*s*; symmetric) and 1583 cm^{-1} (*vs*; asymmetric).[691–2]

Gaseous $CH_3.SiD_3$ shows its fundamentals 433[691, 692] (*m*; SiD_3 rocking), 650 (*s*; largely, a Si—C stretching), 740 (*vs*; mainly, a symmetric SiD_3 deformation), 821 (*vs*; CH_3 rocking), 1264 (*m*; CH_3 symmetric deformation), 1411 (*w*; CH_3 asymmetric deformation), 1558 (*vs*; symmetric Si—D stretching), 1577 (*vs*; asymmetric Si—D stretching), 2923 (*m*; symmetric CH_3 stretching) and 2982 cm^{-1} (*m*; asymmetric CH_3 stretching). The analysis of the fine structure of the 2923 cm^{-1} rotation–vibration band yielded values for a number of its rotational constants.[693]

The calculated vibrational spectrum of deuterated methylsilane was also reported.[694]

The main absorption bands of gaseous ethyl-D_3-silane and $(CH_3)_2SiD_2$ in the 400–3500 cm^{-1} region were observed at[695] 558 (*s*), 703 (*s*), 804 (*vs*), 831 (*vs*), 1260 (*s*), 1554 (*vs*), 1570 (*vs*) and 2975 (*s*) cm^{-1}.† In the case of $(CH_3)_3SiD$ these appear at[695] 520 (*s*), 708 (*s*), 792 (*s*), 852 (*vs*), 1264 (*s*), 1419 (*s*), 1437 (*s*), 1548 (*vs*), 2913 (*m*) and 2969 (*s*) cm^{-1}.

A solution of $(CD_3)_3SiCl$ in CS_2 shows its CD_3 rocking band at 608 (*w*) cm^{-1} while its main CD_3 deformation fundamental is displayed at 1004 cm^{-1} (*vs*) and its C—D stretching modes absorb at 2125 (*w*; symmetrical) and[696] 2218 cm^{-1} (*m*; asymmetrical).

The A_1, E and F_2 normal vibrations of D_{12}-tetramethylsilane have all been calculated.[697] The 670–3350 cm^{-1} spectrum of its monodeuterated isotopomer[226] (in the gas phase) shows its main bands at 690 (*vs*), 1248 (*s*), 1285 (*m*), 1415 (*s*, *b*), 2189 (*s*) and 2955 (*s*) cm^{-1}.

The 24–300 cm^{-1} absorption of gaseous 1,1-D_2-silacyclobutane has also been recorded.[698]

Liquid $SiD_3CH{=}CH_2$[699] exhibits its SiD_3 rocking band at 509 (*m*) cm^{-1} while its asymmetric and symmetric SiD_3 deformation modes absorb at 672 (*m*) and 724 cm^{-1} (*vs*, *b*), correspondingly. The corresponding Si—D stretching absorptions appear at 1575 (*vs*, *b*) and 1553 (*m*) cm^{-1}, respectively.

Gaseous $SiD_3C{\equiv}CH$ which belongs to the C_3 point group shows its fundamentals at[700] 206 (Si—C≡C deformation), 537 (SiD_3 rocking), 672 (asymmetric SiD_3 deformation), 683 (C≡C—H bending), 709 (symmetric SiD_3 deformation), 1595 (Si—D stretching), 2055 (C≡C stretching) and 3315 cm^{-1} (C—H stretching). A number of its molecular constants have also been reported.[701]

The spectrum of gaseous $SiD_3C{\equiv}CCH_3$ shows its main absorption bands[702] at 125 (ν_{16}), 347 (ν_{15}), 534 (SiD_3 rocking), 683 (ν_7), 704 (symmetric SiD_3 deformation), 708 (asymmetric SiD_3 deformation), 1042 (ν_5), 1060 (ν_{12}), 1383 (ν_4), 1415 (ν_{11}), 1589 (symmetric Si—D stretching), 1593 (asymmetric Si—D stretching), 2074 (ν_3), 2940 (ν_1) and 2983 (ν_9) cm^{-1}.

The absorption of $(CH_3)_3SiC{\equiv}CD$ in the 650–5000 cm^{-1} region[703] exhibits its C—D bending fundamental at 540 cm^{-1} (*vs*) and its C—D stretching band at 2590 (*s*) cm^{-1}.

The 650–4000 cm^{-1} spectrum of liquid triphenyldeuterosilane[704] is almost the same as that of its normal isotopomer but for the (Si—H deformation) band at 805 cm^{-1}, which is much weaker, and the (Si—H stretching) 2135 cm^{-1} band which shifts to 1547 cm^{-1}. This shift is analogous to that observed for the Si—H(D) stretching vibration of $SiCl_3H(D)$ which moves upon deuteration from 2258 to 1647 cm^{-1} (the respective Si—D deformation band shifting from 799 to 605 cm^{-1}).

Liquid 88% $(C_2H_5)_3SiOD$[705] shows its bonded O—D stretching absorption at 2450 cm^{-1} (*s*; *b*) while its CCl_4 solution displays the respective free O—D band at 2732 cm^{-1} (sharp).

† The absorption of gaseous ethyl-D_3-silane is reported in ref. 762.

Gaseous perdeutero(methoxysilane)[706] exhibits its main absorption bands in the 280–4000 cm^{-1} region at 545 (SiD_3 rocking), 677 (SiD_3 deformation), 1088 (*s*), 1092, 1138 (*vs*; CD_3 rocking), *ca.* 1570 (*s*; Si—D stretching), 2073 (*m*; C—D stretching) and *ca.* 2230 cm^{-1} (*m*; C—D stretching).

SiD_3OCH_3 (vapour) shows its SiD_3 centred fundamentals[706] at 550 (*m*; rocking), 684 (*m*; deformation), 694 (*s*; deformation) and 1570 cm^{-1} (*s*; Si—D stretching).

The 550–3000 cm^{-1} spectrum of $(CH_3)_3SiOCD_3$[707] displays its principal bands at 700 (*m*), 747 (*s*), 753 (*s*), 841 (*vs*), 859 (*vs*), 924 (*m*), 1080 (*vs*; CD_3 bending), 1127 (*vs*; CD_3 bending), 1252 (*vs*), 1409 (*m*), 1418 (*m*), 1444 (*m*), 2067 (*vs*; symmetrical C—D stretching), 2203 (*s*), 2226 (*s*; asymmetrical C—D stretching), 2252 (*s*), 2899 (*s*) and 2958 cm^{-1} (*vs*).

The 260–5000 cm^{-1} spectra of gaseous $HCOOSiD_3$, $CH_3COOSiD_3$ and $CF_3COOSiD_3$ exhibit their SiD_3 rocking bands at 513 (*m*) and 551 (*m*), 495 (*m*) and 522 (*m*), and 543 (*m*) cm^{-1}, respectively. Their asymmetric bending fundamentals appear at 694 (*s*), 698 (*vs*) and 703 (*vs*), and 702 (*s*) and 710 (*m–s*) cm^{-1} while the corresponding symmetric vibrations give rise to absorptions at 674 (*s*), 675 (*m*; *sh*) and 691 (*vs*) cm^{-1}, respectively. Their Si—D stretching bands are shown at 1587 (*vs*) and 1623 cm^{-1} (*vs*; asymmetric) for the formate, at 1600 (*vs*) and 1620 cm^{-1} (*vs*) for the acetate and at 1596 (*vs*) and 1625 cm^{-1} (*vs*) for the trifluoroacetate.[708]

The Si—N stretching frequency of N-deuterated mono- and disilylamines has also been studied.[709]

Liquid N-deuterated trimethylsilylmethylamine[710] shows its N—D stretching band at 2545 (*w*) cm^{-1} while the respective ethylamine[710] shows it at 2534 cm^{-1}, exhibiting its N—D bending fundamental at 1204 (*s*) cm^{-1}.

In the case of liquid $(CH_3)_3SiNDC_6H_5$[710] the N—D centred vibrations absorb at 337 (*m*, *b*; out-of-plane deformation), 1043 (*vs*; bending), 1114 (*w*; C—N—D deformation) and 2592 cm^{-1} (*vw*; stretching).

4.129 Phosphorus Compounds

The 190–4000 cm^{-1} spectra of gaseous and polycrystalline CD_3PD_2, CH_3PD_2 and CD_3PH_2 have been studied in connection with the assignment of the methylphosphine fundamentals and the calculation of its force constants.[711] Gaseous CD_3PD_2 shows its main bands at 522·9 (*m*; ν_8), 637·9 (*s*; ν_9), 753·3 (*s*; ν_7), 781·2 (*m*; ν_{13}), 789·3 (*s*; ν_6), 1007·8 (*w*; ν_5), 1045·9 (*m*; ν_4), 1056·3 (*m*; ν_{12}), 1673·4 (*vs*; ν_3), 1681·3 (*vs*; ν_{11}), 2141·9 (*m*; ν_2), 2237·6 (*m*; ν_{10}) and 2250·5 cm^{-1} (*m*; ν_1). For CH_3PD_2[711,712] vapour appreciable absorption bands appear at 574 (*m*), 686 (*s*), 790 (*s*), 918 (*m*), 927 (*m*), 1424 (*m*), 1433 (*m*), 1674 (*vs*; symmetrical P—D stretching), 1681 (*vs*; asymmetrical P—D stretching), 2941 (*m*), 2988 (*m*) and 3003 cm^{-1}.

CD_3PH_2[711] exhibits its C—D stretching fundamentals at 2138 (*m*), 2306 (*vs*) and 2309 cm^{-1} (*vs*) while the P—D stretching vibration of CH_3PHD[712] absorbs at 1678 cm^{-1}.

D_9-trimethylphosphine in the gas phase shows its asymmetric D_3C—P stretching band at 649 cm^{-1}, its CD_3 rocking fundamentals at 760 and 791 cm^{-1} and its CD_3 deformation absorption at about 1006, 1026 and 1054 cm^{-1} while the in-phase and out-of-phase CD_3 stretching bands appear at *ca.* 2125 and 2226 cm^{-1}, respectively.[713] The spectra of deuterated phosphiranes have also been studied.[714]

The C—D stretching frequencies of various $[(C_6H_5)_3P^+—CRR_1D]X^-$ (X = Cl, Br or I) salts have been reported.[715] If R=R_1=D these frequencies appear at 2015, 2115, 2200 and 2280 cm^{-1} as compared with the respective $(C_6H_5)_2P—CD_3$ values of 2050, 2110 and 2220 cm^{-1}. In the case of R=R_1=CH_3 one C—D stretching band only is observed, at 2145 cm^{-1}, while for R = D, R_1 = n-C_3H_7, Cl, Br, I, C_6H_5 these frequencies are shown at (*a*) 2085, 2160, (*b*) 2120, 2210, (*c*) 2050, 2120, 2200, (*d*) 2105, 2200 and (*e*) 1990, 2105 and 2175 cm^{-1}, respectively. The location of these bands in the spectra of a number of similar compounds (R = D, R_1 = OCH_3, $COCH_3$, COC_6H_5, CO_2CH_3, SCH_3) has also been determined.[715]

$(C_6H_5)_2POCD_3$ exhibits its CD_3 stretching absorption at[715] 2025, 2125 and 2230 cm^{-1}.

The $CPOCl_2$ moiety fundamentals of CD_3POCl_2[716] appear at 116, 187, 210, 294, 391, 456, 505, 716 and 1049 cm^{-1}.

The P—D stretching band of (O, P-deuterated) deuterophenylphosphinic acid[717] is shown at 1750 cm^{-1} while its sodium salt[718] displays it at 1650 cm^{-1} and its n-propyl ester[719] exhibits it at 1710 cm^{-1}.

The effect of solvents on the infrared absorption of P-deuterated cyclic hydrogen phosphites (and hydrogen thiophosphites) was investigated in a number of cases.[719]

The 400–3600 cm^{-1} spectrum of O-deuterated di-2-ethylhexyl hydrogen phosphate was measured on a 0·25 M solution.[720] The absorption of aqueous K_2DPO_4, anhydrous Na_2DPO_4 and its hydrate, aqueous and anhydrous NaD_2PO_4 and aqueous D_3PO_4 has also been reported[721] as well as the internal fundamentals of RbD_2PO_4 in the non-ferro electric phase.[722]

The absorption of (fluorolube) mulls and solutions (in CCl_4 or CH_2Cl_2) of N-deuterated O-methyl O-(2,4,5-trichlorophenyl)N-alkylphosphoramidates (alkyl = methyl, ethyl, isopropyl, isobutyl, *t*-butyl) was investigated in the 1050–1450 and 2100–2800 cm^{-1} regions.[723] The "N—D bending" frequency of these hydrogen bonded compounds appears at about 1210 cm^{-1}, as compared with the respective "N—H bending' frequency of *ca.* 1440 cm^{-1}, giving a ν_H/ν_D ratio of 1·19. Their monomeric N—D stretching fundamentals is shown in the 3401 (*t*-butyl derivative in CCl_4)-3428 cm^{-1} (isobutyl amidate in CCl_4) zone, yielding a ν_H/ν_D ratio of 1·34.

4.130 Deuterated Sulphur Compounds

The infrared adsorption spectra of CH_3SD have been investigated in the gas, liquid and solid phases.[724]

The absorption of gaseous CF_3SD[724, 725] down to 100 cm^{-1} shows its bands at 125 (*s*; torsion), 308 (*m*; CF_3 rocking), 451 (*m*; symmetric CF_3 deformation), 528 (*m*; asymmetric CF_3 deformation), 555 (*w*), 699 (*s*; C—S—D bending), 763 (*s*; C—S stretching), 1138 and 1146 (*vs*; symmetrical CF_3 stretching) and 1163 and 1172 cm^{-1} (*vs*; asymmetrical CF_3 stretching).

The solution spectrum of CH_3—S—CD_3[726] (in CCl_4 or CS_2) exhibits its CD_3 rocking bands at 725·5 (perpendicular) and 740 cm^{-1} (parallel), its CD_3 bending fundamentals at 961 (symmetric), 1040·5 and 1050·5 cm^{-1} and its C—D stretching bands at 2128 (symmetric), 2231 (asymmetric) and 2240 cm^{-1}.

Gaseous $CH_3SC{\equiv}CD$[727] displays its 300–3500 cm^{-1} fundamentals at 401 (*s*), 421 (*vs*; out-of-plane C—D bending), 531 (*m*; in-plane C—D bending), 962 (*m*), 983 (*m*), 1322 (*m*), 1431 (*m*), 1447 (*m*), 1938 (*m*), 2591 (*vs*; C—D stretching), 2944 (*s*), 3008 (*m*) and 3017 (*m*) cm^{-1}.

The vibrational spectra of gaseous, liquid[728] and crystalline D_4-thiophene[729] have been recorded down to 400 cm^{-1} and their observed absorption bands, as well as those of 2-deutero-, 3-deutero-, 2,3-D_2-, 2,5-D_2-, 3,4-D_2-, 2,3,4-D_3- and 2,3,5-D_3-thiophene, have all been assigned to their respective vibration modes. Table 4.67 reports the infrared active fundamentals observed for gaseous thiophene and D_4-thiophene (point group $\boldsymbol{C_{2v}}$).

Table 4.67 Observed Fundamental Bands (cm^{-1}) of Thiophene and D_4-thiophene[728]

Band	Species	Approximate description	C_4H_4S	C_4D_4S
ν_1	A_1	Symmetrical C—H(D) stretching	3126	2343 (*m*)
ν_2	A_1	Symmetrical C—H(D) stretching	3098	2305 (*m*)
ν_3	A_1	Symmetrical ring stretching	839	731 (*vs*)
ν_4	A_1	Symmetrical ring stretching	1360	1248 (*w*)
ν_5	A_1	Symmetrical ring stretching	1409	1376 (*s*)
ν_6	A_1	Symmetrical C—H(D) deformation	1083	896 (*m*)
ν_7	A_1	Symmetrical C—H(D) deformation	1036	785 (*m*)
ν_8	A_1	Symmetrical ring deformation	608	585 (*vw*)
ν_{13}	B_1	Asymmetrical C—H(D) stretching	3098	2290 (*m*)
ν_{14}	B_1	Asymmetrical ring stretching	—	1459 (*m*)
ν_{15}	B_1	Asymmetrical C—H(D) bending	1256	1034 (*s*)
ν_{16}	B_1	Asymmetrical C—H(D) bending	—	846 (*s*)
ν_{17}	B_1	Ring vibration	872	—
ν_{18}	B_1	Ring deformation	—	*ca.* 712 (*w*)
ν_{19}	B_2	Out-of-plane C—H(D) deformation	712	531 (*vs*)
ν_{20}	B_2	Out-of-plane C—H(D) deformation	—	684 (*vw*)
ν_{21}	B_2	Out-of-plane ring deformation	452	414 (*vw*)

The main absorption bands of liquid D_2-trithiocarbonic acid[730] appear at 467, 487, 605, 633, 635, 905, 1080 (C=S stretching) and *ca.* 1800 cm^{-1} (S—D stretching).

Crystalline perdeuterated dimethyl trithiocarbonate exists in the *cis-cis* rotational isomeric form.[731] In the liquid phase it is present as an equilibrium mixture of this isomer with the *trans-cis* form of this ester. This mixture[731] shows its CD_3 rocking bands at 738 (*s*) and 757 (*w*) cm^{-1}, its symmetric and asymmetric CD_3 deformation fundamentals at 994 (*s*) and 1085 cm^{-1} (*vs*), respectively, and its C—D stretching absorption at 2127 (*m*; symmetric) and 2249 cm^{-1} (*w*; asymmetric).

D_6-dimethyl sulphoxide belongs to point group C_s and as a vapour exhibits its absorption bands in the 700–4000 cm^{-1} region at 750 (*m*), 756 (*m*; *sh*), 803 (*m*; *sh*), 814 (*m*; all four due to CD_3 rocking vibrations), 824 (*m*; *sh*), 1015 (*m*), 1025 (*m*), 1043 (*m*), 1084 (*s*, *sh*; last four due to CD_3 bending modes), 1096 (*vs*; S=O stretching), 2133 (*w*; symmetric C—D stretching), 2185 (*vw*; 2 × 1096) and 2250 cm^{-1} (*m*; degenerate C—D stretching). As a liquid it absorbs in the 250–700 cm^{-1} region at[732] 307 (*s*; asymmetric C—S—O bending), 340 (*s*), 611 (*m*; symmetric C—S stretching) and 619 cm^{-1} (*m*; asymmetric C—S stretching).

The 800–4000 cm^{-1} absorption spectrum of a 10% solution of O-deuterated methanesulphonic acid in dimethyl sulphoxide, which contains a high amount of the $(CH_3)_2SOD$ ions, does not show clear-cut O—D bands, this fact being probably due to the strong hydrogen bonding existing between this group and another solvent molecule.[733]

The complex formed between 3 molecules of D_6-dimethyl sulphoxide and 1 molecule of UCl_4 shows its CD_3 bands[734] at 765 (*m*), 782 (*w*), 820 (*m*; *sh*), 827 (*m*; rocking), 1008 (*sh*), 1015 (*s*; asymmetric deformation), 1040 (*m*; symmetric deformation), 2120 (*w*; symmetric stretching) and 2245 cm^{-1} (*w*; asymmetric stretching). Its S—O stretching frequency is exactly that of the respective normal complex (960 cm^{-1} as compared with the 1096 cm^{-1} value for the free sulphoxide).

The CD_3 frequencies of the similar 5:1 complex are very close to the corresponding 3:1 complex frequencies.[734]

The metal–oxygen stretching vibrations of various D_6-dimethyl sulphoxide complexes in the 200–1200 cm^{-1} region have also been investigated.[735, 736]

D_{10}-diphenyl sulphone[737] exhibits its 400–4000 cm^{-1} fundamentals at 439 (*w*), 535, 553 (*sh*), 564 (*sh*), 589, 624 (*vw*; out-of-plane C—D deformation), 680 (*w*) and 715 (*m*; both due to a mixing of a ring deformation and a C—S stretching), 698 (*sh*), 750 *vw*), 763 (*m*; all three: out-of-plane C—D bending), 797 (*vw*) 821 (*s*; both: in-plane C—D bending), 828 (*s*; out-of-plane C—D bending), 846 (*sh*), 857 (*w*; both: in-plane C—D bending), 956 (*s*), 1032 (*m*) and 1063 (*s*; both due to a mixing of a C—S stretching mode with a ring deformation vibration), 1150 (*vs*; symmetric SO_2 stretching), 1295 (*s*),

1309 (*vs*; asymmetric SO_2 stretching), 1330 (*s*), 1546 (*m*), 2265 (*sh*) and 2280 cm^{-1} (*w*; both: C—D stretching).

D_5-benzenesulphonyl chloride[737] displays its C—D bands at 698 (*sh*), 761 (*vw*), 826 (*m*; out-of-plane deformation), 845 (*vw*), 855 (*vw*; in-plane bending) and 2280 cm^{-1} (*vw*; stretching).

The 800–4000 cm^{-1} absorption of liquid O-deuterated methanesulphonic acid[733] is reproduced in Fig. 4.31.

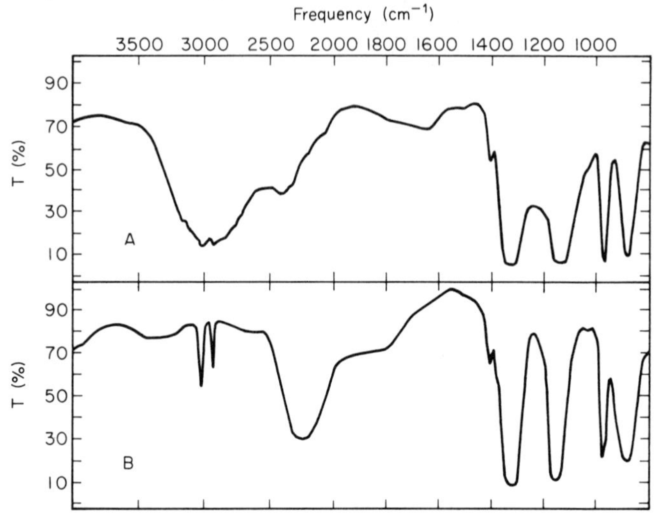

Fig. 4.31 Infrared spectra: A is neat methanesulphonic acid in a capillary film; B is the same, deuterated on oxygen.

The 670–4000 cm^{-1} spectrum of the sodium salts of 2-deutero-1-naphthol-4- and 1-deutero-2-naphthol-6-sulphonic acids, as well as of 1-deutero-2-naphthol-6, 8-disulphonic acid, shows no clear C—D stretching bands.[412]

4.131 Sulphur Compounds which contain Nitrogen or Phosphorus Atoms

Gaseous DNCS shows its fundamental bands[738, 739] at 369 (*s*; ν_5), 586 (*m*; ν_4), 1941 (*vs*; ν_2) and 2641 cm^{-1} (*s*; ν_1). Its rotation spectrum (as well as that of $DCNO_4$) has also been measured.[739]

The fundamentals of liquid D_3-methylthiocyanate[740] appear at 405 (*s*; out-of-plane S—C≡N bending), 639 (*m*; symmetric C—S—C stretching), 686 (*w*; asymmetric C—S—C stretching), 733 (*m*; CD_3 wagging), 780 (*s*; CD_3 rocking), 1006 (*s*; symmetric CD_3 deformation), 1041 (*s*; CD_3 bending), 2140 (*s*; symmetric C—D stretching), 2156 (*vs*; C≡N stretching) and 2267 cm^{-1} (*m*; C—D stretching).

The 400–1600 cm^{-1} spectrum of 2-deuterothiazole[741, 742] and partial spectral data for 2-deutero-4-chlorothiazole and 2,5-D_2-thiazole[742] have been reported.

The 670–1630 cm^{-1} absorption of N-deuterated thiobenzanilide, N-methylthiofuramide and thiofuranilide[743] exhibit their N—D bending ("II′ amide") and C—N stretching ("III′ amide") bands near 1100 and between 1380 and 1410 cm^{-1}, respectively.

D_5-Benzene-sulphonamide[737] shows its C—D bands at 624 (*w*; out-of-plane bending), 821 (*m*; in-plane bending), 830 (*m*; out-of-plane bending), 845 (*w*), 860 (*sh*), 1030 (*sh*; all three: in-plane bending) and 2275 cm^{-1} (*vw*; stretching). The frequency of the respective C—D bands of $C_6D_5SO_2NH$-C_6H_5 is quite similar.[737] The SO_2 group frequencies of $C_6D_5SO_2ND_2$, $C_6H_5SO_2ND_2$, $C_6H_5SO_2NDC_6G_5$ and $C_6D_5SO_2NDC_6H_5$ are also very similar to the corresponding values for the respective normal isotopomers.[737]

The N—D bending and C—N stretching vibrations of N-deuterated-N-methyl-*p*-toluenesulphonamide are coupled and give rise to bands at[743] 918 and 1218 cm^{-1}.

The lattice vibrations of crystalline D_4-thiourea[744] were observed at 42 (*w*), 90 (*sh*), 106 (*vs*) and 163 cm^{-1} (*vs*), in agreement with the values and the intensities calculated for these bands.

The infrared spectra of various inclusion compounds formed with D_4-thiourea as a host molecule have also been reported.[745]

The fundamental absorption bands of liquid D_2-1,2,5-thiadiazole[746] have been observed at 502 (*s*), 654 (*s*; out-of-plane C—D bending), 678 (*w*), 751 (*vs*), 844 (*vs*; in-plane C—D bending), 866 (*vs*), 966 (*w*; in-plane C—D bending), 1181 (*vs*), 1295 (*vs*), 1408 (*m*) and 2301 cm^{-1} (*m*; C—D stretching).

The 650–3600 cm^{-1} absorption of various O-deuterated thiohydroxamic acids, their metal chelates and their S—CH_3 derivatives has also been measured.[747]

A mull of $(CF_3SND)_2CO$ absorbs at[758] 463 (*m*), 500 (*m*; *b*), 529 (*m*), 562 (*w*), 614 (*m*), 750 (*m*), 792 (*m*), 873 (*m*), 993 (*m*), 1125 (*s*; *b*), 1180 (*s*; *b*), 1209 (*m*; N—D bending), 1363 (*w*), 1394 (*s*), 1672 (*s*), 1730 (*w*) and 2420 cm^{-1} (*m*; bonded N—D stretching). Mulled $(CF_3S)_2N$—$COND_2$[758] shows its ND_2 stretching bands at 2430 (*s*) (symmetric) and 2590 cm^{-1} (*s*; asymmetric) while its ND_2 bending fundamental appears at 990 (*m*) cm^{-1} and the respective out-of-plane vibration mode is observed at 437 (*m*) cm^{-1}.

Mulled N-deuterated D_2-*p*-toluenesulphonamide[748] exhibits its S—N stretching band at 860 (*s*) and its ND_2 absorption peaks at 1191 (*s*; bending), 2403 (*m*) and 2492 cm^{-1} (*s*; both due to stretching modes).

A potassium bromide pellet of *p*-$CH_3C_6H_4SO_2NDCH_3$ displays its N—D bending fundamental at 1217 (*w*) cm^{-1} and its N—D stretching band at 2447 (*w*) cm^{-1} while solutions of *p*-$CH_3C_6H_4SO_2N(CD_3)_2$ (in C_2Cl_4 or CS_2) show[749] their CD_3 bands at 1048 (*sh*; symmetric bending), 1055, 1108 (both: *w*; asymmetric bending), 2063 (*w*; symmetric stretching), 2123 and 2207 cm^{-1} (both: *w*; asymmetric stretching).

A KBr pellet of N-deuterated D_3-taurine exhibits its main absorption bands below 1800 cm^{-1} at[641] 446 (*m*), 524 (*s*), 533 (*m–s*), 594 (*s*), 727 (*s*), 773 (*m–s*; ND_3^+ rocking), 830 (*m–s*), 1007 (*m*; ND_3^+ rocking), 1037 (*s*), 1047 (*s*), 1068 (*m*; ND_3^+ bending), 1154 (*s*; asymmetric ND_3^+ bending), 1186 (*s*), 1218 (*vs*), 1278 (*m–s*), 1338 (*m–s*), 1429 (*m*) and 1465 cm^{-1} (*m–s*).

The 300–4000 cm^{-1} spectra of the dipolar N-deuterated *o*-, *m*- and *p*-D_3-aminobenzene sulphonic acids[750] show their ND_3^+ rocking bands at 793 (*w*), 910 (*m*) and 811 (*s*), respectively. Their ND_3^+ deformation fundamentals appear at 1130 (*s*; symmetric), 1168 (*sh*; asymmetric), 1100 (*s*), 1122 (*m*) and 1140 (*vs*) and 1170 cm^{-1} (*vs*), correspondingly, while their respective N—D stretching modes absorb strongly at 2160 (symmetric), 2260 (asymmetric), 2170, 2270 and 2140 and 2290 cm^{-1} (with a shoulder at 2220 cm^{-1}).

The amino group deuteration of the potassium salts of these acids, as well as of the various isomeric aminonaphthalene sulphonic acids, results in 1·32–1·39 ν_H/ν_D ratios for the symmetric amino group stretching frequencies and 1·26–1·42 ratios in the case of the respective asymmetric vibrations.[751]

The infrared spectra of 2-deutero-1-naphthol-4-sulphamide and 1-deutero-2-naphthol-6-sulphamide do not show distinct C—D stretching bands.[412]

Gaseous SiD_3NCS[752] exhibits its fundamental frequencies in the 400–2500 cm^{-1} region at 479 (*m*; Si—N stretching), 541 (*m*; SiD_3 rocking), 701 (*vs*; symmetric SiD_3 bending), 710 (*s*; asymmetric SiD_3 deformation), 1026 and 1034 (both: *m*; C—S stretching), 1584 (*s*; symmetric Si—D stretching), 1610 (*s*; asymmetric Si—D stretching) and 2100 cm^{-1} (*vs*; C—N stretching). The force constants, mean amplitudes, etc. of this silyl isothiocyanate have also been reported.[753]

O,O-dimethyl phosphorochloridothioate in the liquid state exists in an equilibrium between two rotational isomeric configurations.[754] Liquid $(CD_3O)_2P(=S)Cl$ displays its CD_3 fundamentals[754] at *ca.* 895 (*sh*; out-of-plane deformation), 924 (bending), 1105 (bending), 2076 (symmetric stretching; in resonance with an overtone vibration) and 2261 cm^{-1} (*b*; asymmetric stretching).

The 400–3800 cm^{-1} spectra of liquid $CD_3OP(=S)Cl_2$ and its solutions in CCl_4 or CS_2 have been investigated at various temperatures.[755] The CD_3 bands of this compound are located at 916 (deformation), 1095 (symmetric bending), 1100 (asymmetric bending), 2071 (overtone), 2198 (symmetric stretching), 2250 and 2279 cm^{-1} (both due to asymmetric stretching modes).

The 400–3600 cm^{-1} absorption of $CH_2DOP(=S)Cl_2$ and as well as the spectrum of DPS_2F_2 have also been reported.[756, 757]

The free ND_2 stretching fundamentals of CCl_4 solutions of N-deuterated O,O-dimethyl phosphoramidothioate and its diethyl analogue appear at 2490–2491 (symmetric) and 2611–2600 cm^{-1} (asymmetric), respectively.[723] In the case of the corresponding NHD isotopomers, as well as in the case of $(C_6H_5O)_2$ P(=S) $NDCH_3$, the free N—D stretching frequencies are shown at[723] 2530–2539 cm^{-1} for the *trans* bond and at 2553–2562 cm^{-1} for the *cis* bond. The NHD bending vibration of these compounds is exhibited at *ca.* 1384 cm^{-1} (the corresponding ND_2 frequency being 1170–1171 cm^{-1}) while the respective twisting fundamental is displayed at 938 cm^{-1} (718–720 cm^{-1} for the parallel ND_2 groups).

4.132 Deuterated Ferrocenes and Ferricinium Compounds

The infrared absorption of perdeuterated ferrocene has been investigated on a Nujol mull,[758] in a potassium bromide pellet[759] and in solution in various solvents[758] down to 400 cm^{-1}. Its bands (in KBr) appear at 447 (*vs*; ring-Fe stretching), 478 (*vs*; ring-Fe bending), 637 (*s*; symmetric C—D deformation), 682 (*w*; asymmetric C—D deformation), 740 (*w*), 772 (*m–s*; asymmetric C—D deformation), 826 (*w*), 853 (*w*), 907 (*w*), 1004 (*vw*), 1046 (*s*; ring breathing), 1063 (*w*), 1099 (*vw*), 1285 (*w*, *b*; C—C stretching), 1360 (*w*; *b*), 1389 (*vw*), 1407 (*w*), 1443 (*w*) and 2360 cm^{-1} (*m–s*, *b*; C—D stretching).

The 750–2000 cm^{-1} spectrum of deuteroacetyl-ferrocene was also reported.[760]

The infrared absorption of ferricinium salts is generally quite similar to that of ferrocene itself. Thus, the main absorption bands of ferricinium tetrachloroferrate[759] are displayed at 467 (*m–s*; ring-Fe bending), 670 (*s*; asymmetric C—D deformation), 781 (*m–s*; asymmetric C—D deformation), 1304 (*m–s*; asymmetric C—C stretching) and 2360 cm^{-1} (*m–s*, *b*; C—D stretching). The 400–3200 cm^{-1} spectra of ferricinium salts containing the $AlCl_4^-$, HgI_3^- and $Cr(NH_3)_2(SCN)_4^-$ anions have similarly been measured.

4.133 Germanium Compounds

Gaseous D_3-methyl-germane shows its 11 infrared active fundamentals[761] at 452 (*s*), 550 (*s*), 713 (*s*; CD_3 rocking), 844 (*vs*), 904 (*s*), 977·5 (*s*; CD_3 deformation), 1032 (*m*; CD_3 deformation), 2073 (*s*; C—D stretching), 2088·5 (*vs*), 2090 (*vs*) and 2247 cm^{-1} (*s*; C—D stretching). The resolution of the rotational structure of a number of these bands yielded values for some of its rotational constants.

The absorption spectrum of its isotopomer, methyl-D_3-germane,[761] exhibits its GeD_3 rocking band at 382 (*m*) cm^{-1}, the GeD_3 deformation fundamentals at 627 (*m*) and 658 cm^{-1} (*m–s*) and the Ge—D stretching absorption at 1496 (*vs*) and 1500 (*s*) cm^{-1}. Gaseous ethyl-D_3-germane displays these bands[762] at 384 (*m–w*), 602 (*vs*), 633 (*m*), 1489 (*vs*) and 1496 cm^{-1} (*vs*), respectively.

The 140–4000 cm^{-1} absorption of $GeD_3C{\equiv}CH$,[763] which is assumed to belong to point group C_{3v}, shows its fundamentals at 203, 484 (GeD_3 rocking), 518, 608 (symmetric GeD_3 deformation), 643 (asymmetric GeD_3 deformation), 673, 1525 (GeD stretching), 2052·5 and 3316·5 cm^{-1}.

The 85–3400 cm^{-1} spectrum of $(CH_3)_3GeC{\equiv}CD$ has also been reported and interpreted.[764]

The absorption bands of liquid $C_6D_5GeCl_3$ in the 33–4000 cm^{-1} region were all interpreted in detail, assuming C_{2v} symmetry.[765,766] Its C—D stretching frequencies give rise to fundamentals at 2270 (*m–w–*; *sh*), 2287 (*m*) and 2298 (*m*) cm^{-1}. The infrared spectrum of the corresponding bromide has equally been recorded and analyzed.[767]

The 33–4000 cm^{-1} spectra of crystalline, vitreous or gaseous GeH_3OCD_3 have been measured and have been interpreted on basis of a bent molecular skeleton.[768]

The force constants and rotation distortion constants of D_3-germyl cyanide have been calculated.[769] D_3-germyl isocyanate vapour[770] shows its GeD_3 bands at 495 (*s*; rocking), 622 (*vs*; bending), 647 (*m–s*; asymmetric bending), 1523 (*vs*; symmetric stretching) and 1550 cm^{-1} (*s*; asymmetric stretching).

The 100–4000 cm^{-1} absorption of gaseous GeD_3NCS[771] shows its main bands at 391 (*m–s*), 501 (*m*), 610 (*s*), 624 (*m*), 1047 (*vvs*), 1524 (*s*), 1551 (*m*) and 2119 (*m*) cm^{-1}.

4.134 Different Other Labelled Compounds

The C—D stretching frequency of various deuterated organomagnesium compounds[772] depends upon the substitution on this carbon atom and in the case of solutions was observed to vary between 2072 (for $[(CH_3)_2CD]_2Mg$) and 2130 cm^{-1} (for CH_2DMgBr).

Fully deuterated dimeric trimethylaluminium shows its infrared active fundamentals[773] at 315, 435, 472, 579, 677 (all three: CD_3 rocking), 955, 1036 (both: symmetric CD_3 bending), 1140 (asymmetric CD_3 bending), 2110, 2170 and 2200 cm^{-1} (all three: C—D stretching).

The CD_3 group bands of *o*-$C_6H_4(As(CD_3)_2)_2$ appear at[774] 639, 654 (both: *s*; rocking), 965 (*m*; symmetric bending), 1036 (*s*; asymmetric deformation), 2119 (*s*; symmetric CD_3 stretching) and 2209 cm^{-1} (*s*; asymmetric CD_3 stretching).

The absorption spectra of mulls of $(CD_3)_3AsCl_2$, $(CD_3)_3AsBr_2$, and R_3AsODX (R = CH_3, C_6H_5) compounds were also investigated.[775]

Liquid D_6-dimethyl selenide shows its 400–4000 cm^{-1} fundamental bands[776] at 548 (*w*), 564 (*w*), 658 (*vw*), 696 (*m*), 744 (*m*; all three: CD_3 rocking), 968 (*m*), 983 (*w*; symmetric CD_3 bending), 1038 (*vs*; asymmetric C—D bending), 2134 (*m*; symmetric C—D stretching) and 2256 cm^{-1} (*m–w*; asymmetric C—D stretching).

The characteristic CD_3 frequencies calculated by factoring them from the secular determinant of dimethyl selenide agree well with the values observed for them[777]

The out-of-plane deformation fundamentals of crystalline D_4-selenophene were observed at 475 and 748 cm^{-1} as compared with the corresponding 541 and 905 cm^{-1} frequencies of normal selenophene.[729] The out-of-plane normal vibrations of 2-deutero-, 3-deutero- and 2,5-D_2-selenophene have also been reported.[729]

The infrared absorption of deuterated methylsilver was recorded.[778]

Gaseous CD_3SnH_3 exhibits its 250–4000 cm^{-1} fundamental frequencies[779] at 392 (*w*), 478 (*m*), 628 (*s*), 703·5 (*s*), 738 (*m*; CD_3 rocking), 920 (*m*; CD_3 symmetric deformation), 1017 (*w*; CD_3 asymmetric deformation), 1889 (*s*), 2144 (*m*; CD_3 symmetric stretching) and 2254·5 cm^{-1} (*m*; asymmetric CD_3 stretching).

CH_3SnD_3 vapour shows its SnD_3 group absorption bands[779] at 317 (*m*; rocking), 493 (*s*; C—SnD_3 stretching), 502·5 (*s*; asymmetric bending), 509 (*m*; symmetric bending), 612 (*w*; 2 × 316), 1012 (*w*), and 1352 cm^{-1} (*s*; Sn—D stretching).

The 400–4000 cm^{-1} spectrum of liquid D_6-dimethylmercury displays bands at[780] 491 (C—Hg—C stretching), 598 (CD_3 rocking), 669 (*sh*; CD_3 rocking), 931, 1030 (both: CD_3 bending), 2039 (combination), 2114 and 2224 cm^{-1} (both: C—D stretching). Nujol or hexachlorobutadiene mulls of D_3-methyl mercuric bromide[781] show its corresponding absorption bands at 495, 597, —, 916, 1027, —, 2126 and 2262 cm^{-1}, respectively.

The 650–4000 cm^{-1} absorption of $CH_3OCD_2CD_2HgI$[782] exhibits its CD_2 bands at 710 (rocking), 735, 920 (both: torsion), 960, 1071 (both: out-of-plane deformation), 1137, 1220 (both: bending), 2070 (symmetric stretching) and 2170 cm^{-1} (asymmetric stretching).

4.135 Deuterated Metal Complexes

Solid symmetrically monodeuterated sodium acetylacetonate (see Fig. 4.32) absorbs below 2000 cm^{-1} at[783] 405 (*m*), 462 (*w*), 512 (*m*), 519 (*m*), 526

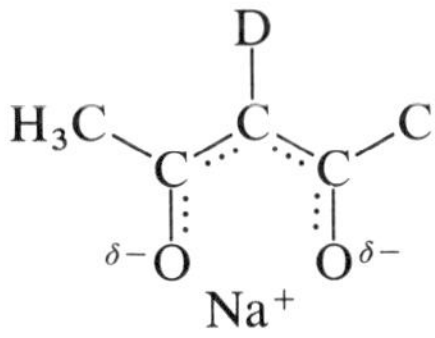

Fig. 4.32 The structure of sodium acetylacetonate.

(*m*), 560 (*w*), 569 (*w*; both: out-of-plane C—D deformation), 591 (*w*), 621 (*w*), 650 (*m*), 910 (*m*), 959 (*w*), 1012 (*m*), 1044 (*w*; C—D bending), 1232 (*m*), 1250 (*m*), 1283 (*m*), 1358 (*s*), 1410 (*vs*), 1455 (*vs*), 1510 (*vs*) and 1595 cm^{-1} (*vs*). In an aqueous solution the acetylacetonate anion probably loses its planarity and shows its bands in the 800–2000 cm^{-1} region[783] at 910 (*w*), 966 (*m*), 996 (*m*), 1060 (*w*), 1418 (*s*), 1439 (*s*), 1475 (*vs*) and 1555 (*m*) cm^{-1}.

The out-of-plane C—D deformation frequencies of the corresponding solid lithium complex (in a KBr pellet) are observed[784] at 570 and 594 cm^{-1} while those of the beryllium and palladium acetylacetonates appear at[784] 581, 590, 597 cm^{-1} and 580, 586 cm^{-1}, respectively. The in-plane C—D bending vibration of the solid 3-deuterated copper(II) complex (in KBr) appears[784, 785] at 918 cm^{-1} as compared with the 1031, 1061 and 1088 cm^{-1} values for the CD_3 deformation frequencies of the 1,5-D_6-copper(II) acetylacetonate.[784]

The infrared spectra of 3-deuterated platinum(II) acetylacetonate,[786] three deuterated derivatives of potassium dichloro(acetylacetonato)platinate-(II),[786, 787] two deuterated sodium dichlorobis(γ-acetylacetonato)platinates-(II),[787] deuterated disodium dichlorobis(acetylacetonato)platinate(II),[786] its bromo-analogue,[786] deuterated potassium chlorobis(acetylacetonato)-plate[786] and tris(acetylacetonato)platinate,[786] DPt(acetuacetonyl)$_2$Cl[786] and other acetylacetonato platinum complexes[788, 789] have also been investigated.

The 250–4000 cm^{-1} spectrum of solid $Co(CD_2ODCO_2)_2$[790] shows its C—D stretching bands at 2087 and 2160 cm^{-1} and its O—D stretching fundamental at 2300 cm^{-1}. $Co(CD_2OHCO_2)_2$ displays only the 2160 cm^{-1} C—D stretching band while $Co(CH_2ODCO_2)_2$ exhibits its O—D stretching absorption[790] at 2298 cm^{-1}.

The infrared absorption of various labelled metal biuret complexes has been investigated[791] down to 180 cm^{-1}.

4.136 Deuterated Steroids

The exact C—D stretching frequency of steroids monodeuterated in a cyclohexane ring depends upon the orientation of this bond, it being usually higher when this bond is equatorial than when it is axial. The C—D stretching bands of 3β-acetoxy-6α-deuterocholestan-7-one appear thus at 2137 (*s*) and 2181 (*m*) cm^{-1} as compared with the respective values of 2134 (*m*) and 2164 (*s*) cm^{-1} for the corresponding axial 6β-deutero- isomer and the 2131 and 2220 cm^{-1} frequencies in the case of the 6-D_2-isotopomer.[792]

This trend is characteristic of cyclohexane derivatives in general as can be seen from the fact that the equatorial *trans*-1-deutero-4-phenylcyclohexane, e.g. also shows its C—D stretching fundamental at 2171 cm^{-1} as compared with the corresponding 2158 cm^{-1} vibration observed for the respective (axial) *cis* isomer[792] (their D_2-isotopomer absorbs at 2106, 2160, 2189 and 2201 cm^{-1}). Some α-deuterated steroid alcohols and ketones, however, do not follow this tendency.[793]

The C—D valence vibration of CCl_4 or $CHCl_3$ solutions of 3α- and 3β-deuterocholestane, 2β-deuterocholestan-3α-ol, 3α-deuterocholestan-3β-ol, 2β-deuterocholestan-3-one, 6β-deuterocholestan-7β:7α-diol 3-acetate, 5α-

or 7α-deutero-3β-acetoxycholestan-6-one[792] and other monodeuterated steroids[792, 793] was observed between 2090 and 2356 cm^{-1}.

Gem-D_2-steroids show a doublet band in this region,[792, 793] which in the case of α,α-D_2-3- or 17-ketones[794] appears at 2140 and 2215 cm^{-1} with an additional band at 2105 cm^{-1} for the D_2-3-one steroids (the absorption of 6-D_2-7-cholestanone was also investigated).[794] D_3-21-one-20 steroids exhibit their C—D stretching bands at[794] 2220 and 2255 cm^{-1}.

α-Dideuteration decreases the C=O valence vibration of keto-steroids by[794] 1–5 cm^{-1}.

The 1350–1500 cm^{-1} spectra of the D_3-acetates of various steroid carbinols (as well as of cyclohexanol) were investigated in order to enable the identification of the different methyl and methylene deformation bands.[794] The absorption of a number of α,α′-D_4-keto-steroids[794, 238b] and 2,4,16-D_6-3,17-androstanedione[238b] has also been measured for the same reason.

The intensity of the C—D stretching bands of the CD_3COO group was observed to be very low.[794]

4.137 Deuterated Xanthines

The infrared spectra of various N-deuterated caffeine salts were investigated in connection with the assignment of the NH bands of the normal salts.[795] Their N^+—D valency fundamentals appear between 1932 (for the hydrobromide) and 2365 cm^{-1} (for the perchlorate) while the corresponding in-plane bending frequencies are displayed at 950 and 932 cm^{-1}, respectively. Their out-of-plane N^+—D deformation bands are observed in the 700–850 cm^{-1} region.[795]

In the theobromine case the valence N—D frequency of the free base appears at 2310 cm^{-1} while the respective bending vibration is exhibiting at 1378 cm^{-1}. The theobromine salts show both N—D and N^+—D bands;[796] for they hydroiodide, e.g. they are found at 972 (N^+—D bending), 1388 (N—D bending), 2245 and 2250 cm^{-1}. The N—D stretching frequency of these salts is in certain cases[796] as high as 2466 cm^{-1} ($HAsF_6$ salt).

4.138 Deuterated Polymers

Partly deuterated polythene (film) shows its C—D stretching fundamentals as a doublet at about 2080 and 2180 cm^{-1} and its C—D deformation absorption near[797] 1080 cm^{-1}. The far infrared bands of crystalline perdeuterated polyethylene have also been reported and assigned.[798]

Polymerizing *cis*-D_2-ethylene with the help of the $Al(C_2H_5)_3$-$TiCl_4$ catalyst gives a product which exhibits its CHD rocking bands[799] at 590 (*m*) and 597 (*m*) cm^{-1}, its characteristic CHD deformation fundamental at[800] 1306 (*m*) cm^{-1} and its C—D valence vibrations at[799] 2137 (*vs*) and 2155 cm^{-1}

(*sh*). The *trans*-D_2-ethylene yields under these conditions a polymer which absorbs[799,800] at 586, 594 (both: CHD rocking), 1288, 1335 (both are characteristic poly-*trans* CHD deformation bands), 2133 (*vs*) and 2152 cm^{-1} (*sh*).

Calculation of the infrared active CHD deformation fundamentals of the various $(CHD)_n$ polymers yielded in the case of the *erythro*-diisotactic configuration a frequency of 1310 cm^{-1}. For the *threo*-diisotactic form such vibrations were computed to absorb[801] at 1281 and 1329 cm^{-1}. It is therefore concluded that the poly-*cis* polymer exists mainly as the *erythro*-configurations while the poly-*trans* product is mainly in the *threo*-forms.[800,801]

The frequency of the rocking bands characteristic of the various polythene CXD (X = H, D) groups[802] is given in Table 4.68.

TABLE 4.68 The Frequency (cm^{-1}) of Various CXD Rocking Bands[727]

Type of group	Frequency	Type of group	Frequency
$(CHD)CD_2(CHD)$	581	$(CH_2)CD_2CD_2CD_2(CHD)$	561
$(CHD)CD_2CD_2(CHD)$	566	$(CH_2)CD_2(CD_2)_nCD_2(CHD)$, $n = 3 - 6$	~530
$(CHD)CD_2CD_2CD_2(CHD)$	545	$(CH_2)CHD(CH_2)$	660
$(CHD)CD_2(CD_2)_4CD_2(CHD)$	528	$(CH_2)CHDCHD(CH_2)$	630
$(CHD)CD_2(CD_2)_nCD_2(CHD)$, $n = 7, 8, \ldots$	522	$(CH_2)CHD(CHD)_2CHD(CH_2)$	612
$(CH_2)CD_2(CHD)$	618	$(CH_2)CHD(CHD)_8CHD(CH_2)$	590

The observed polarized light spectra of stretched (highly crystalline) perdeuterated polypropylene, in the 80–4000 cm^{-1} region, agree well with its calculated normal frequencies.[803] Their main bands appear at 786 (*m*; CD_3 rocking), 1034 (*m*; CD_2 wagging), 1050, 1057, 1062 (all three: *s*; CD_3 deformation), 1098 (*m*; CD_2 scissoring), 1224 (*m*), 2069 (*s*; symmetric CD_3 stretching[804]), 2078 (*s*), 2086 (symmetric CD_2 stretching), 2132 (*s*; $R_1R_2R_3C—D$ stretching), 2180 (*vs*; asymmetric CD_2 stretching), 2209 and 2217 cm^{-1} (both: *vs*; asymmetric CD_3 stretching).

Poly-*cis*-1-deuteropropylene[803,805] shows its CHD rocking bands at 736 (*m*) and 746 (*m*) cm^{-1} while its CHD deformation fundamental is exhibited at 1286 cm^{-1}. Since its vibrational spectrum is very similar to that calculated for the *erythro*-di-isotactic poly-1-deuteropropylene it is concluded that it also exists in this form.[803] Poly-*trans*-1-deuteropropylene, on the other hand, displays an infrared absorption which agrees with it being in the respective *threo*-configuration.[803] Its CHD rocking bands appear at 719 (*m*) and 749 (*m*) cm^{-1} while its CHD deformation vibrations give rise to absorptions near 1252 (*m*) and 1290 (*s*) cm^{-1}.

Poly-2-deuteropropylene[803,806] shows its C—D deformation bands at 781 (*w*), 790 (*s*), 821 (*m*) and 890 cm^{-1} (*vw*). Its C—D stretching absorption appears near[807] 2160 cm^{-1}.

Poly-1,1-D_2-propylene shows its CD_2 bands at[806, 807] 691, 715 (both: *m*; rocking), 764, 811 (both: *w*; twisting), 1051 (*m*; deformation), 1058 (*s*; scissoring), 1110 (*m*; wagging), 2092 (*m*; symmetric stretching), 2169 (*w*) and 2171 cm^{-1} (*s*; asymmetric stretching).

Poly-1,1,2-D_3-propylene exhibits its CD_2 wagging frequencies at[803] 1037 (*m*) and 1229 (*m*) cm^{-1} while its CD_2 deformation bands appear at 1076 (*m*) and 1080 (*m*) cm^{-1}. The CD_3 stretching vibrations of poly-3,3,3-D_3-propylene[807] are located at 2066 (*m*), 2067 (*m*), 2204 (*s*) and 2215 cm^{-1} (*vs*).

Poly-D_6-butadiene[808] shows its C—D bending fundamentals as weak bands near 1100 cm^{-1}. Its *cis* form shows out-of-plane C—D deformation bands at 482 and 593 cm^{-1} while its *trans* form shows only one such band, at 712 cm^{-1}.

cis-Poly-1-deuterobutadiene[809] exhibits its CHD deformation absorption at 1300 (*m*) cm^{-1} while its highly *trans* form displays it at 1310 (*m*) cm^{-1}. Both forms, however, show their C—D stretching fundamental[809] at 2155 (*m*) cm^{-1}.

The out-of-plane C—D deformation bands of *cis*-poly-2-deuterobutadiene appear at[808] 580 and 658 cm^{-1} while the respective *trans*-form band is observed as high as 892 cm^{-1}. The spectra of *cis*- and *trans*-poly-2,3-D_2-butadiene[808] and *cis*- and *trans*-poly-1,1,4,4-D_4-butadiene[809] have also been reported.

The infrared-active fundamentals of poly-2,3-D_2-, 1,1,4,4-D_4- and perdeuterobutadiene have been calculated.[810]

Poly-D_3-vinyl chloride[811, 812] shows its C—D stretching bands at 2110 (*m–w*: symmetric; CD_2), 2148 (*m–w*; asymmetric; CD_2) and 2222 cm^{-1} (*m*; C—D). Poly-α-deutero[811, 812] and β-deuterovinyl chloride[811] exhibit their C—D valence vibration at 2207 (*w*) and 2170 (*w*) cm^{-1}, respectively, while the CD_2 bands of poly-α,β-D_2[811] and β,β-D_2-vinyl chloride[812] appear at 2180 (*w*), 2208 (*w*) and 2112 (*m*), 2210 (*w*) cm^{-1}, correspondingly. The other CD_2 characteristic fundamentals of the β,β-D_2- isomer are observed at[812] 709 (*sh*), 900 (*s*; both: rocking) and 1020 cm^{-1} (*w*; bending).

The spectra of various deuterated vinylidene chloride and vinyl chloride copolymers have also been measured.[813]

Poly-α-deuterostyrene[814] shows its C—D wagging vibration at 960 (*w*) cm^{-1}, its C—D bending fundamental at 1275 (*w*) cm^{-1} and its C—D stretching band at 2134 (*m*) cm^{-1}. In the case of the respective β-isomer the absorption in the region of the last band appears at[815] 2102 (*w*; *sh*), 2150 (*m*) and 2183 (*m*) cm^{-1}, the spectra of its α,β-D_2- and β,β-D_2-isotopomers in this vicinity being similar to it.[815]

The 400–3500 cm^{-1} spectra of poly-*p*-deutero- and poly-α,β,β-D_3-styrene[816] as well as the 900–1400 cm^{-1} absorption of poly-D_5- styrene[817] have also been reported.

4.139 Labelled Oxygen-containing Polymers

O-deuterated polyhydroxymethylene films[818] show their O—D valency fundamental at 2475 (*s*) cm^{-1} and their O—D deformation absorption in the vicinity of 1000 cm^{-1} (*vs*).

The 100–4000 cm^{-1} spectra of oriented $(CD_2O)_n$ films exhibit their CD_2 stretching bands[819] at 2105 (*m*; symmetric) and 2235 cm^{-1} (*m*; asymmetric).

Poly-D_8-tetrahydrofuran[820] absorbs at 2180 and 2220 cm^{-1} (both: *s*; asymmetric CD_2 stretching) while poly-3,3,4,4-D_4-tetrahydrofuran[820] displays its CD_2 valence vibrations at 2100 (*m*; symmetric) and 2202 cm^{-1} (*m*; asymmetric).

The 800–3500 cm^{-1} spectrum of O-deuterated polyvinyl alcohol[821] shows its O—D deformation band at 977 cm^{-1}.

Poly-D_4-ethylene oxide[822] exhibits its CD_2 characteristic bands at 703 (rocking), 996 (*m*; torsion), 1087 (*sh*; scissoring), 2030 (*sh*), 2072 (*s*), 2082 (*s*), 2164 (*s*), 2171 (*s*) and 2238 (*w*) cm^{-1}.

The absorption spectra of deuterated starch[823] and various celluloses[824–826] have also been recorded.

Quenched (amorphous) samples of poly-D_4-ethylene terephthalate display their C—D stretching bands at 2130, 2200 and 2260 cm^{-1}; heating(=crystallizing) them gives rise to new bands at[827] 2175 and 2275 cm^{-1}.

The O—D valence vibrations of D_2O-hydrated polystyrene-sulphonic acid films absorb,[828] at room temperature, at 1805 (*w*) and 2254 cm^{-1} (*b*; *s*). The O—D stretching absorption of a few other, similar, acids has also been reported.[828]

4.140 Deuterated Nitrogen-containing Polymers

Poly-α-deuteroacrylonitrile[829] shows its C—D out-of-plane deformation vibration at 765 (*m*) cm^{-1}, its bending frequency at 889 (*m*) cm^{-1} and its stretching band at 2170 (*w*) cm^{-1}. The CD_2 characteristic absorption of poly-β,β-D_2-acrylonitrile[829] appears at 702 (*m*; rocking), 810 (*vw*; wagging), 1043 (*m*; rocking), 1094 (*s*; bending), 2130 (*w*; symmetric stretching) and 2180 cm^{-1} (*w*; asymmetric stretching). In the case of poly-D_3-acrylonitrile[829] the C—D bands are observed at 807 (*vw*; wagging), 873 (*m*; bending), 942 (*m*; wagging), 1085 (*vs*; bending), 2130 (*w*; symmetric CD_2 stretching), 2170 (*vw*; C—D stretching) and 2180 cm^{-1} (*w*; asymmetric CD_2 valence vibration).

The N-deuterated *A* (extended) form poly-6-hexanolactam[830, 831] shows its N—D stretching bands at 2410 (*s*) and 2461 cm^{-1} (*vs*) while its *B* (pleated) form exhibits them at 2410 (*vs*) and 2471 (*s*) cm^{-1}.

The 400–3500 cm^{-1} spectrum of N-deuterated poly-ε-methylcaprolactam was also recorded.[831]

The spectra of various kinds of N-deuterated nylon fibres were investigated extensively[832–834] as were the normal vibrations of N-deuterated polyglycine[835] and the N—D stretching bands of deuterated glycyl clycine.[836]

4.141 Labelled Natural Products

Deuterated α and β-keratin show a characteristic absorption band at 1575 cm^{-1} which has tentatively been assigned to the absorption of a deuterated guanidinium group in an arginine moiety.[837] Deuterated α-keratin also shows a characteristic (N—D stretching) overtone band at[838] 4900 cm^{-1}.

The change of the infrared absorption spectrum of silk fibres, brought about by their deuteration, was investigated[833] as was the 1000–5000 cm^{-1} spectrum of deuterated gelatin solutions.[839]

Carboxyhaemoglobin films cast from D_2O solutions show bands[840] at about 2410 (N—D stretching?) and 2440 cm^{-1} (O—D stretching ?). Its solution in D_2O shows a strong and broad band centred at about[840] 2100 cm^{-1} (C—D stretching ?) and a band at about[840] 1450 cm^{-1}, probably due to deuterated peptide links, exhibited at every *p*D.

The 160–667 cm^{-1} spectra of Nujol mulls of fully deuterated chlorophylls a and b as well as their 160–400 cm^{-1} absorption in solution are generally quite similar to the corresponding normal spectra except for a 5–40 cm^{-1} downward shift of most of their bands.[841]

The infrared spectrum of deuterated deoxyribonucleic acid was also investigated.[842]

References

1. Church, C. H. (1959). *Diss. Abs.* **20**, 1398.
2. Reddy, S. P. and Cho, C. W. (1965). *Canad. J. Phys.* **43**, 793.
3. Pai, S. T., Reddy, S. P. and Cho, C. W. (1966). *Canad. J. Phys.* **44**, 2893.
4. Kolos, W. and Wolniewicz, L. (1966). *J. Chem. Phys.* **45**, 944.
5. Durie, R. A. and Herzberg, G. (1960). *Canad. J. Phys.* **38**, 806.
6. Trefler, M. and Gush, H. P. (1968). *Phys. Rev. Letters* **20**, 703.
7. (a) Adams, R. M. and Katz, J. J. (1956). *J. Opt. Soc. Amer.* **46**, 895.
   (b) Revich, V. E. and Stankiewich, S. A. (1966). *Doklady Akad. Nauk S.S.S.R.* **170**, 1376.
   (c) Spanbauer, R. N., Rao, K. N. and Jones, L. H. (1965). *J. Mol. Spectroscopy* **16**, 100.
   (d) Perkins, A. J. (1968). *Spectrochim. Acta* **24A**, 285.
8. Jones, L. H. and Goldblatt, M. (1957). *J. Mol. Spectroscopy* **1**, 43.
9. Rank, D. H., Eastman, D. P., Rao, B. S. and Wiggins, T. A. (1962). *J. Opt. Soc. Amer.* **52**, 1.
10. Rao, B. S. (1965). *Diss. Abs.* **25**, 4972.
11. Jones, L. H. and Robinson, E. S. (1956). *J. Chem. Phys.* **24**, 1246.
12. Bulanin, M. O., Orlova, N. D. and Shchepkin, D. N. (1965). *Optics and Spectroscopy* **19**, 406.
13. Datta, P. and Barrow, G. M. (1965). *J. Chem. Phys.* **43**, 2137.
14. Rank, D. H., Glickman, W. A. and Wiggins, T. A. (1965). *J. Chem. Phys.* **43**, 1304.
15. Keller, F. L. and Nielsen, A. H. (1954). *J. Chem. Phys.* **22**, 294.

16. Palik, E. D. (1955). *J. Chem. Phys.* **23**, 217.
17. Benedict, W. S., Gailar, N. M. and Plyler, E. K. (1956). *J. Chem. Phys.* **24**, 1139.
18. Staschewski, D. (1965). *Berichte Buns, Gesell, Phys. Chem.* **69**, 426.
19. Dickey, F. P. and Hoffman, J. M. (1955). *J. Chem. Phys.* **23**, 1718.
20. Gailar, N. M. and Dickey, F. P. (1960). *J. Mol. Spectroscopy* **4**, 1.
21. Libby, W. F. (1943). *J. Chem. Phys.* **11**, 101.
22. Posener, D. W. (1953). "Microwave Spectrum of the Water Molecule", MIT Research Laboratory of Electronics Tech. Rept. No. 225.
23. Morozov, V. P., Kvasa, N. T., Caune, A. Ja. and Lisovenko, V. A. (1963). *Optics and Spectroscopy* **15**, 617.
24. Khachkuruzov, G. A. (1959). *Trudy Gosuddarst. Inst. Priklad Khim.* **42**, 109.
25. Allavena, M. and Bratoz, S. (1963). *J. Chim. Phys.* **60**, 1199.
26. Slone, H. J. Jr. (1965). *Diss. Abs.* **25**, 7325.
27. (a) Bayly, J. G., Kartha, V. B. and Stevens, W. H. (1963). *Infra-red Phys.* **3**, 211.
    (b) Simon, M. N. (1968). *Diss. Abs.* **28B**, 4536.
28. Giguère, P. A. and Harvey, K. B. (1956). *Canad. J. Chem.* **34**, 798.
29. Thompson, W. K. (1965). *Trans. Faraday Soc.* **61**, 2635.
30. Saumagne, P. and Josien, M. L. (1958). *Bull. Soc. chim. France* (*5*) **25**, 813.
31. Mecke, R. (1961). *Current Science* (*India*) **30**, 43.
32. Waldron, R. D. (1957). *J. Chem. Phys.* **26**, 809.
33. Thomas, M. R., Scheraga, H. A. and Schrier, E. E. (1965). *J. Phys. Chem.* **69**, 3722.
34. Falk, M. and Giguère, P. A. (1957). *Canad. J. Chem.* **35**, 1195.
35. Draegert, D. A. and Williams, D. (1968). *J. Chem. Phys.* **48**, 401.
36. (a) Nibler, J. W. and Pimentel, G. C. (1965). *Spectrochim. Acta* **21**, 877.
    (b) Venkatesh, G. M. and Neelakantan, P. (1966). *Proc. Indian Acad. Sci. Sect.* **A 64**, 36.
    (c) Romo, L. A. (1956). *J. Phys. Chem.* **60**, 987.
    (d) Bernard, M. P. (1965). *Annls. Phys.* **10**, 57.
37. Postmus, C. and Ferraro, J. R. (1968). *J. Chem. Phys.* **48**, 3605.
38. Staats, P. A., Morgan, H. W. and Goldstein, J. H. (1956). *J. Chem. Phys.* **24**, 916.
39. (a) Hedberg, K. and Badger, R. M. (1951). *J. Chem. Phys.* **19**, 508.
    (b) Rochkind, M. M. (1966). *Diss. Abs.* **27**, 785B.
40. Buchanan, R. A., Kinsey, E. L. and Caspers, H. H. (1962). *J. Chem. Phys.* **36**, 2665.
41. Busing, W. R. (1955). *J. Chem. Phys.* **23**, 933.
42. Oehler, O. and Guenthard, H. H. (1968). *J. Chem. Phys.* **48**, 2036.
43. Cummins, R. A. (1965). *Aust. J. Chem.* **18**, 98.
44. Miller, R. E. (1966). *Diss. Abs.* **27B**, 438.
45. Miller, R. E. and Eggers, D. F. Jr. (1966). *J. Chem. Phys.* **45**, 3028.
46. Miller, R. E. , Leroi, G. E. and Eggers, D. F. Jr. (1967). *J. Chem. Phys.* **46**, 2292.
47. Allen, H. C. Jr., Plyler, E. K. and Blaine, L. R. (1957). *J. Res. Nat. Bur. Stand.* **59**, 211.
48. Allen, H. C. Jr., Naylor, R. E. and Plyler, E. K. (1954). *J. Res. Nat. Bur. Stand.* **53**, 321.
49. Morozov, V. P., Kvasa, N. T., Caune, A. J. and Lisovenko, V. A. (1963). *Optics and Spectroscopy* **15**, 617.
50. Venkateswarlu, K. and Mariam, S. (1965). *Indian. J. Pure Appl. Phys.* **3**, 117.
51. Venkateswarlu, K. and Mariam, S. (1965). *Proc. Indian Academy Sci. Sect A* **61**, 260.
52. Crosby, G. A., Bair, E. J. and Cross, P. C. (1955). *J. Chem. Phys.* **23**, 1660.

53. Herzberg, G. (1945). "Infrared and Raman Spectra of Polyatomic Molecules". p. 283. D. Van Nostrand Company, Princeton, U.S.A.
54. (a) Saumagne, P., Lascombe, J. and Devaure, J. (1961). *Compt. rend.* **253**, 632.
(b) Winnewisser, B. P. and Winnewisser, M. (1966). *Z. Naturforsch.* **23A**, 832.
55. Lynds, L. (1965). *J. Chem. Phys.* **42**, 1124.
56. Bass, C. D., Lynds, L., Wolfram, T. and DeWames, R. E. (1964). *J. Chem. Phys.* **40**, 3611.
57. Lynds, L. and Bass, C. D. (1964). *J. Chem. Phys.* **41**, 3165.
58. Lynds, L. and Bass, C. D. (1965). *J. Chem. Phys.* **43**, 4357.
59. Perec, M. and Becka, L. N. (1965). *J. Chem. Phys.* **43**, 721.
60. Lynds, L., Wolfram, T. and Bass, C. D. (1965). *J. Chem. Phys.* **43**, 3775.
61. Lynds, L. and Bass, C. D. (1964). *J. Chem. Phys.* **40**, 1590.
62. Corset, J., Guillermet, J. and Lascombe, J. (1966). *Bull. Soc. chim. France*, 1231.
63. Benedict, W. S. and Plyler, E. K. (1957). *Canad. J. Phys.* **35**, 1235.
64. Rao, K. N., Brim, W. W. and Hoffman, J. M. (1961). *J. Mol. Spectroscopy* **7**, 362. See also Levine, D. M. and Dows, D. A. (1967). *J. Chem. Phys.* **46**, 1168.
65. Jones, L. H., Brim, W. W. and Rao, K. N. (1963). *J. Mol. Spectroscopy*, **11**, 389.
66. Sundaram, S., Suszek, F. and Cleveland, F. F. (1960). *J. Chem. Phys.* **32**, 251.
67. Sundaram, S. and Cleveland, F. F. (1960). *J. Mol. Spectroscopy* **5**, 61.
68. McDowell, R. S. (1966). *J. Mol. Spectroscopy* **19**, 239.
69. Palik, E. D. and Bell, E. E. (1957). *J. Chem. Phys.* **26**, 1095.
70. Buerger, H., Sawodny, H. and Hoefler, F. (1965). *Monatsh.* **96**, 1437.
71. Powell, D. B. and Sheppard, N. (1956). *J. Chem. Soc.* 3108.
72. Barrow, G. M., Krueger, R. H. and Basolo, F. (1956). *J. Inorg. Nuclear Chem.* **2**, 341.
73. Mueller, A. and Krebs, B. (1966). *Z. anorg. Chem.* **345**, 165.
74. Dows, D. A. and Pimentel, G. C. (1955). *J. Chem. Phys.* **23**, 1258.
75. Moore, C. B. and Rosengren, K. (1966). *J. Chem. Phys.* **44**, 4108.
76. Levine, D. M. and Dows, D. A. (1967). *J. Chem. Phys.* **46**, 1168.
77. Krakow, B., Lord, R. C. and Neely, G. O. (1968). *J. Mol. Spectroscopy* **27**, 148.
78. McGraw, G. E., Bernitt, D. L. and Hisatsune, I. C. (1966). *J. Chem. Phys.* **45**, 1392.
79. (a) Bain, O. and Giguère, P. A. (1955). *Canad. J. Chem.* **33**, 527.
(b) Hunt, R. H. and Leacock, R. A. (1956). *J. Chem. Phys.* **45**, 3141.
80. Benedict, W. S., Gailar, N. and Plyler, E. K. (1953). *J. Chem. Phys.* **21**, 1302.
81. Ward, W. M. (1958). *Diss. Abs.* **18**, 1823.
82. Stroup, R. E., Oetjen, R. A. and Bell, E. E. (1953). *J. Opt. Soc. Amer.* **43**, 1096.
83. Tamaru, K. (1955). *J. Phys. Chem.* **59**, 777.
84. Haynie, W. H. and Nielsen, H. H. (1953). *J. Chem. Phys.* **21**, 1840.
85. Nixon, R. (1956). *J. Phys. Chem.* **60**, 1055.
86. Sundaram, S. (1961). *Canad. J. Phys.* **39**, 370.
87. Daykin, P. N. and Sundaram, S. (1962). *Z. phys. Chem.* (*Frankfurt*) **32**, 222.
88. Flotow, H. E., Osborne, D. W., Otto, K. and Abraham, B. M. (1963). *J. Chem. Phys.* **38**, 2620.
89. Roszinski, H., Dautel, R. and Zeil, W. (1963). *Z. phys. Chem.* (*Frankfurt*) **36**, 26.
90. Greenwood, N. N., Storr, A. and Wallbridge, M. G. H. (1963). *Inorg. Chem.* **2**, 1036.
91. Rush, J. J. and Ferraro, J. R. (1966). *J. Chem. Phys.* **44**, 2497.
92. Schwarzmann, E. and Lange, L. (1968). *Z. Naturforsch.* **23b**, 874.
93. Benoit, A. (1963). *Spectrochim. Acta* **19**, 2011.
94. Cornelis-Benoit, A. (1965). *Spectrochim. Acta* **21**, 623.

95. Nightingale, R. E. and Wagner, E. L. (1954). *J. Chem. Phys.* **22**, 205.
96. McGraw, G. E., Bernitt, D. L. and Hisatsune, I. C. (1965). *J. Chem. Phys.* **42**, 237.
97. Meal, J. H. and Wilson, M. K. (1956). *J. Chem. Phys.* **24**, 385.
98. Boyd, D. R. J. (1955). *J. Chem. Phys.* **23**, 922.
99. Polo, S. R. and Wilson, M. K. (1954). *J. Chem. Phys.* **22**, 1559.
100. Venkateswarlu, K. and Devi, V. M. (1965). *Indian J. pure appl. Phys.* **3**, 120.
101. Kovalchuk, D. S. and Morozov, V. P. (1964). *Spekroskopiya, Metody i Primenic. Akad. Nauk S.S.R., Sibersk. Otd.* 113.
102. Freeman, D. E. and Wilson, M. K. (1965). *Spectrochim. Acta* **21**, 1825.
103. Thyagarajan, G., Herranz, J. and Cleveland, F. F. (1961). *J. Mol. Spectroscopy* **7**, 154.
104. Venkateswarlu, K. and Devi, V. M. (1965). *Indian J. pure appl. Phys.* **3**, 195.
105. Kovalchuk, D. S. and Morozov, V. P. (1964). *Ukr. phys. J.* **9**, 876; *C. A.* **62**, 119h (1965).
106. Newman, C., Polo, S. R. and Wilson, M. K. (1959). *Spectrochim. Acta* **15**, 793.
107. Campbell-Ferguson, H. J. and Ebsworth, E. A. V. (1967). *J. Chem. Soc.* (*A*). 705.
108. Linton, H. R. and Nixon, E. R. (1958). *Spectrochim. Acta* **12**, 41.
109. Linton, H. R. and Nixon, E. R. (1958). *Spectrochim. Acta* **10**, 299.
110. Pillai, M. G. K. and Ramaswamy, A. (1964). *J. Chim. phys.* **61**, 1082.
111. Lindeman, L. P. and Wilson, M. K. (1956). *Z. Phys. Chem.* (*Frankfurt*) **9**, 29.
112. McKean, D. C. and Chalmers, A. A. (1967). *Spectrochim. Acta* **23A**, 777.
113. Lindeman, L. P. and Wilson, M. K. (1954). *J. Chem. Phys.* **22**, 1723.
114. Lindeman, L. P. and Wilson, M. K. (1957). *Spectrochim. Acta* **9**, 47.
115. Beach, A. L. and Griffiths, J. E. (1966). *Canad. J. Chem.* **44**, 743.
116. Griffiths, J. E. (1967). *Canad. J. Chem.* **45**, 2639.
117. Drake, J. E., Riddle, C. and Rogers, D. E. (1969). *J. Chem. Soc.* (*A*). 910.
118. Levin, I. W. and Ziffer, H. (1965). *J. Chem. Phys.* **43**, 4023.
119. Bethke, G. W. and Wilson, M. K. (1957). *J. Chem. Phys.* **26**, 1121.
120. Venkateswarlu, K. and Thanalakshmi, R. (1962). *Proc. Indian. Acad. Sci. A* **56**, 247.
121. Wagner, E. L. and Hornig, D. F. (1950). *J. Chem. Phys.* **18**, 305.
122. Plumb, R. C. and Hornig, D. F. (1955). *J. Chem. Phys.* **23**, 947.
123. Wagner, E. L. and Hornig, D. F. (1950). *J. Chem. Phys.* **18**, 296.
124. Morgan, H. W., Staats, P. A. and Goldstein, J. H. (1957). *J. Chem. Phys.* **27**, 1212.
125. Rocchiccioli, C. (1961). *Compt. rend.* **253**, 838.
126. Charitonov, Ju. Ja. and Saruchanov, M. A. (1968). *Zhur. neorg. Khim.* **13**, 71.
127. Durig, J. R., Bush, S. F. and Mercer, E. E. (1966). *J. Chem. Phys.* **44**, 4238.
128. Decius, J. C. and Pearson, D. P. (1953). *J. Amer. Chem. Soc.* **75**, 2437.
129. Giguère, P. A. and Savoie, R. (1962). *Canad. J. Chem.* **40**, 495.
130. Martinez, J. V. and Wagner, E. L. (1957). *J. Chem. Phys.* **27**, 1110.
131. Durig, J. R., Antion, D. J. and Baglin, F. G. (1968). *J. Chem. Phys.* **49**, 666.
132. Chackalackal, S. M. and Stafford, F. E. (1966). *J. Amer. Chem. Soc.* **88**, 723.
133. Giguère, P. A. and Savoie, R. (1963). *J. Amer. Chem. Soc.* **85**, 287.
134. Stopperka, K. (1966). *Z. anorg. Chem.* **344**, 263.
135. Lord, R. C. and Nielsen, E. (1951). *J. Chem. Phys.* **19**, 1.
136. Lehmann, W. J., Ditter, J. F. and Shapiro, I. (1958). *J. Chem. Phys.* **29**, 1248.
137. Venkateswarlu, K. and Radhakrishnan, M. (1962). *Proc. Indian Acad. Sci.* **A55**, 161.
138. Kapshtal, V. N. and Sverdlov, L. M. (1963). *J. Phys. Chem. Moscow* **37**, 1275.

139. Norman, A. D. and Schaeffer, R. (1966). *J. Amer. Chem. Soc.* **88**, 1143.
140. Hrostowski, H. J. and Pimentel, G. C. (1954). *J. Amer. Chem. Soc.* **76**, 998.
141. Coriell, S. R. (1962). *Diss. Abs.* **22**, 3865.
142. Kaufman, J. J. and Koski, W. S. (1956). *J. Amer. Chem. Soc.* **78**, 5774.
143. Bethke, G. W. and Wilson, M. K. (1957). *J. Chem. Phys.* **26**, 1107.
144. Ponomarenko, V. A. and Egorov, P. Ju. (1960). *Bull. Acad. Sci. U.S.S.R.* 1133.
145. Lord, R. C., Robinson, D. W. and Schumb, W. C. (1956). *J. Amer. Chem. Soc.* **78**, 1327.
146. Ebsworth, E. A. V., Taylor, R. and Woodward, L. A. (1959). *Trans. Faraday Soc.* **55**, 211.
147. Robinson, D. W. (1958). *J. Amer. Chem. Soc.* **80**, 5924.
148. Ebsworth, E. A. V., Hall, J. R., Mackillop, M. J., McKean, D. C., Sheppard, N. and Woodward, L. A. (1958). *Spectrochim. Acta* **13**, 202.
149. Davidson, G., Ebsworth, E. A. V., Sheldrick, G. M. and Woodward, L. A. (1966). *Spectrochim. Acta* **22**, 67.
150. Aylett, B. J., Hall, J. R., Mckean, D. C., Taylor, R. and Woodward, L. A. (1960). *Spectrochim. Acta* **16**, 747.
151. Dupuis, T. (1956). *Compt. rend.* **243**, 1621.
152. Vuagnat, A. M. and Wagner, E. L. (1957). *J. Chem. Phys.* **26**, 77.
153. Uno, T., Machida, K. and Hanai, K. (1966). *Spectrochim. Acta* **22**, 2065.
154. Thomas, J. P. (1965). *Diss. Abs.* **26**, 1942.
155. Cotton, F. A., Down, J. L. and Wilkinson, G. (1959). *J. Chem. Soc.* 833.
156. (a) Watanabe, H., Totani, T., Nakagawa, T. and Kubo, M. (1960). *Spectrochim. Acta* **16**, 1076.
(b) Kartha, V. B., Krishnamachari, S. L. N. G. and Subramaniam, C. R. (1967). *J. Mol. Spectroscopy*, **23**, 149
(c) Vaska, L. and Catone, D. L. (1966). *J. Amer. Chem. Soc.* **88**, 5324; 5325.
(d) Pliskin, W. A. and Eischens, R. P. (1960). *Z. phys. Chem.* (Frankfurt) **24**, 11.
157. (a) Wilmshurst, J. K. and Bernstein, H. J. (1957). *Canad. J. Chem.* **35**, 226.
(b) Kaylor, H. M. and Nielsen, A. H. (1955). *J. Chem. Phys.* **23**, 2139.
(c) Fox, K., Hecht, K. T., Meredith, R. E. and Peters, C. W. (1962). *J. Chem. Phys.* **36**, 3135.
158. Herranz, J., Morcillo, J. and Gómez, A. (1966). *J. Mol. Spectroscopy* **19**, 266.
159. Jones, L. H. and McDowell, R. S. (1959). *J. Mol. Spectroscopy* **3**, 632.
160. Fox, K. (1962). *J. Mol. Spectroscopy* **9**, 381.
161. Orlova, N. D. (1964). *Bull. Leningr. Iniv. Ser. Phys. et Chim.* **19**, No. 16 (3), 30.
162. Andersen, F. A. (1963). *K. Danske Vidensk. Selsk. Mat.-Fys. Medd.* **33** (12), 1.
163. Mears, G. E. F., Parkyn, D. G. and Leisegang, E. C. (1966). *J.S. African Chem. Inst.* **19**, 93, 130; (1967). **20**, 16.
164. (a) Allen, Jr., H. C. and Plyler, E. K. (1959). *J. Res. Nat. Bur. Stand.* **63A**, 145.
(b) Betrencourt-Stirnemann, C., Alamichel, C. and Graner, G. (1967). *Compt. rend.* **265B**, 549.
165. Moritz, A. G. (1966). *Spectrochim. Acta* **22**, 385.
166. Leisegang, E. C. and Parkyn, D. G. (1968). *J.S. African Chem. Inst.* **21**, 64.
167. Olson, W. B., Allen H. C. Jr. and Plyler, E. K. (1963). *J. Res. Nat. Bur. Stand.* **67A**, 27.
168. Dowling, J. Jr. (1965). *Diss. Abs.* **25**, 4198.
169. Rea, D. G. and Thompson, H. W. (1956). *Trans. Faraday Soc.* **52**, 1304.
170. (a) Wiggins, T. A., Shull, E. R., Bennett, J. M. and Rank, D. H. (1953). *J. Chem. Phys.* **21**, 1940.
(b) Blass, W. E. and Edwards, T. H. (1967). *J. Mol. Spectroscopy* **24**, 116.

171. Jones, L. H. and Goldblatt, M. (1958). *J. Mol. Spectroscopy* **2**, 103.
172. Jones, L. H. (1960). *J. Mol. Spectroscopy* **4**, 84.
173. (a) Edgell, W. F. and May, C. E. (1954). *J. Chem. Phys.* **22**, 1808.
(b) Polo, S. R. and Wilson, M. K. (1953). *J. Chem. Phys.* **21**, 1129.
174. Galasso, V., de Alti, G. and Costa, G. (1965). *Spectrochim. Acta* **21**, 669.
175. Shimanouchi, T. and Suzuki, I. (1961). *J. Mol. Spectroscopy* **6**, 277.
176. Meister, A. G., Rosser, S. E. and Cleveland, F. F. (1950). *J. Chem. Phys.* **18**, 346.
177. Maklakov, L. I. and Nikitin, V. N. (1963). *Optics and Spectroscopy* **15**, 445.
178. Long, D. A., Gravenor, R. B. and Jones, D. T. L. (1964). *Trans. Faraday Soc.* **60**, 1509.
179. Sushchinskii, M. M. and Muldakhmetov, Z. M. (1964). *Optika i Spektroskopia* **17**, 45.
180. Forel, M. T., Leicknam, J. P. and Josien, M. L. (1960). *J. Chim. phys.* **57**, 1103.
181. Lascombe, J., Devaure, J. and Josien, M. L. (1964). *J. Chim. phys.* **61**, 1271.
182. Lord, R. C., Nolin, B. and Stidham, H. D. (1955). *J. Amer. Chem. Soc.* **77**, 1365.
183. Kagarise, R. E. (1963). *Spectrochim. Acta* **19**, 629.
184. Polo, S. R., Palm, A., Voelz, F. L., Cleveland, F. F., Meister, A. G., Bernstein, R. B. and Sherman, R. H. (1955). *J. Chem. Phys.* **23**, 833.
185. Weissman, H. B., Meister, A. G. and Cleveland, F. F. (1958). *J. Chem. Phys.* **29**,
186. Palm, A., Voelz, F. L. and Meister, A. G. (1955). *J. Chem. Phys.* **23**, 726.
187. Palma, F. E., Piotrowski, E. A., Sundaram, S. and Cleveland, F. F. (1964). *J. Mol. Spectroscopy* **13**, 119.
188. Takata, A. N., Narasimham, K. V., Meister, A. G., Dowling, J. M., Cleveland, F. F., Sundaram, S., Piotrowski, E. A., Bernstein, R. B. and Miller, S. I. (1965). *J. Mol. Spectroscopy* **15**, 319.
189. Riter, J. R. and Egger, D. F. Jr. (1966), *J. Chem. Phys.* **44**, 745.
190. Ditzel, E. F., Meister, A. G., Piotrowski, E. A., Cleveland, F. F., Sarna, Y. A. and Sundaram, S. (1964). *Can. J. Chem.* **42**, 2841.
191. Herzberg, G. (1945). "Infrared and Raman Spectra of Polyatomic Molecules", p. 317. D. Van Nostrand, Princeton, U. S. A.
192. Jones, E. W., Popplewell, R. J. L. and Thompson, H. W. (1966). *Proc. Roy. Soc.* **A290**, 490.
193. Mears, G. E. F., Parkyn, D. G. and Leisegang, E. C. (1966). *J. S. African Chem. Inst.* **19**, 131.
194. Jones, E. W., Popplewell, R. J. L. and Thompson, H. W. (1966). *Spectrochim. Acta* **22**, 659.
195. Weissman, H. B., Bernstein, R. B., Rosser, S. E., Meister, A. G. and Cleveland, F. F. (1955). *J. Chem. Phys.* **23**, 544.
196. (a) Jones, E. W., Popplewell, R. J. L. and Thompson, H. W. (1965). *Spectrochim. Acta* **22**, 639.
196. (b) Jones, E. W., Popplewell, R. J. L. and Thompson, H. W. (1965). *Proc. Roy. Soc. A.* **288**, 39.
197. Joffrin, C., Van Than, N. and Barchewitz, P. (1966). *Canad. J. Phys.* **44**, 2695.
198. Joffrin, C., Van Than, N. and Barchewitz, P. (1966). *J. Phys.* (*Paris*) **27**, 15.
199. Palik, E. D. and Narahari, Rao, K. (1957). *J. Chem. Phys.* **26**, 1402.
200. Andersen, F. A., Bak, B. and Brodersen, S. (1956). *J. Chem. Phys.* **24**, 989.
201. Van Riet, R. (1957). *Ann. Soc. sci. Bruxelles.* **71**, 102.
202. Herzberg, G. (1945). "Infrared Spectra of Polyatomic Molecules", pp. 372–346. D. Van Nostrand, Princeton, U.S.A.
203. Thompson, S. O., Turkevich, J. and Irsa, A. P. (1952). *J. Phys. Chem.* **56**, 243.
204. Tejada, S. B. (1964). *Diss. Abs.*, **25**, 876.

205. Weiss, S. and Leroi, G. E. (1968). *J. Chem. Phys.*, **48**, 962.
206. Lysne, P. C. (1967). *Diss. Abs.* **27B**, 3218.
207. Bak, B., Christensen, D., Møller, J. and Detoni, S. (1958). *Acta Chem. Scand.* **12**, 2021.
208. Bernstein, H. J., Pullin, A. D. E., Rabinovitch, B. S. and Larson, N. R. (1952). *J. Chem. Phys.* **20**, 1229.
209. Lafon, B. and Nielsen, J. R. (1966). *J. Mol. Spectroscopy* **21**, 175.
210. Neu, J. T. and Gwinn, W. D. (1950). *J. Chem. Phys.* **18**, 1642.
211. Kukina, V. S. and Sverdlov, L. M. (1966). *Zhur. Fiz. Khim.* **40**, 2837.
212. Gayles, J. N. Jr. and King, W. T. (1965). *Spectrochim. Acta* **21**, 543.
213. (a) McMurry, H. L., Thornton, V. and Condon, F. E. (1949). *J. Chem. Phys.* **17**, 919.
(b) Friedman, L. and Turkevich, J. (1949). *J. Chem. Phys.* **17**, 1012.
(c) Documentation of Molecular Spectroscopy Catalogue, Spectral Card No. 4016. Butterworths Scientific Publications, London.
(d) Spectral Card No. 4017. Butterworths Scientific Publications, London.
214. McMurry, H. L. and Thornton, V. (1951). *J. Chem. Phys.* **19**, 1014.
215. Ogawa, T., Hara, J. and Hisota, K. (1968). *J. Chem. Soc., Japan* **89**, 19.
216. McMurry, H. L. and Speas, D. (1965). *Spectrochim. Acta* **21**, 2105.
217. (a) Condon, F. E., McMurry, H. L. and Thornton, V. (1951). *J. Chem. Phys.* **19**, 1010.
(b) Frejaville, G., Gounelle, Y., Julien, J. and Paillous, A. (1963). *Bull. Soc. chim. France*, 2171.
218. Thompson, O., Turkevich, J. and Irsa, A. P. (1952). *J. Phys. Chem.* **56**, 248.
219. Helmkamp, G. K. and Rickborn, B. F. (1957). *J. Org. Chem.* **22**, 479.
220. Wilmshurst, J. K. and Bernstein, H. J. (1957). *Canad. J. Chem.* **35**, 969.
221. Evans, J. C. and Bernstein, H. J. (1956). *Canad. J. Chem.* **34**, 1037.
222. Evans, J. C. and L. G. Y. S. (1966). *J. Amer. Chem. Soc.* **88**, 2118.
223. Zeil, W., Buchert, H., Heel, H. and Pfoertner, H. (1960). *Z. Elektrochem.* **64**, 769.
224. Shull, E. R., Oakwood, T. S. and Rank, D. H. (1953). *J. Chem. Phys.* **21**, 2024.
225. Shimizu, K. and Murata, H. (1957). *Bull. Chem. Soc. Japan* **30**, 487.
226. Rank, D. H., Saksena, B. D. and Shull, E. R. (1950). *Discuss. Faraday Soc.* **9**, 187, 219.
227. Baker, A. W. and Lord, R. C. (1955). *J. Chem. Phys.* **23**, 1636.
228. Guenthard, Hs. H., Lord, R. C. and McCubbin, T. K. Jr. (1956). *J. Chem. Phys.* **25**, 768.
229. Duncan, J. L. (1968). *J. Mol. Spectroscopy* **23**, 451.
230. Duncan, J. L. and McKean, D. C. (1968). *J. Mol. Spectroscopy* **27**, 117.
231. Lord, R. C. and Nakagawa, I. (1963). *J. Chem. Phys.* **39**, 2951.
232. Alexanyan, V. T., Kuzyants, G. M., Lukina, M. Yu., Zotova, S. V. and Vostokova, E. I. (1968). *Zhur. strukt. Khim*, 141.
233. (a) Durig, J. R. and Morrissey, A. C. (1967). *J. Chem. Phys.* **46**, 4854.
(b) Durig, J. R. and Green, W. H. (1967). *J. Chem. Phys.* **47**, 673.
234. Sverdlov, L. M. and Prokofeva, N. I. (1959). *Optics and Spectroscopy* **7**, 588.
235. Miller, F. A. (1950). *J. Chem. Phys.* **18**, 1522.
236. (a) Miller, F. A. and Golob, H. R. (1964). *Spectrochim. Acta* **20**, 1517.
(b) Abramowitz, S. and Bauman, R. P. (1963). *J. Chem. Phys.* **39**, 2757.
237. Fuhrer, H. and Guenthard, Hs. H. (1965). *Helv. Chim. Acta* **48**, 236.
238. (a) Lalonde, R. T., Ding, J. and Tobias, M. A. (1967). *J. Amer. Chem. Soc.* **89**, 6651.

(b) Jones, R. N., Nolin, B. and Roberts, G. (1955). *J. Amer. Chem. Soc.* **77**, 6334.
239. Charette, J. and De Hemptinne, M. (1951). *Bull. classe Sci.* **37**, 436.
240. Crawford, B. L. Jr., Lancaster, J. E. and Inskeep, R. G. (1953). *J. Chem. Phys.* **21**, 678.
241. Sverdlov, L. M. and Pakhomova, N. L. (1954). *Zhur. Eksper. Teoret. Fiz.* **26**, 64.
242. Arnelt, R. L. and Crawford, B. L. Jr. (1950). *J. Chem. Phys.* **18**, 118.
243. Callear, A. B. and Cvetanovich, R. J. (1956). *J. Chem. Phys.* **24**, 873.
244. Grogan, M. J. and Nakamoto, K. (1966). *J. Amer. Chem. Soc.* **88**, 5454.
245. (a) Mann, D. E., Acquista, N. and Plyler, E. K. (1954). *J. Chem. Phys.* **22**, 1586.
(b) Ref. 213b, Card No. 462.
246. Houser, T. J., Bernstein, R. B., Miekka, R. G. and Angus, J. C. (1955). *J. Amer. Chem. Soc.* **77**, 6201.
247. Scherer, J. A., Evans, J. C. and Overend, J. (1960). *J. Chem. Phys.* **33**, 314.
248. Klochkovskii, Yu. V., Kukina, V. S. and Sverdlov, L. M. (1965). *Zhur. fiz. Khim.* **39**, 1912.
249. Kukina, V. S. and Sverdlov, L. M. (1966). *Zhur. Fiz. Khim.* **40**, 1613.
250. Craig, N. C., Lo, G. Y., Needham, C. D. and Overend, J. (1964). *J. Amer. Chem. Soc.* **86**, 3232.
251. Craig, N. C. and Lo, G. Y. (1967). *J. Mol. Spectroscopy.* **23**, 307.
252. Craig, N. C. and Entemann, E. A. (1962). *J. Chem. Phys.* **36**, 243.
253. Ramavataram, K. and Rahman, A. (1956). *Ann. Soc. sci. Bruxelles.* **70**, 194.
254. De Croes, M., Perlinghi, M. and Van Riet, R. (1956), *Bull. Acad. Belg. Cl. Sci.* **42**, 379.
255. Edgell, W. F. and Ultee, C. J. (1954). *J. Chem. Phys.* **22**, 1983.
256. Bak, B. and Christensen, D. (1958). *Spectrochim. Acta* **12**, 355.
257. Enomoto, S. and Asahina, M. (1966). *J. Mol. Spectroscopy* **19**, 117.
258. Kukina, V. S. and Sverdlov, L. M. (1966). *Zhur. Fiz. Khim.* **40**, 2837.
259. Sverdlov, L. M., Klochkovskij, Ju. V., Kukina, V. S. and Mezhueva, T. D. (1960). *Optics and Spectroscopy* **9**, 728.
260. Vandenhaute, R. (1967). *Ann. Soc. sci. Bruxelles.* **81**, 163.
261. Van Riet, R. and De Hemptinne, M. (1961). *Bull. Classe. sci. Acad. Roy. Belg.* **47**, 1128.
262. Enomoto, S. and Ichinohe, S. (1958). *J. Chem. Soc. Japan* **79**, 1343.
263. Dowling, J. M., Puranik, P. G., Meister, A. G. and Miller, S. I. (1957). *J. Chem. Phys.* **26**, 233.
264. Krupp, R. H., Piotrowski, E. A., Cleveland, F. F. and Miller, S. I. (1963). *Dev. Appl. Spectroscopy* **2**, 52.
265. De Hemptinne, M., Germain-Lefevre, G., Van Riet, R. and Lenaerts, D. (1960). *Bull. Acad. Belg. Clin. Sci.* **46**, 310.
266. Narita, Sh., Ichinohe, S. and Enomoto, S. (1959). *J. Chem. Phys.* **31**, 1151.
267. Lord, R. C. and Venkateswarlu, P. (1953). *J. Opt. Soc. Amer.* **43**, 1079.
268. Sverdlov, L. M. (1956). *C.R. Acad. Sci. U.R.S.S.* **106**, 80.
269. Adams, D. M. and Chatt, J. (1962). *J. Chem. Soc.* 2821.
270. Farina, M. and Peraldo, M. (1960). *Gazzetta* **90**, 973.
271. Hunziker, H. and Guenthard, Hs. H. (1965). *Spectrochim. Acta* **21**, 51.
272. Meyer, R. and Guenthard, Hs. H. (1967). *Spectrochim. Acta* **23A**, 2341.
273. Bernstein, H. J. and Powling, J. (1951). *J. Amer. Chem. Soc.* **73**, 1843.
274. Purlee, E. L. and Taft, R. W. Jr. (1956). *J. Amer. Chem. Soc.* **78**, 5807.
275. Pathak, C. M. (1968). *Diss. Abs.* **29B**, 144.
276. Hoffmann, E. G. (1958). *Annalen* **618**, 276.

277. Schroetter, H. W. and Hoffmann, E. G. (1964). *Annalen* **672**, 44.
278. Meyer, R., Hunziker, H. and Guenthard, Hs. H. (1969). *Spectrochim. Acta* **25A**, 295.
279. London, L. H. and Moeller, K. D. (1968). *J. Mol. Structure* **2**, 493.
280. Lord, R. C. and Rea, D. G. (1957). *J. Amer. Chem. Soc.* **79**, 2401.
281. Mitchell, R. W., Dorko, E. A. and Merritt, J. A. (1968). *J. Mol. Spectroscopy* **26**, 197.
282. Neto, N., Di Lauro, C., Castellucci, E. and Califano, S. (1967). *Spectrochim Acta.* **23A**, 1763.
283. Overend, J. and Crawford, B. Jr. (1958). *J. Chem. Phys.* **29**, 1002.
284. Overend, J. and Thompson, H. W. (1956). *Trans. Faraday. Soc.* **52**, 1295.
285. Van Than, N. and Joffrin, C. (1965). *J. de Phys.* **26**, 809.
286. Maki, A. G. and Toth, R. A. (1965). *J. Mol. Spectroscopy* **17**, 136.
287. Evans, J. C., Wilmshurst, J. K. and Bernstein, H. J. (1956). *Canad. J. Chem.* **34**, 1139.
288. Shuler, W. E. and Fletcher, W. H. (1957). *J. Mol. Spectroscopy* **1**, 95.
289. Eaton, D. R. and Thompson, H. W. (1959). *Proc. Roy. Soc.* **A250**, 39.
290. Sverdlov, L. M. and Borisov, M. G. (1960). *Optics and Spectroscopy* **9**, 432.
291. Tarasova, N. V. and Sverdlov, L. M. (1965). *Optics and Spectroscopy* **18**, 336.
292. Panchenko, Yu. N., Pentin, Yu. A., Tyulin, V. I. and Tatevskii, V. M. (1964). *Optics and Spectroscopy*, **16**, 536.
293. Koptev, G. S., Panchenko, Yu, N., Tyulin, V. I. and Tatevskii, V. M. (1965). *Optics and Spectroscopy* **19**, 108.
294. Borshagovskaya, I. S., Panchenko, Yu. N. and Pentin, Yu. A. (1967). *Optics and Spectroscopy* **22**, 194.
295. Panchenko, Yu. N., Choan, T. S. and Pentin, Yu. A. (1967). *Vestnik Moscov. Yuniversit.* **6**, 32.
296. Sobolev, E. V., Alexanyan, V. T. and Mironov, V. A. (1963). *Proc. Acad. Sci. (U.S.S.R.)* **152**, 892.
297. Stidham, H. D. (1965). *Spectrochim. Acta* **21**, 23.
298. Documentation of Molecular Spectroscopy Catalogue, Spectral Cards 13266 and 13267.
299. Lippincott, E. R., Lord, R. C. and McDonald, R. S. (1951). *J. Amer. Chem. Soc.* **73**, 3370.
300. Van Tets, A. and Guenthard, Hs. H. (1963). *Spectrochim. Acta.* **19**, 1495.
301. Steele, D. (1965). *J. Mol. Spectroscopy* **15**, 333.
302. Talley, R. M. and Nielsen, A. H. (1945). *J. Chem. Phys.* **22**, 2030.
303. Overend, J. and Thompson, H. W. (1955). *Proc. Roy. Soc.* **232**, 291.
304. Plyler, E. K. and Gailar, N. M. (1952). *J. Res. Nat. Bur. Stand.* **47**, 249.
305. Ghersetti, S. and Rao, K. N. (1968). *J. Mol. Spectroscopy* **28**, 273.
306. Saksena, B. D. (1952). *J. Chem. Phys.* **20**, 95.
307. Allen, H. C. Jr., Blaine, I. R. and Plyler, E. K. (1956). *J. Res. Nat. Bur. Stand.* **56**, 279.
308. Jones, L. H., Goldblatt, M., McDowell, R. S., Armstrong, D. E., Scott, J. F., Williamson, J. G. and Rao, K. N. (1967). *J. Mol. Spectroscopy* **23**, 9.
309. Lafferty, W. J., Plyler, E. K. and Tidwell, E. D. (1962). *J. Chem. Phys.* **37**, 1981.
310. Overend, J. and Thompson, H. W. (1956). *Proc. Roy. Soc.* **234A**, 306.
311. Allen, H. C. Jr., Tidwell, E. D. and Plyler, E. K. (1956). *J. Amer. Chem. Soc.* **78**, 3034.
312. (a) Grisenthwaite, R. J. and Thompson, H. W. (1954). *Trans. Faraday Soc.* **50**, 212.

(b) Christensen, M. T. and Thompson, H. W. (1956). *Trans. Faraday. Soc.* **52**, 1439.
(c) Thomas, R. K. and Thompson, H. W. (1968). *Spectrochim. Acta* **24A**, 1353.
313. Jones, A. V. (1952). *J. Chem. Phys.* **20**, 861.
314. Berney, C. V., Cousins, L. R. and Miller, F. A. (1963). *Spectrochim. Acta* **19**, 2019.
315. Galasso, V. and Bigotto, A. (1965). *Spectrochim. Acta* **21**, 2085.
316. Davidson, D. W. and Bernstein, H. J. (1955). *Canad. J. Chem.* **33**, 1226.
317 Romanet, R. and Wojtkowiak, B. (1960). *Compt. rend.* **251**, 364.
318. Wojtkowiak, B. and Romanet, R. (1960). *Compt. rend.* **251**, 62.
319. Pham-Van-Huong and Lascombe, J. (1962). *Compt. rend.* **254**, 2543.
320. (a) Miller, F. A. (1956). *J. Chem. Phys.* **24**, 996.
(b) Ref. 53, p. 365.
321. Brodersen, S. and Langseth, A. (1956). *Mat. Fys. Skr. Dan. Selsk.* **1**, 3.
322. Danti, A. and Lord, R. C. (1958). *Spectrochim. Acta* **13**, 180.
323. Neto, N., Sorocco, M. and Califano, S. (1966). *Spectrochim. Acta* **22**, 1981.
324. Albrecht, A. C. (1960). *J. Mol. Spectroscopy* **5**, 236.
325. Duinker, J. C. and Mills, I. M. (1968). *Spectrochim. Acta* **24A**, 417.
326. Kakiuti, Y. (1959). *J. Chem. Soc. Japan* **80**, 21.
327. Ferguson, E. E. (1957). *J. Chem. Phys.* **26**, 1357.
328. Fritz, H. P. and Manchot, J. (1962). *Z. Naturforsch.* **17b**, 711.
329. Brodersen, S. and Langseth, A. (1959). *Mat. Fys. Skr. Danske Vid. Selsk.* **1**, 7.
330. Brodersen, S. and Langseth, A. (1958). *Mat. Fys. Skr. Danske Vid. Selsk.* **1**, 1.
331. Garrigou-Lagrange, C., Le Calve, N. and Vignalou, C. (1966). *J. Chem. Phys.* **63**, 1454.
332. Documentation of Molecular Spectroscopy Catalogue, Spectral Cards Nos. 5574, 5575, 11263, 11267.
333. Lucazean, G. and Lebas, J. M. (1967). *J. Chim. Phys.* **64**, 1473.
334. Marcou, A. and Pinton, M. M. (1968). *Compt. rend.* **266B**, 917.
335. Kakiuti, Y. and Ogura, J. (1961). *Bull. Chem. Soc. Japan* **34**, 899.
336. Saeki, S. (1961). *Bull. Chem. Soc. Japan* **34**, 1552.
337. Saeki, S. (1962). *Bull. Chem. Soc. Japan* **35**, 442.
338. Saeki, S. (1960). *Bull. Chem. Soc. Japan* **33**, 1024.
339. Steele, D., Kynaston, W. and Gebbie, H. A. (1963). *Spectrochim. Acta* **19**, 785.
340. Gates, P. N., Radcliffe, K. and Steele, D. (1969). *Spectrochim. Acta* **25A**, 507.
341. Saeki, S. (1962). *Bull. Chem. Soc. Japan* **35**, 326.
342. Saeki, S. (1959). *Compt. rend.* **249**, 2537.
343. Steele, D., Lippincott, E. R. and Xavier, J. (1960). *J. Chem. Phys.* **33**, 1242.
344. Nanney, T. R., Bailey, R. T. and Lippincott, E. R. (1965). *Spectrochim. Acta* **21**, 1495.
345. Nanney, T. R., Lippincott, E. R. and Hamer, J. C. (1966). *Spectrochim. Acta* **22**, 737.
346. Fuson, N., Garrigou-Lagrange, C. and Josien, M. L. (1960). *Spectrochim. Acta* **16**, 106.
347. Turkevich, J., McKenzie, H. A., Friedman, L. and Spurr, R. (1949). *J. Amer. Chem. Soc.* **71**, 4045.
348. Ref. 332, Spectral Cards Nos. 11264, 11271.
349. Evans, J. C. and Nyquist, R. A. (1960). *Spectrochim. Acta* **16**, 918.
350. Gruger, A. and Lebas, J. M. (1965). *J. Chim. Phys.* **62**, 1318.
351. Weston, R. E. Jr., Tsukamoto, A. and Litchin, M. N. (1966). *Spectrochim. Acta* **22**, 433.

352. Sandroni, S. and Geiss, F. (1966). *Spectrochim. Acta* **22**, 235.
353. Katon, J. E. and McDevitt, N. T. (1964). *J. Mol. Spectroscopy* **14**, 308.
354. Peregudov, G. V. (1960). *Optics and Spectroscopy* **9**, 295.
355. Zerbi, G. (1968). *Spectrochim. Acta* **24A**, 511.
356. Person, W. B., Pimentel, G. C. and Schnepp, O. (1955). *J. Chem. Phys.* **23**, 230.
357. Lippincott, E. R. and O'Reilly, Jr, E. J. (1955). *J. Chem. Phys.* **23**, 238.
358. Chantry, G. W., Anderson, A. and Gebbie, H. A. (1964). *Spectrochim. Acta* **20**, 1465.
359. Mitra, S. S. and Bernstein, H. J. (1959). *Canad. J. Chem.* **37**, 553.
360. Krainov, E. P. (1964). *Optika i Spektroskopiya* **16**, 763.
361. Peregudov, G. V. (1961). *Optics and Spectroscopy* **11**, 397.
362. Califano, S. (1962). *J. Chem. Phys.* **36**, 903.
363. Colombo, L. (1964). *Spectrochim. Acta* **20**, 547.
364. Krainov, E. P. (1964). *Optika i Spektroskopiya* **16**, 984.
365. Bouas-Laurent, H., Calas, R. and Josien, M. L. (1961). *Compt. rend.* **252**, 285.
366. Chafik, A. and Mecke, R. (1968). *Z. Naturforsch.* **23a**, 716.
367. Schettino, V. (1967). *J. Chem. Phys.* **46**, 302.
368. Califano, S. and Abbondanza, G. (1963). *J. Chem. Phys.* **39**, 1016.
369. Falk, M. and Whalley, E. (1961). *J. Chem. Phys.* **34**, 1554.
370. Cardinaud, R. (1960). *Bull. Soc. chim. Fr.* **1960**, 629.
371. Cardinaud, R. (1959). *Compt. rend.* **249**, 1641.
372. Maclou, M. and Henry, L. (1957). *Compt. rend.* **244**, 1494.
373. Halford, J. O., Anderson, L. C. and Kissin, G. H. (1937). *J. Chem. Phys.* **5**, 927.
374. Takenaka, T. (1962). *J. Chem. Soc. Japan* **83**, 267.
375. Perchard, J. P. and Jpsien, M. L. (1968). *J. Chim. Phys.* **65**, 1834; 1856.
376. Margottin-Maclou, M. and Corval, M. (1963). *J. Chim. Phys.* **60**, 1044.
377. Hadži, D. and Jeramic, D. (1957). *Spectrochim. Acta* **9**, 263.
378. Tarte, P. and Deponthière, R. (1957). *Bull. Soc. chim. Belg.* **66**, 525.
379. Quinan, J. R. and Wiberley, S. E. (1954). *Analt. Chem*, **26**, 1762.
380. Fukushima, K. and Zurolinski, B. J. (1968). *J. Mol. Spectroscopy* **26**, 368.
381. Tanaka, C. (1962). *J. Chem. Soc. Japan* **83**, 521.
382. Hirsjaervi, P. (1956). *Acta. Chem. Fenn.* **29**, 200.
383. Tanaka, C. (1962). *J. Chem. Soc. Japan* **83**, 792.
384. Tanaka, C. (1962). *J. Chem. Soc. Japan* **83**, 661.
385. Seubold, F. H. Jr. (1956). *J. Org. Chem.* **21**, 156.
386. Streitwieser, A. Jr. (1953). *J. Amer. Chem. Soc.* **75**, 5015.
387. Zeil, W., Buchert, H., Heel, H. and Pfoertner, H. (1960). *Z. Elektrochem.* **64**, 769.
388. Tanaka, C. (1960). *J. Chem. Soc. Japan* **81**, 1042.
389. Tanaka, C. (1962). *J. Chem. Soc. Japan* **83**, 398.
390. Pritchard, J. G. and Nelson, H. M. (1960). *J. Phys. Chem.* **64**, 795.
391. Cannell, L. G. and Taft, R. W. Jr. (1956). *J. Amer. Chem. Soc.* **78**, 5812.
392. Stuart, A. V. and Sutherland, G. B. B. M. (1956). *J. Chem. Phys.* **24**, 566.
393. Karabatsos, G. J. (1960). *J. Org. Chem.* **25**, 1409.
394. Geiseler, G., Fruwert, J. and Huettig, R. (1966). *Chem. Ber.* **99**, 1594.
395. Cook, J. S. and Reece, I. H. (1961). *Austral. J. Chem.* **14**, 211.
396. (a) Sawodny, W., Niedenzu, K. and Dawson, J. W. (1967). *Spectrochim. Acta* **23A**, 799.
(b) Matsuura, H. and Miyazawa, T. (1967). *Spectrochim. Acta* **23A**, 2433.
(c) Durig, J. R. and Green, W. H. (1969). *Spectrochim. Acta* **25A**, 849.
397. Prelog, V. and Kueng, W. (1956). *Helv. Chim. Acta* **39**, 1394.
398. Curtin, D. Y. and Kellom, D. B. (1953). *J. Amer. Chem. Soc.* **75**, 6011.

399. Evans, J. C. (1960). *Spectrochim. Acta* **16**, 1382.
400. Kovner, M. A., Davydova, N. I. and Zhigunova, I. A. (1965). *Optika i Spektroskopiya* **18**, 152.
401. Leicknam, J. P., Lascombe, J., Fuson, N. and Josien, M. L. (1959). *Bull. Soc. chim. France*, 1516.
402. Pham Van Huong, Couzi, M. and Lascombe, J. (1968). *Compt. rend.* **B266**, 172.
403. Singh, S. and Rao, C. N. R. (1966). *Can. J. Chem.* **44**, 2611.
404. Ginn, S. G. W. and Wood, J. L. (1965). 628; *Spectrochim. Acta* **23A**, 611, (1967).
405. Lin, T. and Fishman, E. (1967). *Spectrochim. Acta* **23A**, 491.
406. Mecke, R. and Rossmy, G. (1955). *Z. Elektrochem.* **59**, 866.
407. Jakobsen, R. J. (1965). *Spectrochim. Acta* **21**, 433.
408. Bist, H. D., Brand, J. C. D. and Williams, D. R. (1967). *J. Mol. Spectroscopy* **24**, 402.
409. Davydova, N. I., Zhigunova, I. A., Ignateva, L. A. and Kovner, M. A. (1965). *Optika i Spektroskopiya* **18**, 1077.
410. Jakobsen, R. J. and Brasch, J. W. (1965). *Spectrochim. Acta* **21**, 1753.
411. Steger, E., Stahlberg, U. and Diêu, N. Q. (1968). *Spectrochim. Acta* **24A**, 1023.
412. Koller, E. J. and Zollinger, H. (1956). *Helv. Chim. Acta* **39**, 1610.
413. Anderson, D. M. W., Bellamy, L. J. and Williams, R. L. (1958). *Spectrochim. Acta* **12**, 233.
414. Lord, R. C. and Nolin, B. (1956). *J. Chem. Phys.* **24**, 656.
415. Wray, J. H. (1968). *J. Phys.* **1B**, 485.
416. Kanazawa, Y. and Nukada, K. (1962). *Bull. Chem. Soc. Japan* **35**, 612.
417. Perchard, J. P., Forel, M. T. and Josien, M. L. (1964). *J. Chim. phys.* **61**, 632.
418. Lacarbe, P., Forel, M. T. and Bessis, G. (1968). *Spectrochim. Acta* **24A**, 2165.
419. Le Calvé, J., Grange, P. and Lascombe, J. (1965). *Compt. rend.* **260**, 6065.
420. Wieser, H., Laidlaw, W. G., Krueger, P. J. and Fuhrer, H. (1968). *Spectrochim. Acta* **24A**, 1055.
421. Le Calvé, J., Derouault, J., Forel, M. T. and Lascombe, J. (1967). *Compt. rend.* **B264**, 448.
422. Palm, A. and Bissell, E. R. (1960). *Spectrochim. Acta* **16**, 459.
423. Dalton, F., Hill, R. D. and Meakins, G. D. (1960). *J. Chem. Soc.* 2927.
424. Bak, B., Brodersen, S. and Hansen, L. (1955). *Acta Chem. Scand.* **9**, 749.
425. Rico, M., Barrachina, M. and Orza, J. M. (1967). *J. Mol. Spectroscopy* **24**, 133.
426. Forel, M. T., Garrigou-Lagrange, C., Gemin, J. and Josien, M. L. (1963). *J. Chim. phys. et Phys. Chim. biol.* **60**, 1047.
427. Moravec, J., Kouřím, P. and Horák, M. (1955). *Coll. Czech. Chem. Comm.* **30**, 2298.
428. Nukada, K. (1962). *Spectrochim. Acta* **18**, 745.
429. Nukada, K. (1961). *Bull. Chem. Soc. Japan* **34**, 1624.
430. Kobayashi, M., Iwamoto, R. and Tadokoro, H. (1966). *J. Chem. Phys.* **44**, 922.
431. Ebers, E. S. and Nielsen, H. H. (1938). *J. Chem. Phys.* **6**, 311.
432. Davidson, D. W., Stoicheff, B. P. and Bernstein, H. J. (1954). *J. Chem. Phys.* **22**,
433. Beckmann, L., Gutjahr, L. and Mecke, R. (1965). *Spectrochim. Acta* **21**, 307.
434. Budzinski, J. (1965). *Bull. Acad. polon. Sci., Ser. math. astr. phys.* **13**, 183.
435. Evans, J. C. and Bernstein, H. J. (1956). *Canad. J. Chem.* **34**, 1083.
436. Worden, E. F. Jr. (1962). *Spectrochim. Acta* **18**, 1121.
437. Berney, C. V. (1969). *Spectrochim. Acta* **25A**, 793.
438. Eggers, D. F. Jr. and Lingren, W. E. (1956). *Anal. Chem.* **28**, 1328.
439. Brand, J. C. D. and Minkoff, G. J. (1954). *J. Chem. Soc.* 2970.
440. George, W. O. and Mansell, V. G. (1968). *Spectrochim. Acta* **24A**, 145.

441. Johansen, H. (1964). *Z. phys. Chemie (Leipzig)* **227**, 305.
442. Wiberg, K. B. (1954). *J. Amer. Chem. Soc.* **76**, 5371.
443. Pinchas, S. (1957). *Anal. Chem.* **29**, 334.
444. Dellepiane, G. and Overend, J. (1966). *Spectrochim. Acta* **22**, 593.
445. Forel, M. T. and Fouassier, M. (1967). *Spectrochim. Acta* **23A**, 1977.
446. Nolin, B. and Jones, R. N. (1953). *J. Amer. Chem. Soc.* **75**, 5627.
447. Karabatsos, G. J. (1960). *J. Org. Chem.* **25**, 315.
448. (a) Frei, K. and Guenthard, Hs. H. (1960). *J. Mol. Spectroscopy* **5**, 218.
(b) Durig, J. R. and Green, W. H. (1968). *J. Mol. Spectroscopy* **27**, 95.
448. (c) Durig, J. R. and Morrissey, A. C. (1967). *J. Chem. Phys.* **47**, 4455.
449. Durig, J. R. and Lord, R. C. (1966). *J. Chem. Phys.* **45**, 61.
450. Reisse, J. Peters, P. A., Ottinger, R., Bervelt, J. P. and Chiurdoglu, G. (1966). *Tetrahedron Letters*, 2511.
451. Bervelt, J. P., Ottinger, R., Peters, P. A., Reisse, J. and Chiurdoglu, G. (1968). *Spectrochim Acta* **24A**, 1411.
452. Eglinton, G., Martin, J. and Parker, W. (1965). *J. Chem. Soc.* 1243.
453. Dabrowski, J. and Terpiński, J. (1966). *J. Org. Chem.* **31**, 2159.
454. Dabrowski, J. and Kamieńska-Trela, K. (1966). *Bull. Chem. Soc. Japan* **39**, 2565.
455. Krebs, A. and Schrader, B. (1966). *Z. Naturforsch.* **21B**, 194.
456. Krebs, A. and Schrader, B. (1967). *Annalen* **709**, 46.
457. Jenny, E. F. and Roberts, J. D. (1956). *J. Amer. Chem. Soc.* **78**, 2005.
458. Noack, K. and Jones, R. N. (1960). *Z. Elektrochem.* **64**, 707.
459. Bhatt, M. V., Srinivasan, G. and Neelakantan, P. (1965). *Tetrahedron* **21**, 291.
460. Moore, C. B. and Pimentel, G. C. (1963). *J. Chem. Phys.* **38**, 2816.
461. Durig, J. R. and Willis, J. N. Jr. (1966). *Spectrochim. Acta* **22**, 1299.
462. Charney, E. and Becker, E. D. (1965). *J. Chem. Phys.* **42**, 910.
463. Ziffer, H., Charney, E. and Becker, E. D. (1965). *J. Chem. Phys.* **42**, 914.
464. Becker, E. D., Charney, E. and Anno, T. (1965). *J. Chem. Phys.* **42**, 942.
465. Pecile, C. and Lunelli, B. (1967). *J. Chem. Phys.* **46**, 2109.
466. Gazis, E. and Heim, P. (1967). *Tetrahedron Letters* **13**, 1185.
467. Cook, D. (1966). *Spectrochim. Acta* **22**, 415.
468. Ikegami, Y. (1961). *Bull. Chem. Soc. Japan* **34**, 94.
469. Barker, S. A., Moore, R. H., Stacey, M. and Whiffen, D. H. (1960). *Nature* **186**, 307.
470. Stratton, R. F. and Nielsen, A. H. (1960). *J. Mol. Spectroscopy* **4**, 373.
471. Venkateswarlu, K., Jagatheesan, S. and Rajalakshmi, K. V. (1963). *Proc. Indian Acad. Sci. A.* **58**, 373.
472. Ramsey, J. A. and Ladd, J. A. (1968). *J. Chem. Soc. (B)*. 118.
473. Hall, L. C. and Overend, J. (1967). *Spectrochim. Acta* **23A**, 2535.
474. Wilmshurst, J. K. (1965). *J. Chem. Phys.* **25**, 478.
475. Millikan, R. C. and Pitzer, K. S. (1958). *J. Amer. Chem. Soc.* **80**, 3515.
476. Mikawa, Y., Brasch, J. W. and Jakobsen, R. J. (1967). *J. Mol. Spectroscopy* **24**, 314.
477. Carlson, G. L., Witkowski, R. E. and Fateley, W. G. (1966). *Spectrochim. Acta* **22**, 1117.
478. Bellamy, L. J., Osborn, A. K. and Pace, R. J. (1963). *J. Chem. Soc.* 3749.
479. Haas, C. M. (1966). *Diss. Abs.* **26**, 5741.
480. (a) Haurie, M. and Novak, A. (1965). *J. Chim. phys.* **62**, 137.
(b) Statz, G. and Lippert, E. (1967). *Ber. Bunsenges. phys. Chem.* **71**, 673.
481. Haurie, M. and Novak, A. (1965). *J. Chim. phys.* **62**, 146.
482. Stanevich, A. E. (1964). *Optika i Spektroskopiya* **16**, 446.

483. Haurie, M. and Novak, A. (1966). *J. Chim. phys.* **63**, 1584.
484. Bratož, S., Hadži, D. and Sheppard, N. (1956). *Spectrochim. Acta* **8**, 249.
485. Nakamura, K. (1959). *J. Chem. Soc. Japan* **80**, 225.
486. Hadži, D. and Sheppard, N. (1953). *Proc. Roy. Soc.* **216A**, 247.
487. Tsuboi, M., Takenishi, T. and Nakamura, A. (1963). *Spectrochim. Acta* **19**, 284.
488. Eliel, E. L., Price, C. C., Convery, R. J. and Prosser, T. J. (1958). *Spectrochim. Acta* **10**, 423.
489. Takenishi, T. (1961). *J. Chem. Soc. Japan* **82**, 319.
490. Susi, H. (1959). *Spectrochim. Acta* **15**, 1063.
491. Pava, B. M. and Stafford, F. E. (1968). *J. Phys. Chem.* **72**, 4628.
492. (a) Cardwell, H. M. E., Dunitz, J. D. and Orgel, L. E. (1953). *J. Chem. Soc.* 3741.
(b) Craig, D. and Shipman, J. J. (1952). *J. Amer. Chem. Soc.* **74**, 2905.
493. Katon, J. E. and McDevitt, N. T. (1965). *Spectrochim. Acta* **21**, 1717.
494. Spinner, E. (1967). *J. Chem. Soc.* B. 879.
495. Nakamura, K. (1958). *J. Chem. Soc. Japan* **79**, 1411.
496. Jones, L. H. and McLaren, E. (1954). *J. Chem. Phys.* **22**, 1796.
497. Nakamura, K. (1958). *J. Chem. Soc. Japan* **79**, 1420.
498. Scott, F. A., Goldenson, J., Wiberley, S. E. and Bauer, W. H. (1954). *J. Phys. Chem.* **58**, 61.
499. Nakamoto, K., Sarna, Y. A. and Behnke, G. T. (1965). *J. Chem. Phys.* **42**, 1662.
500. Bernard, M. P. (1959). *Compt. rend.* **248**, 1153.
501. (a) Susi, H. and Zell, T. (1963). *Spectrochim. Acta* **19**, 1933.
(b) Susi, H. and Scherer, J. R. (1969). *Spectrochim. Acta* **25A**, 1243.
502. Corval, M. and Lecomte, J. (1955). *Mikrochim, Acta*, 25.
503. Nolin, B. and Jones, R. N. (1956). *Canad. J. Chem.* **34**, 1382; 1392.
504. Jones, R. N. (1962). *Canad. J. Chem.* **40**, 301.
505. Dalton, F., Ellington, P. S. and Meakins, G. D. (1960). *J. Chem. Soc.* 3681.
506. Roberts, J. D., Regan, C. M. and Allen, I. (1952). *J. Amer. Chem. Soc.* **74**, 3679.
507. Adembri, G., Sbrana, G. and Califano, S. (1963). *Ricerca sci.* **A3**, 431.
508. Di Lauro, C., Califano, S. and Adembri, G. (1968). *J. Molec. Struct.* **2**, 173.
509. Gray, A. P. and Lord, R. C. (1957). *J. Chem. Phys.* **26**, 690.
510. Wolff, H. and Wolff, E. (1965). *Ber. Bunsen Gesellschaft Phys. Chem.* **69**, 467.
511. Tamagake, K., Tsuboi, M. and Hirakawa, A. Y. (1968). *J. Chem. Phys.* **48**, 5536.
512. Tsuboi, M., Hirakawa, A. Y. and Tamagake, K. (1967). *J. Mol. Spectroscopy* **22**, 272.
513. Perchard, J. P., Forel, M. T. and Josien, M. L. (1964). *J. Chim. Phys.* **61**, 652.
514. Buttler, M. J. and McKean, D. C. (1965). *Spectrochim. Acta* **21**, 465.
515. Dellepiane, G. and Zerbi, G. (1968). *J. Chem. Phys.* **48**, 3573.
516. Gayles, J. N. (1967). *Spectrochim. Acta* **23A**, 1521.
517. Sabatini, A. and Califano, S. (1960). *Spectrochim. Acta* **16**, 677.
518. Hughes, M. N. and McWhinnie, W. R. (1967). *J. Chem. Soc. A*. 592.
519. Ostrowska, B. and Tramer, A. (1967). *Acta Phys. Pol.* **32**, 111.
520. (a) Califano, S. and Moccia, R. (1957). *Gazzeta* **87**, 805.
(b) Moritz, A. G. (1962). *Spectrochim. Acta* **18**, 671.
521. Lady, J. H. and Whetsel, K. B. (1965). *Spectrochim. Acta* **21**, 1669.
522. Hadži, D. and Škrbljak, M. (1957). *J. Chem. Soc.* 843.
523. Kaplan, E. D. and Thornton, E. R. (1967). *J. Amer. Chem. Soc.* **89**, 6644.
524. Krueger, P. J. (1967). *Can. J. Chem.* **45**, 2135.
525. Theoret, A. and Sandorfy, C. (1967). *Spectrochim. Acta* **23A**, 519.
526. Evans, J. C. and Lo, G. Y. S. (1966). *J. Phys. Chem.* **70**, 11.
527. Evans, J. C. and Lo, G. Y. S. (1966). *J. Phys. Chem.* **70**, 20.

528. Evans, J. C. and Lo, G. Y. S. (1967). *J. Phys. Chem.* **71**, 3942.
529. Salthouse, J. A. and Waddington, T. C. (1966). *J. Chem. Soc.* (*A*). 28.
530. Gaver, R. W. and Murmann, R. K. (1964). *J. Inorg. Nuclear Chem.* **26**, 881.
531. Anisimova, O. S., Kost, A. N., Pentin, Yu. A. and Yudin, L. G. (1966). *Vest. mosk. gos. Univ. Ser. II khim.* **21**, 19.
532. Titova, T. F., Anisimova, O. S. and Pentin, Yu. A. (1967). *Optika i Spektrosk.* **23**, 905.
533. Levi, G. (1965). *Compt. rend.* **261**, 4007.
534. Heacock, R. A. and Marion, L. (1956). *Canad. J. Chem.* **34**, 1782.
535. Cope, A. C., Cotter, R. J. and Roller, G. G. (1955). *J. Amer. Chem. Soc.* **77**, 3590.
536. Margerum, J. D. and Sousa, J. A. (1965). *Appl. Spectroscopy* **19**, 91.
537. Checkland, P. B. and Thompson, H. W. (1955). *Trans. Faraday Soc.* **51**, 1.
538. Allen, H. C. Jr., Tidwell, E. D. and Plyler, E. K. (1956). *J. Chem. Phys.* **25**, 302.
539. Caldow, G. L. and Thompson, H. W. (1960). *Proc. Roy. Soc.* **A254**, 1.
540. Maki, A. G., Plyler, E. K. and Thibault, R. (1964). *J. Opt. Soc. Amer.* **54**, 869.
541. Mann, R. H. and Horrocks, W. D. Jr. (1966). *J. Chem. Phys.* **45**, 1278.
542. Staats, P. A., Morgan, H. W. and Goldstein, J. H. (1956). *J. Chem. Phys.* **25**, 582.
543. Morino, Y. and Nakagawa, T. (1966). *J. Chem. Phys.* **44**, 841.
544. (a) Fletcher, W. H. and Shoup, C. S. (1963). *J. Mol. Spectroscopy* **10**, 300.
(b) Pace, E. L. and Noe, L. J. *J. Chem. Phys.* **49**, 5317; 5325.
545. Evans, J. C. and Lo, G. Y. S. (1965). *Spectrochim. Acta* **21**, 1033.
546. Janz, G. J., Tait, M. J. and Meier, J. (1967). *J. Phys. Chem.* **71**, 963.
547. Fujiyama, T. and Shimanouchi, T. (1964). *Spectrochim. Acta* **20**, 829.
548. Jakobsen, R. J. (1965). *Spectrochim. Acta* **21**, 127.
549. Bak, B. and Nielsen, J. T. (1960). *Z. Elektrochem.* **64**, 560.
550. Kuthan, J., Jehlička, V. and Hakr, E. (1967). *Colln. Czech. Chem. Comm.* **32**, 4309.
551. Mottern, J. G. and Fletcher, W. H. (1962). *Spectrochim. Acta* **18**, 995.
552. Morcillo, J. and Orza, J. M. (1960). *Anales real Soc. españ. Fis. Quim.* **56B**, 253.
553. Lord, R. C. and Miller, F. A. (1942). *J. Chem. Phys.* **10**, 328.
554. Corsin, L., Fax, B. J. and Lord, R. C. (1953). *J. Chem. Phys.* **21**, 1170.
555. Stănoiu, I. L. and Balaban, A. T. (1968). *Revue roum. Chim.* **13**, 127.
556. Pratesi, F. and Freymann, R. (1967). *Compt. rend.* **B265**, 804.
557. Long, D. A., Murfin, F. S. and Thomas, E. L. (1963). *Trans. Faraday Soc.* **59**, 12.
558. Andersen, F. A., Bak, B., Brodersen, S. and Rastrup-Andersen, J. (1955). *J. Chem. Phys.* **23**, 1047.
559. Wilmshurst, J. K. and Bernstein, H. J. (1957). *Canad. J. Chem.* **35**, 1183.
560. Kovner, M. A., Korostelev, Yu. S. and Berezin, V. I. (1961). *Optics and Spectroscopy* **10**, 233.
561. Berezin, V. I. (1963). *Optika i Spektroskopia* **15**, 310.
562. Long, D. A. and Thomas, E. L. (1963). *Trans. Faraday Soc.* **59**, 783.
563. Abramovitch, R. A., Kroeger, D. J. and Staskun, B. (1962). *Canad. J. Chem.* **40**, 2030.
564. Long, D. A. and George, W. O. (1963). *Spectrochim. Acta* **19**, 1777.
565. Spinner, E. (1967). *Aust. J. Chem.* **20**, 1805.
566. Fritzsche, H. (1966). *Spectrochim. Acta* **22**, 1139.
567. Hinman, R. L. and Bauman, C. P. (1964). *J. Org. Chem.* **29**, 2437.
568. Moore, C. B. and Pimentel, G. C. (1964). *J. Chem. Phys*, **40**, 329, 342.
569. Nemes, L., Duncan, J. L. and Mills, I. M. *Spectrochim. Acta* (in press).
570. Fletcher, W. H. and Brown, F. B. (1963). *J. Chem. Phys.* **39**, 2478.

571. Durig, J. R., Walker, M. and Baglin, F. G. (1968). *J. Chem. Phys.* **48**, 4675.
572. Long, D. A. and Jones, D. T. L. (1963). *Trans. Faraday Soc.* **59**, 1033.
573. Zecchina, A., Cerruti, L., Coluccia, S. and Borello, E. (1967). *J. Chem. Soc.* (*B*), 1363.
574. Perchard, C. and Josien, M. L. (1965). *J. Chim. phys.* **62**, 423.
575. Bellocq, A. M., Perchard, C., Novak, A. and Josien, M. L. (1965). *J. Chim. phys.* **62**, 1334, 1344.
576. Cordes, M. and Walter, J. L. (1968). *Spectrochim. Acta* **24A**, 237.
577. Mitchell, R. W. and Merritt, J. A. (1968). *J. Mol. Spectroscopy* **27**, 197.
578. Perchard, C. and Novak, A. (1967). *Spectrochim. Acta* **23A**, 1953.
579. Stidham, H. D. and Tucci, J. V. (1967). *Spectrochim. Acta* **23A**, 2233.
580. Sbrana, G., Adembri, G. and Califano, S. (1966). *Spectrochim. Acta* **22**, 1831.
581. Califano, S., Adembri, G. and Sbrana, G. (1964). *Spectrochim. Acta* **20**, 385.
582. Simmons, J. D., Innes, K. K. and Begun, G. M. (1964). *J. Mol. Spectroscopy* **14**, 190.
583. Stidham, H. D. and Chandler, J. A. (1965). *J. Inorg. Nuclear Chem.* **27**, 397.
584. Berezin, V. I. (1964). *Optika i Spektroskopiya* **16**, 240.
585. Neto, N., Ambrosino, F. and Califano, S. (1964). *Spectrochim. Acta* **20**, 1503.
586. Mecke, R. and Kutzelnigg, W. (1960). *Spectrochim. Acta* **16**, 1216.
587. Yamaguchi, A. (1957). *J. Chem. Soc. Japan* **78**, 672; 694.
588. Miller, F. A. and Bassi, D. (1963). *Spectrochim. Acta* **19**, 565.
589. Schmidt, A. (1967). *Chem. Ber.* **100**, 3319.
590. Lancaster, J. E., Stamm, R. F. and Colthup, N. B. (1961). *Spectrochim. Acta* **17**, 155.
591. Shurvell, H. F., Faniran, J. A. and Dodsworth, H. (1967). *Spectrochim. Acta* **A23**, 1313.
592. Franks, L. A., Merer, A. J. and Innes, K. K. (1968). *J. Mol. Spectroscopy* **26**, 458.
593. Brownlee, R. T. C., Katritzky, A. R. and Topsom, R. D. (1966). *J. Chem. Soc.* (*B*), 726.
594. Novak, A. and Lautie, A. (1967). *Nature* **216**, 1202.
595. Lautie, A. and Novak, A. (1968). *J. Chim. phys.* **65**, 1359.
596. Kyogoku, Y., Higuchi, S. and Tsuboi, M. (1967). *Spectrochim. Acta* **23A**, 969.
597. Kyogoku, Y., Lord, R. C. and Rich, A. (1967). *J. Amer. Chem. Soc.* **89**, 496.
598. Onishi, T., Uyematsu, T., Watanabe, H. and Tamaru, K. (1967). *Spectrochim. Acta* **23A**, 730.
599. Sidorov, A. N. and Kotlyar, I. P. (1961). *Optics and Spectroscopy* **11**, 92.
600. Sutton, L. E. and Kenney, M. E. (1967). *Inorg. Chem.* **6**, 1869.
601. Esposito, J. N., Sutton, L. E. and Kenney, M. E. (1967). *Inorg. Chem.* **6**, 1116.
602. Suzuki, I. (1960). *Bull. Chem. Soc. Japan* **33**, 1359.
603. Suzuki, I. (1962). *Bull. Chem. Soc. Japan* **35**, 540.
604. Miyazawa, T., Shimanouchi, T. and Mizushima, S. I. (1956). *J. Chem. Phys.* **24**, 409.
605. Suzuki, I. (1962). *Bull. Chem. Soc. Japan* **35**, 1279.
606. Schneider, B., Hoření, A., Pivcovă, H. and Honzl, J. (1965). *Coll. Czech. Chem. Comm.* **30**, 2196.
607. Pivcová, H., Schneider, B. and Štokr, J. (1965). *Coll. Czech. Chem. Comm.* **30**, 2215.
608. Jakeš, J. and Schneider, B. (1968). *Spectrochim. Acta* **24A**, 286.
609. Lenormant, H. (1953). *Bull. Soc. Chim. France* (5) **20**, 216.
610. Dorris, K. L., Siddall, T. H., Stewart, W. E. and Good, M. L. (1967), *Spectrochim. Acta* **23A**, 1657.

611. Sirdeshmukh, L. (1967). *Current Sci.* **36**, 391.
612. Ellul, B. M. J. and Moodie, R. B. (1967). *J. Chem. Soc.* (*B*) 253.
613. Chalapathi, V. V. and Ramiah, K. V. (1968). *Proc. Indian Acad. Sci.* **67A**, 184.
614. Mirek, J. and Pasko, J. (1967). *Roczniki Chem.* **41**, 2083.
615. Allenstein, E. and Beyl, V. (1967). *Chem. Ber.* **100**, 3551.
616. Rey-Lafon, M., Forel, M. T. and Lascombe, J. (1967). *J. Chim. phys.* **64**, 1435.
617. Coburn, R. A. and Dudek, G. O. (1968). *J. Phys. Chem.* **72**, 1177; 3681.
618. Levin, E. S. and Rodinova, G. N. (1967). *Dokl. Akad. Nauk SSSR* **172**, 607.
619. Sohár, P. and Farkas, L. (1967). *Acta Chim. Hung.* **54**, 79; 91.
620. Hadži, D. and Premru, L. (1967). *Spectrochim. Acta* **23A**, 35.
621. Ramaswamy, K. K., Jose, C. I. and Sen, D. N. (1967). *Indian J. Chem.* **5**, 156.
622. Kimura, M., Kuroda, Y., Yamamoto, O. and Kubo, M. (1961). *Bull. Chem. Soc. Japan* **34**, 1081.
623. Kuroda, Y. and Kimura, M. (1963). *Bull. Chem. Soc. Japan* **36**, 464.
624. Hadži, D. (1956). *J. Chem. Soc.* 2725.
625. Voter, R. C., Banks, C. V. Fassel, V. A. and Kehres, P. W. (1951). *Anal. Chem.* **23**, 1731.
626. Caton, J. E. Jr. and Banks, C. V. (1967). *Inorg. Chem.* **6**, 1670.
627. Ashby, R. A. and Werner, R. L. (1966). *Spectrochim. Acta* **22**, 1345.
628. Adembri, G., Speroni, G. and Califano, S. (1963). *Spectrochim. Acta* **19**, 1145.
629. Frasco, D. L. (1964). *J. Chem. Phys.* **41**, 2134.
630. Suzuki, S., Shimanouchi, T. and Tsuboi, M. (1963). *Spectrochim. Acta* **19**, 1195.
631. Khanna, R. K., Horak, M. and Lippincot, E. R. (1966). *Spectrochim. Acta* **22**, 1759.
632. Tsuboi, M., Onishi, T., Nakagawa, I., Shimanouchi, T. and Mizushima, S. I. (1958). *Spectrochim. Acta* **12**, 253.
633. Callahan, M. V. and May, L. (1968). *J. Mol. Struct.* **2**, 154.
634. Dupuy, B. and Garrigou-Lagrange, C. (1967). *J. Chim. phys.* **64**, 1509.
635. Laulicht, I., Pinchas, S., Samuel, D. and Wasserman, I. (1966). *J. Phys. Chem.* **70**, 2719.
636. Nakamura, K. (1959). *J. Chem. Soc. Japan* **80**, 113.
637. Ogiso, T. (1967). *J. Pharm. Soc. Japan* **87**, 617.
638. Shindo, H. and Brown, Th. H. (1965). *J. Amer. Chem. Soc.* **87**, 1904.
639. Dupuy, B., Castinel, C. and Garrigou-Lagrange, C. (1969). *Spectrochim. Acta* **25A**, 571.
640. Ablov, A. V., Čapurina, L. F. and Proskina, N. N. (1967) *Zhur. neorg. Khim.* **12**, 499.
641. Stahlberg, U. and Steger, E. (1967). *Spectrochim. Acta* **23A**, 475.
642. Cotrait, M. and Novak, A. (1967). *Compt. rend.* **264B**, 1659.
643. Callahan, M. V. and May, L. (1968). *J. Mol. Structure* **2**, 154.
644. Oshima, T. and Tamiya, N. (1961). *Spectrochim. Acta* **17**, 384.
645. Tsuboi, M., Takenishi, T. and Nakamura, A. (1963). *Spectrochim. Acta* **19**, 271.
646. Sultanov, A. and Ilbiagimov, A. P. (1965). *Chem. Abs.* **62**, 7250.
647. Tamiya, N. and Shimanouchi, T. (1962). *Spectrochim. Acta* **18**, 895.
648. Gopal, L., Jose, C. I. and Biswas, A. B. (1967). *Spectrochim. Acta* **23A**, 513.
649. Wilson, T. P. (1943). *J. Chem. Phys.* **11**, 361.
650. Gluziński, P. and Eckstein, Z. (1968). *Spectrochim. Acta* **24A**, 1777.
651. Brookes, M. J. and Jonathan, N. (1968). *J. Chem. Soc.* (*A*) 1529.
652. Kamlet, M. J., Oesterling, R. E. and Adolph, H. G. (1965). *J. Chem. Soc.* 5838.
653. Popov, E. M. (1967). *Zh. strukt. Khim.* **8**, 1033.
654. Loos, K. R. and Guenthard, H. H. (1967). *J. Chem. Phys.* **46**, 1200.

655. Kross, R. D. and Fassel, V. A. (1956). *J. Amer. Chem. Soc.* **78**, 4225.
656. Shurvell, H. F., Faniran, J. A., Symons, E. A. and Buncel, E. (1967). *Can. J. Chem.* **35**, 117.
657. Cardinaud, R. (1960). *Bull. Soc. chim. Fr.* 634.
658. Avakjan, V. G., Gosteeva, G. N. and Čekrygin, V. A. (1968). *Izv. Akad. Nauk SSSR Ser. khim.* **2**, 432.
659. Cook, D. (1966). *Spectrochim. Acta* **22**, 419.
660. Prutkova, N. M., Martynenko, L. I., Grigorev, A. I. and Mitrofanova, M. D. (1967). *Zh. neorg. Khim.* **12**, 126.
661. Paranik, P. G. and Sirdeshmukh, L. (1968). *Proc. Indian Acad. Sci.* **67A**, 99.
662. Jose, C. I. (1969). *Spectrochim. Acta* **25A**, 111.
663. Shigorin, D. N., Pavlenishvili, I. Ya. and Kokachashvili, V. I. (1966). *Zh. Fiz. Khim.* **40**, 3006.
664. Fukushima, K., Ideguchi, Y. and Miyazawa, T. (1964). *Bull. Chem. Soc. Japan* **37**, 349.
665. Foucaud, A. and Person, H. (1964). *Compt. rend.* **258**, 523.
666. Kyogoku, Y., Higuchi, S. and Tsuboi, M. (1967). *Spectrochim. Acta* **23A**, 969.
667. Kyogoku, Y., Lord, R. C. and Rich, A. (1967). *J. Amer. Chem. Soc.* **89**, 496.
668. Brownlee, R. T. C., Katritzky, A. R. and Topsom, R. D. (1966). *J. Chem. Soc.* (*B*), 726.
669. Alemagna, A. and Lorenzelli, V. (1964). *J. Chim. Phys.* **61**, 884.
670. Sato, Y., Kotera, K., Takahashi, T. and Meshi, T. (1960). *J. Pharm. Soc. Japan* **80**, 976.
671. Bacaloglu, R., Catalina, E. and Ostrogovich, G. (1967). *Spectrochim. Acta* **23A**, 1571.
672. Newman, R. and Badger, R. M. (1952). *J. Amer. Chem. Soc.* **74**, 3545.
673. Bell, N. A. and Coates, G. E. (1965). *J. Chem. Soc.* 692.
674. Bell, N. A., Coates, G. E. and Emsley, J. W. (1966). *J. Chem. Soc.* (*A*) 49.
675. Coates, G. E., Roberts, P. D. and Downs, A. J. (1967). *J. Chem. Soc.* (*A*) 1085.
676. Lehmann, W. J., Wilson, C. O. Jr. and Shapiro, I. (1959). *J. Chem. Phys.* **31**, 1071.
677. Kapshtal, V. N. and Sverdlov, L. M. (1966). *Zhur. Fiz. Khim.* **40**, 2818.
678. Lehmann, W. J., Onak, T. P. and Shapiro, I. (1959). *J. Chem. Phys.* **30**, 1215.
679. Faniran, J. A. and Shurvell, H. F. (1968). *Canad. J. Chem.* **46**, 2089.
680. Lehmann, W. J., Weiss, H. G. and Shapiro, I. (1959). *J. Chem. Phys.* **30**, 1222.
681. Meller, A. and Schaschel, E. (1967). *Monatsh.* **98**, 390.
682. Meller, A. and Marecek, H. (1967). *Monatsh.* **98**, 2336.
683. Becher, H. J. (1963). *Spectrochim. Acta* **19**, 575.
684. Rice, B., Galliano, R. J. and Lehmann, W. J. (1957). *J. Phys. Chem.* **61**, 1222.
685. Mann, D. E. (1954). *J. Phys. Chem.* **22**, 71.
686. Kapshtal, V. N. and Sverdlov, L. M. (1968). *Zhur. Fiz. Khim.* **42**, 2154.
687. Meller, A. and Wechsberg, M. (1967). *Monatsh.* **98**, 513.
688. Porter, R. F. and Yeung, E. S. (1968). *Inorg. Chem.* **7**, 1306.
689. Totani, T., Watanabe, H. and Kubo, M. (1969). *Spectrochim. Acta* **25A**, 585.
690. Meller, A. and Schaschel, E. (1967). *Monatsh.* **98**, 1358.
691. Ball, D. F., Carter, T., McKean, D. C. and Woodward, L. A. (1964). *Spectrochim. Acta* **20**, 1721.
692. Wilde, R. E. (1968). *J. Mol. Spectroscopy* **25**, 291.
693. Wilde, R. E. (1962). *J. Mol. Spectroscopy* **8**, 427.
694. Ponomorev, Yu. I., Kovalev, I. F. and Orlov, V. A. (1967). *Optics and Spectroscopy* **23**, 258.

695. Ball, D. F., Goggin, P. L., McKean, D. C. and Woodward, L. A. (1960). *Spectrochim. Acta* **16**, 1358.
696. Lazarev, A. N., Poiker, K. and Shchukovskaya, L. L. (1967). *Izv. Akad. Nauk. SSSR. nerog. Mater.* **3**, 2021.
697. Kovalev, I. F. (1959). *Optics and Spectroscopy* **6**, 387.
698. Laane, J. and Lord, R. C. (1968). *J. Chem. Phys.* **48**, 1508.
699. Frankiss, S. G. (1966). *Spectrochim. Acta* **22**, 295.
700. Reeves, R. B., Wilde, R. E. and Robinson, D. W. (1964). *J. Chem. Phys.* **40**, 125.
701. Pillai, M. G. K., Ramaswamy, K. K. and Gnanadesikan, S. G. (1966). *Czech. J. Phys.* **16B**, 150.
702. Reeves, R. B. and Robinson, D. W. (1964). *J. Chem. Phys.* **41**, 1699.
703. Buchert, H. and Zeil, W. (1961). *Z. Phys. Chem.* (*Frankfurt*) **29**, 317.
704. Kaplan, L. (1954). *J. Amer. Chem. Soc.* **76**, 5880.
705. Ryskin, Ya. I., Voronkov, M. G. and Shabrova, Z. I. (1959). *Izvest. Akad. Nauk. SSSR. Otdel. Khim. Nauk.* 1019.
706. Weiss, G. S. and Nixon, E. R. (1965). *Spectrochim. Acta* **21**, 903.
707. Lazarev, A. N., Poiker, K. and Kucharskaya, E. V. (1967). *Izv. Akad. Nauk SSSR. neorg. Mater* **3**, 2029.
708. Robiette, A. G. and Thompson, J. C. (1965). *Spectrochim. Acta* **21**, 2023.
709. Buerger, H. (1965). *Inorg. Nucl. Chem. Lett.* **1**, 11.
710. Buerger, H. and Goetze, U. (1968). *Monatsh.* **99**, 155.
711. Lannon, J. A. and Nixon, E. R. (1967). *Spectrochim. Acta* **23A**, 2713.
612. Moritz, A. G. (1966). *Spectrochim. Acta* **22**, 1015.
713. Park, P. J. D. and Hendra, P. J. (1968). *Spectrochim. Acta* **24A**, 2081.
714. Mitchell, R. W., Kuma, L. J., Pirkle, R. J. and Merritt, J. A. (1969). *Spectrochim. Acta* **25A**, 819.
715. Schlosser, M. (1964). *Chem. Ber.* **97**, 3219.
716. Durig, J. R., Clark, J. W. and DiYorio, J. S. (1968). *J. Chem. Phys.* **48**, 4154.
717. Daasch, L. W. and Smith, D. C. (1951). *Analyt. Chem.* **23**, 853.
718. Chittenden, R. A. and Thomas, L. C. (1965). *Spectrochim. Acta* **21**, 861.
719. Zwierzak, A. (1967). *Can. J. Chem.* **45**, 2501.
720. Vorsina, I. A. and Levin, I. S. (1967). *Zh. neorg. Khim.* **12**, 3046.
721. Chapman, A. C. and Thirlwell, L. E. (1964). *Spectrochim. Acta* **20**, 937.
722. Ratajczak, H. and Mielke, Z. (1968). *J. Mol. Struct.* **1**, 397.
723. Nyquist, R. A., Blair, E. H. and Osborne, D. W. (1967). *Spectrochim. Acta* **23A**, 2505.
724. (a) May, I. W. (1968). *Diss. Abs.* **28B**, 3671.
(b) May, I. W. and Pace, E. L. (1968). *Spectrochim. Acta* **24A**, 1605.
725. Redington, R. L. (1962). *J. Mol. Spectroscopy* **9**, 469.
726. Perchard, J. P., Forel, M. T. and Josien, M. L. (1964). *J. Chim. Phys.* **61**, 645.
727. Moritz, A. G. (1967). *Spectrochim. Acta* **23A**, 167.
728. Rico, M., Orza, J. M. and Morcillo, J. (1965). *Spectrochim. Acta* **21**, 689.
729. Aleksanian, V. T., Kimelfeld, Ya. M., Magdesieva, N. N. and Yurev, K. Yu. (1967). *Optika i Spektroskopiya* **22**, 216.
730. Mueller, A., Krebs, B. and Gattow, G. (1967). *Z. anorg. Chem.* **346**, 74.
731. Rosmus, P., Mayer, R., Herzog, K. and Steger, E. (1967). *Tetrahedron Letters* 4495.
732. Horrocks, W. D. Jr. and Cotton, F. A. (1961). *Spectrochim. Acta* **17**, 134.
733. Williams, J. M. and Kreevoy, M. (1967). *J. Amer. Chem. Soc.* **89**, 5499.
734. Bagnall, K. W., Brown, D., Holah, D. H. and Lux, F. (1968). *J. Chem. Soc.* (*A*). 465.

735. Berney, C. V. and Weber, J. H. (1968). *Inorg. Chem.* **7**, 283.
736. Olejnik, S., Aylmore, L. A. G., Posner, A. M. and Quirk, J. P. (1968). *J. Phys. Chem.* **72**, 241.
737. Uno, T., Machida, K. and Hanai, K. (1968). *Spectrochim. Acta* **24A**, 1705.
738. Durig, J. R. and Wertz, D. W. (1967). *J. Chem. Phys.* **46**, 3069.
739. Krakow, B., Lord, R. C. and Neely, G. O. (1968). *J. Mol. Spectroscopy* **27**, 148.
740. Moritz, A. G. (1966). *Spectrochim. Acta* **22**, 1021.
741. Davidovics, G., Roepstorff, P., Chouteau, J. and Garrigou-Lagrange, C. (1967). *Spectrochim. Acta* **23A**, 2669.
742. Davidovics, G., Garrigou-Lagrange, C., Chouteau, J. and Metzger, J. (1967). *Spectrochim. Acta* **23A**, 1477.
743. Hadži, D. (1957). *J. Chem. Soc.* 847.
744. Takahashi, H., Schrader, B., Meier, W. and Gottlieb, K. (1967). *J. Chem. Phys.* **47**, 3842.
745. Nishikawa, M. and Takabatake, A. (1964). *Chem. Pharm. Bull.* **12**, 291.
746. Šoptrajanov, A. and Ewing, G. E. (1966). *Spectrochim. Acta* **22**, 1417.
747. Nagata, K. and Mizukami, S. (1966). *Chem. Pharm. Bull.* **14**, 1255; **15**, 70.
748. Haas, A. and Sheppard, N. (1965). *J. Chem. Soc.* **6613**.
749. Rademacher, P., Wiegraebe, W. and Luettke, W. (1967). *Chem. Ber.* **100**, 1213.
750. Ganguly, L., Jose, C. I. and Biswas, A. B. (1968). *Spectrochim. Acta* **24A**, 215.
751. Caullet, C., Virey, A., Gendre, J. C., Kirsche, C. and Virey, J. (1967). *Compt. rend.* **264*C***, 1800.
752. Ebsworth, E. A. V., Mould, R., Taylor, R., Wilkinson, G. R. and Woodward, L. R. (1962). *Trans. Faraday Soc.* **58**, 1069.
753. Sathianandan, K. and Margrave, J. L. (1963). *J. Mol. Spectroscopy* **10**, 442.
754. Nyquist, R. A. and Muelder, W. W. (1968). *Spectrochim. Acta* **24A**, 187.
755. Nyquist, R. A. and Muelder, W. W. (1966). *Spectrochim. Acta* **22**, 1563.
756. Vasilev, A. F. and Tulyakova, T. F. (1968). *Shur. priklad. Spektrosk.* **8**, 102.
757. Mitchell, R. W., Lustig, M., Ruff, J. K. and Merritt, J. A. (1968). *J. Amer. Chem. Soc.* **90**, 6329.
758. Pavlík, I. and Šubrt, J. (1967). *Coll. Czech. Chem. Comm.* **32**, 76.
759. Pavlík, I. and Plecháček, V. (1966). *Coll. Czech. Chem. Comm.* **31**, 2083.
760. Rubalcava, H. E. and Thomson, J. B. (1963). *J. Phys. Chem.* **67**, 310.
761. Griffiths, J. E. (1963). *J. Chem. Phys.* **38**, 2879.
762. Mackay, K. M. and Watt, R. (1967). *Spectrochim. Acta* **23A**, 2761.
763. Lovejoy, R. W. and Baker, D. R. (1967). *J. Chem. Phys.* **46**, 658.
764. Steingross, W. and Zeil, W. (1966). *J. Organomentall. Chem.* **6**, 464.
765. Durig, J. R. and Sink, C. W. (1968). *Spectrochim. Acta* **24A**, 575.
766. Durig, J. R., Sink, C. W. and Turner, J. B. (1969). *Spectrochim. Acta* **25A**, 629.
767. Durig, J. R., Gibson, B. M. and Sink, C. W. (1968). *J. Mol. Structure* **2**, 1.
768. Cradock, S. (1968). *J. Chem. Soc.* (*A*) 1426.
769. Pillai, M. G. K. and Perumal, A. (1964). *Bull. Soc. chim. belges.* **73**, 29.
770. Griffiths, J. E. (1968). *J. Chem. Phys.* **48**, 278.
771. Davidson, G., Woodward, L. A., Mackay, K. M. and Robinson, P. (1967). *Spectrochim. Acta* **23A**, 2383.
772. Jullien, J. and Paillous, A. (1965). *Bull. Soc. Chim. France*, 2236.
773. Ogawa, T. (1968). *Spectrochim. Acta* **24A**, 15.
774. Green, J. H. S., Kynaston, W. and Rodley, G. A. (1968). *Spectrochim. Acta* **24A**, 853.
775. O'Brien, M. H., Doak, G. O. and Long, G. G. (1967). *Inorg. Chem. Acta* **1**, 34.
776. Allkins, J. R. and Hendra, P. J. (1967). *Spectrochim. Acta* **23A**, 1671.

777. Freeman, J. M. and Henshall, T. (1968). *Canad. J. Chem.* **46**, 2175.
778. Costa, G. and DeAlti, G. (1960). *Atti Accad. naz. Lincei, Rend. Classe Sci. fis. mat. nat.* **28**, 627.
779. Kimmel, H. and Dillard, C. R. (1968). *Spectrochim. Acta* **24A**, 909.
780. Bribes, J. L. and Gaufres, R. (1968). *Compt. rend.* **266C**, 584.
781. Meič, Z. and Randič, M. (1968). *Trans. Faraday Soc.* **64**, 1438.
782. Kreevoy, M. M. and Ditsch, L. T. (1960). *J. Amer. Chem. Soc.* **82**, 6124.
783. George, W. O. and Robinson, F. V. (1968). *J. Chem. Soc.* (*A*) 1950.
784. Junge, H. and Musso, H. (1968). *Spectrochim. Acta* **24A**, 1219.
785. Mikami, M., Nakagawa, I. and Shimanouchi, T. (1967). *Spectrochim. Acta* **23A**, 1037.
786. Lewis, J., Long, R. F. and Oldham, C. (1965). *J. Chem. Soc.* 6740.
787. Behnke, G. T. and Nakamoto, K. (1967). *Inorg. Chem.* **6**, 433.
788. Gibson, D., Lewis, J. and Oldham, C. (1967). *J. Chem. Soc.* (*A*) 72.
789. Behnke, G. T. and Nakamoto, K. (1968). *Inorg. Chem.* **7**, 330.
790. Nakamoto, K., McCarthy, P. J. and Miniatas, B. (1965). *Spectrochim. Acta* **21**, 379.
791. Kedzia, B. B., Armendarez, P. X. and Nakamoto, K. (1968). *J. Inorg. Nuclear Chem.* **30**, 849.
792. Corey, E. J., Howell, M. G., Boston, A., Young, R. L. and Sneen, R. A. (1956). *J. Amer. Chem. Soc.* **78**, 5036.
793. Hodoson, F. and Mantsch, H. (1966). *Studii si Cercetari Chim.* **14**, 15.
794. Jones, R. N., Cole, A. R. H. and Nolin, B. (1952). *J. Amer. Chem. Soc.* **74**, 5662.
795. Cook, D. and Regnier, Z. R. (1967). *Canad. J. Chem.* **45**, 2895.
796. Cook, D. and Regnier, Z. R. (1967). *Canad. J. Chem.* **45**, 2899.
797. Jones, R. N. and Sandorfy, C. (1956). In "Chemical Applications of Spectroscopy". West, ed., p. 408. Interscience Publishers. Inc., New York, U.S.A.
798. Tasumi, M. and Krimm, S. (1967). *J. Chem. Phys.* **46**, 755.
799. Ikeda, S., Yamamoto, A. and Tanaka, H. (1963). *J. Polymer. Sci.* **A1**, 2925.
800. Tasumi, M., Shimanouchi, T., Tanaka, H. and Ikeda, S. (1964). *J. Polymer. Sci.* **A2**, 1607.
801. Tasumi, M., Shimanouchi, T. and Ikeda, S. (1966). *J. Polymer. Sci.* **A1 4**, 1023.
802. Tasumi, M., Shimanouchi, T., Kenjo, H. and Ikdeda, S. (1966). *J. Polymer. Sci.* **A1 4**, 1011.
803. Tadokoro, H., Kobayashi, M., Ukita, M., Yasufuku, K., Murahashi, S. and Torii, T. (1965). *G. Chem. Phys.* **42**, 1432.
804. Liang, C. Y. and Lytton, M. R. (1962) *J. Polymer. Sci.* S45.
805. Tadokoro, H., Ukita, M., Kobayashi, M. and Murahashi, S. (1963). *J. Polymer. Sci.* **B1**, 405.
806. Miyazawa, T. and Ideguchi, Y. (1963). *Bull. Chem. Soc. Japan* **36**, 1125.
807. Miyazawa, T. and Ideguchi, Y. (1964). *Bull. Chem. Soc. Japan* **37**, 1065.
808. Golub, M. A. and Shipman, J. J. (1960). *Spectrochim. Acta* **16**, 1165.
809. Golub, M. A. and Shipman, J. J. (1964). *Spectrochim. Acta* **20**, 701.
810. Neto, N. and Di Lauro, C. (1967). *Eur. polymer. J.* **3**, 645.
811. Krimm, S., Folt, V. L., Shipman, J. J. and Berens, A. R. (1963). *J. Polymer. Sci.* **A1**, 2621.
812. Enomoto, S., Asahina, M. and Satoh, S. (1966). *J. Polymer. Sci.* **A1 4**, 1373.
813. Enomoto, S. and Satoh, S. (1967). *Kolloid Z.* **219**, 12.
814. Kobayashi, M. (1960). *Bull. Chem. Soc. Japan* **33**, 1416.
815. Kobayashi, M. (1961). *Bull. Chem. Soc. Japan* **34**, 560.

816. Tadokoro, H., Nishiyama, Y. S., Nozakura, Sh. and Murahashi, Sh. (1961). *Bull. Chem. Soc. Japan* **34**, 381.
817. Kobayashi, M. (1961). *Bull. Chem. Soc. Japan* **34**, 1045.
818. Shaefgen, J. R. and Zbinden, R. (1964). *J. Polymer. Sci.* **A2**, 4865.
819. Tadokoro, H., Kobayashi, M., Kawaguchi, Y., Kobayashi, A. and Murahashim, S. (1963). *J. Chem. Phys.* **38**, 703.
820. Imada, K., Tadokoro, H., Umehara, A. and Murahashim, S. (1965). *J. Chem. Phys.* **42**, 2807.
821. Tadokoro, H. (1959). *Bull. Chem. Soc. Japan* **32**, 1252.
822. Yoshihara, T., Tadokoro, H. and Murahashi, S. (1964). *J. Chem. Phys.* **41**, 2902.
823. Morsi, M. K. S. and Sterling, C. (1966). *Carbohydrate Res.* **3**, 97.
824. Sobue, H. and Fukuhara, S. (1955). *J. Chem. Soc. Japan Ind. Chem. Sect.* **58**, 946.
825. Jeffries, R. (1963). *Polymer. Lond.* **4**, 375.
826. Correns, E. and Dechant, J. (1967). *Faserforsch. Text. Tech.* **18**, 590.
827. Miyake, A. (1960). *J. Phys. Chem.* **64**, 510.
828. Zundel, G. and Metzger, H. (1967). *Spectrochim. Acta* **23A**, 759.
829. Yamadera, R., Tadokoro, H. and Murahashi, S. (1964). *J. Chem. Phys.* **41**, 1233.
830. Schmidt, P., Schneider, B. (1963). *Coll. Czech. Chem. Comm.* **28**, 2685.
831. Schmidt, P., Schneider, B. and Čefelin, P. (1967). *J. Polymer. Sci. C. Symp.* **16**, 229.
832. Larose, P. (1957). *Canad. J. Chem.* **35**, 1242.
833. Sobue, H. and Fukuhara, S. (1957). *J. Chem. Soc. Japan, Ind. Chem. Sect.* **60**, 86.
834. Koshimo, A. (1965). *J. Appl. Polymer. Sci.* **9**, 55.
835. Fukushima, K., Ideguchi, Y. and Miyazawa, T. (1963). *Bull. Chem. Soc. Japan* **36**, 1301.
836. Sultanov. A. and Ibragimov, A. P. (1964). *Radiats Effektg Biol. Akad. Nauk. U. USSR, Inst. Yadern. Fiz*, 107.
837. Bendit, E. G. (1966). *Biopolymers.* **4**, 561.
838. Fraser, R. D. B. and MacRae, T. P. (1959). *J. Chem. Phys.* **31**, 122.
839. Milch, R. A. (1964). *Nature*, **202**, 84.
840. Larson, D. L. and Mclaughlin, J. (1955). *J. Amer. Chem. Soc.* **77**, 1355; 1359.
841. Boucher, L. J., Strain, H. H. and Katz, J. J. (1966). *J. Amer. Chem. Soc.* **88**, 1341.
842. Lenormant, H. and De Loze, C. (1955). *Bull. Soc. Chim. France.* 1501.

CHAPTER 5

Carbon Labelled Compounds

5.1 INTRODUCTION

Most of the carbon labelled compounds the infrared spectra of which were investigated so far contained the stable carbon-13 isotope; in a few cases, however, the compounds studied were labelled by the radioactive carbon-14 modification.

The effect of ^{13}C-labelling on the various molecular absorption frequencies was measured mainly in the cases where these are characteristic for specific CX groups, this happening (see Section 3.2) when X has either a very light mass (X = H) or else when the CX bond is double or triple (X = O, N).

An application of Hooke's law (eqn. 2.17) to the case of the stretching frequency of the diatomic ^{13}C—H oscillator (and the R—^{13}C—H groups—see Section 3.2) yields a ^{12}C—H/^{13}C—H frequency ratio of 0·997 which gives for the observed ^{12}C—H value of *ca.* 3000 cm^{-1} an expected ^{13}C isotopic shift of about 9 cm^{-1}. For ^{17}C—H groups the isotopic shift can similarly be calculated to be about 17 cm^{-1}.

The isotopic shifts for the stretching frequencies of the ^{13}CO and ^{14}CO carbon monoxide molecules are estimated likewise to be about 48 and 89 cm^{-1}, respectively.

5.2 ^{13}C-CARBON MONOXIDE

High resolution of the vibration–rotation absorption band of a natural abundance sample of gaseous carbon monoxide[1] yielded the value of 2096·07 cm^{-1} (in vacuum) for the centre of the ^{13}CO stretching band, as compared with the respective 2143 cm^{-1} value for ^{12}CO. Some of its other constants are given in Table 5.1.

The ^{13}CO vibration–rotation absorption lines were also observed in the infrared solar spectrum[3] (measured below the atmosphere of the earth). The isotopic shift of the first overtone of the ^{13}CO stretching vibration was observed to amount to 93·27 ± 0·02 cm^{-1}, in good agreement with the respective calculated value.[2]

TABLE 5.1 The Main Vibration–Rotation Constants (cm^{-1}) of $^{13}C^{16}O$

	Observed[1]	Calculated[2]
ω_e	2121·41	2121·44
B_e	1·8464 ± 0·0002	1·84615
α_e	0·01645 ± 0·00003	0·01640
D	$5{\cdot}4 \times 10^{-6}$	$5{\cdot}44 \times 10^{-6}$

The ^{13}CO stretching frequency of an aqueous solution of a 53 atom % ^{13}C–carbon monoxide addition compound with human hemoglobin appears at 1906·8 cm^{-1} as compared with the value of 1950·3 cm^{-1} observed for the corresponding ^{12}CO vibration.[4]

5.3 LABELLED CARBON DIOXIDE

The linear and symmetrical $^{13}C^{16}O_2$ molecule was observed[5] to show the centre of its ν_3, antisymmetric stretching band, in the gas phase, at 2284·5 cm^{-1}, as compared with the corresponding value of 2349 cm^{-1} in the case of $^{12}C^{16}O_2$. An application of eqn. 3.11 to the respective $^{13}CO_2$ and $^{12}CO_2$ *zero-order* frequencies yields an isotopic frequency ratio of 0·97154 which after taking into account the $^{13}CO_2$ anharmonicity (see end of Section 3.8) results in a calculated ν_3 value for $^{13}CO_2$ which is within 0·1 cm^{-1} (2284·4 cm^{-1}) of the measured value.[5]

The observed *zero-order* fundamentals of $^{13}CO_2$ and $^{12}CO_2$ are given in Table 5.2.

TABLE 5.2 The Zero-Order Fundamentals[6] (cm^{-1}) of $^{13}CO_2$ and $^{12}CO_2$

Fundamental	Description	$^{13}CO_2$	$^{12}C^{16}O_2$
ω_1	Symmetric stretching	1354·91	1354·91
ω_2	Bending	653·83	673·00
ω_3	Antisymmetric stretching	2328·22	2396·49

The $^{13}CO_2$ infrared absorption lines were also observed in the solar spectrum.[3]

The fine structure of the $2\nu_2$, $3\nu_2 + \nu_3$ (observed[7] at 4147·24 cm^{-1}) and other overtone and combination bands of $^{13}CO_2$ has been analysed, studied carefully, and interpreted.[8,9]

The cubic and quartic anharmonic force constants of $^{13}CO_2$ were calculated from its observed vibrational energy levels and rotational constants[6] and its rotational constant D was determined.[10]

The infrared absorption spectrum of $^{14}CO_2$ has also been measured.[11] The centre of its ν_3, antisymmetrical stretching band was found to be near 2226 cm^{-1}, thus showing an isotopic shift of *ca.* 123 cm^{-1} as compared with the respective $^{13}CO_2$ shift of *ca.* 64 cm^{-1}.

5.4 ^{13}C-Carbonyl Sulphide

Gaseous $O{=}^{13}C{=}S$ shows its ν_3, out-of-phase stretching, fundamental band[12] at 2009·2 cm^{-1} and its $\nu_1 + \nu_3$ combination band at 2861 cm^{-1}. A polycrystalline film of 64 atom % ^{13}C-carbonyl sulphide (at 80°K) exhibits its very strong ν_1, ν_2 and ν_3 fundamentals at 853·1, 503·3 and 1958 cm^{-1}, respectively, as compared with the corresponding 857·8, 518·9 and 2004·3 cm^{-1} values in the case of the normal compound.[13]

The high resolution and analysis of the 3650–7000 cm^{-1} (overtone and combination) absorption bands of a sample of 51·8 atom % ^{13}COS yielded the values of its B_0 and D_0 rotational constants as 0·20220 and 4·3 × 10^{-8} cm^{-1}, respectively.[14]

5.5 Labelled Nickel Tetracarbonyl

The tetrahedral isotopically homogeneous $Ni(CO)_4$ molecule, which belongs to the $\boldsymbol{T_d}$ point group, can be expected to show only four infrared active fundamentals, the other five normal vibrations being infrared inactive. The 400–4200 cm^{-1} absorption spectrum of a 93 molecule % sample of $Ni(^{13}CO)_4$ exhibited[15] three of these bands at the positions reported in Table 5.3.

TABLE 5.3 The Fundamental Infrared Absorption Bands (cm^{-1}) Observed[15] for $Ni(^{12}CO)_4$ and $Ni(^{13}CO)_4$

		$Ni(^{12}CO)_4$		$Ni(^{13}CO)_4$	
Band	Approximate description	gas	solution[a]	gas	solution
ν_5	Asymmetric CO stretching	2057·8	2044·5	2011·2	1998·4
ν_6	Ni—C—O bending	458·9	455	442·7	445
ν_7	Ni—C stretching	423·1	422·5	417·3	418
ν_8	C—Ni—C bending	80	91	[b]	

[a] In CCl_4.
[d] Not observed directly.

The anharmonicity constants of $Ni(^{13}CO)_4$ were calculated.[15]

Gaseous $Ni(^{12}CO)_3$ ^{13}CO no longer possess tetrahedral symmetry and so also show the in-phase CO stretching band,[16] at 2120 cm^{-1}, besides the out-of-phase, ^{13}CO stretching fundamental at 2018 cm^{-1} and the degenerate ^{12}CO stretching absorption at 2058 cm^{-1}. The respective calculated values are[16] 2123·7, 2020 and 2058 cm^{-1}. In a cyclohexane solution, only the two first bands were observed, at 2117 and 2007 cm^{-1}, respectively.[17]

A cyclohexane solution of $Ni(^{12}CO)_2(^{13}CO)_2$ shows only one CO stretching band,[17] at 2014 cm^{-1} (calculated:[17] 2016 cm^{-1}), while gaseous $Ni(^{12}CO)(^{13}CO)_3$ was observed[15] to absorb at 2103 cm^{-1}.

The C–O stretching frequencies of isotopically substituted metal carbonyls in general can now be calculated very accurately.[18]

5.6 OTHER ^{13}C-LABELLED METAL CARBONYLS

Molybdenum hexacarbonyl with ^{13}C in natural abundance contains about 6% of $Mo(^{12}CO)_5(^{13}CO)$ (point group $\boldsymbol{C_{4v}}$) which in solution in cyclohexane gives rise to weak absorption bands at[19] 1956·6 (axial ^{13}CO group stretching), 2009·3 (axial ^{12}CO stretching) and 2108·7 cm^{-1} (in-phase CO stretching) besides the only $Mo(^{12}CO)_6$ (triply degenerate) band at 1988·4 cm^{-1} (see Fig. 5.1).

Using these data the main force constants of this molecule have been evaluated and used to calculate all the CO stretching frequencies of the various bis- and tris-^{13}CO $Mo(CO)_6$ modifications in order to be able to identify such isotopomers in a ^{13}C-enriched sample.[19]

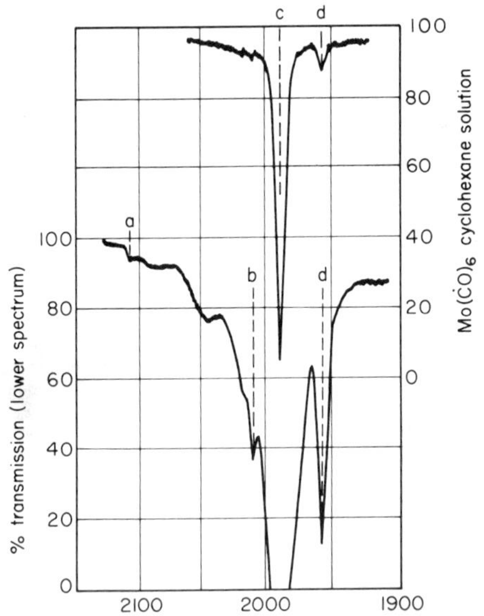

Fig. 5.1 The spectrum of $Mo(CO)_6$ containing ^{13}C in natural abundance.

The absorption spectra of mono-^{13}CO $Cr(CO)_6$, $Mn_2(CO)_{10}$, $Fe(CO)_5$ and $W(CO)_6$ were observed to be quite similar to the respective spectrum of molybdenum carbonyl.[17] The characteristic absorption of ^{13}C-labelled $Re_2(CO)_{10}$ in the 1900–2200 cm^{-1} region was also reported.[20]

Cyclohexane solutions of natural abundance manganese ^{13}C-pentacarbonyl hydride and deuteride show their C=O stretching bands at about 1965, 1982, 2035, and 2110 cm^{-1} while the corresponding $HRe(CO)_5$ fundamentals were observed at 1967, 1982, 2042 and 2123 cm^{-1}, respectively.[21] These bands are due to *ca.* 4% of the *cis*-mono-^{13}CO-form and 1% of the respective *trans* modification, each of which exhibits its own characteristic bands.

Cyclohexane solutions of manganese and rhenium pentacarbonyl halides (X = Cl, Br, I) also show CO stretching bands due to their natural abundance content of 4% of axially-labelled and 1% of radially-labelled mono-^{13}CO-metal pentacarbonyl halides.[22] The former absorb near 1960 cm^{-1} in the case of the manganese compounds and in the vicinity of 1940 cm^{-1} in case of the rhenium halides while the radial ^{13}CO-manganese carbonyl bromide e.g. exhibits bands at 2020, 2073 and 2127 cm^{-1} and the corresponding rhenium carbonyl bromide shows them at 2015, 2073 and 2143 cm^{-1}, respectively.

$Co(^{12}CO)_2(^{13}CO)NO$ in solution in cyclohexane was observed[17] to display its CO stretching bands at 2010 and 2098 cm^{-1} as compared with the 2086 cm^{-1} absorption found for $Co(^{12}CO)(^{13}CO)_2NO$. These frequencies agree well with the values calculated for them.[17]

The frequency of the ^{13}CO stretching vibrations in other carbonyl–nitrosyl complexes has also been investigated.[23]

Hydrocarbon solutions of *ca.* 50% *cis*-^{13}CO methylmanganese pentacarbonyl show its CO stretching bands at (the numbers in parentheses give the molecular extinction coefficients of the respective bands): 1975 (9000), 1992 (7000), 2011 (5000), 2033 (1600) and 2102 (600) cm^{-1} as compared with the 1949, 2011 and 2110 cm^{-1} fundamentals of the 50% *trans*-^{13}CO compound and the 1990 (5600), 2011 (26,000) and 2110 (350) cm^{-1} frequencies of the pure ^{12}CO methylmanganese carbonyl.[24]

A cyclohexane solution of cyclopentadienylmanganese carbonyl shows its ^{13}C=O stretching bands at 1912 and 2018 cm^{-1}, in good agreement with the values calculated for them.[17]

A mixture of *cis*- and *trans*-$CH_3COMn(^{12}CO)_4{}^{13}CO$ exhibits bands[24] at: 1962 (*trans*), 1969 (*cis*) and 2044 cm^{-1} (*cis*) besides a $CH_3{}^{13}COMn(^{12}CO)_5$ ^{13}CO stretching fundamental which appears at 1626 cm^{-1} (the respective ^{12}CO band being at 1661 cm^{-1}).

The natural abundance of various $^{13}CO—[R_3P, Mn(CO)_4]_2$ (R = C_2H_5, n-C_3H_7, n-C_4H_9, C_6H_{11}, Ph, CH_3O, C_2H_5O, PhO) and similar compounds gives rise to series of weak CO stretching bands which were assigned to their respective modes of vibration in each case, these assignments being supported by an agreement with the frequencies calculated for them.[25]

The CO stretching frequencies of the natural abundance content of $^{13}CO[M(CO)_3SR]_4$ (M = Mn, Re; R = C_2H_5, C_7H_7) compounds were also reported.[26]

5.7 LABELLED SALTS

Potassium and sodium cyanates which contain about 60 atom % of ^{13}C were investigated in a dilute solid solution in potassium bromide, in the 600–5000 cm^{-1} region, at various temperatures starting from 150°K and onwards.[27] The main absorption bands of the linear $^{13}C^{14}N^{16}O$-ions were thus observed at: 612 (bending), 1191, 1272 (both due to a Fermi resonance between the in-phase stretching and the bending overtone vibrations) and 2113 cm^{-1} (out-of-phase stretching). The respective normal cyanate ion bands appear at 629, 1206, 1293 and 2170 cm^{-1}, correspondingly. The anharmonicity constants of the $^{13}CNO^-$ ions have also been evaluated.[27]

Potassium thiocyanate labelled by about 60 atom % of ^{13}C shows its fundamentals at[28] 457·5 and 470 (branches of the ν_2, bending vibration of crystalline $^{14}N^{13}C^{32}S^-$), 742·6 (out-of-phase stretching) and 2006 cm^{-1} (in-phase stretching), as compared with the respective $^{14}N^{12}C^{32}S^-$ values of 470, 484, 749 and 2053 cm^{-1}. The zero-point vibration frequencies of the $N^{13}CS^-$ ions were also reported.[28]

Natural calcite, which contains about 1% of $Ca^{13}CO_3$, shows a weak ν_2, symmetrical bending, absorption band due to the $^{13}CO_3^{2-}$ ions, at 850 cm^{-1}, near its strong $^{12}CO_3^{2-}$ ν_2 band which appears at 877 cm^{-1} (see Fig. 5.2). The

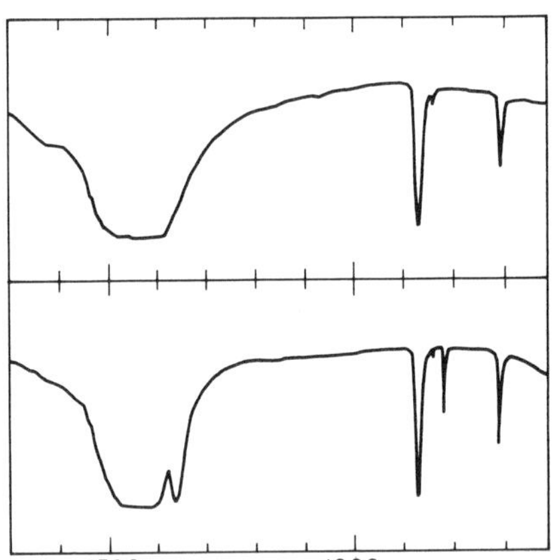

Fig. 5.2 The infrared spectra of natural (above) and 10% ^{14}C-labelled calcite (below).

isotopic shift of 27 cm^{-1} thus observed for the ν_2 fundamental of ^{13}C-calcite agrees very well with the value of 27·5 cm^{-1} calculated for this vibration in the case of the free $^{13}CO_3^{2-}$ ion. In the case of aragonite, strontium carbonate and barium carbonate, however, the isotopic shift of this frequency is much lower (11, 14 and 18 cm^{-1}, respectively) as a result of an interaction between neighbouring CO_3^{2-} ions in these crystals.[29]

^{14}C-labelled calcite (10 atoms %) shows its ν_2 fundamental at 826 cm^{-1} (see Fig. 5.2), i.e. shifted by 51 cm^{-1} relative to the $^{12}CO_3^{2-}$ ν_2 frequency, as compared with the shift of 52 cm^{-1} calculated for the isolated $^{14}CO_3^{2-}$ ion. In this case the ν_3 out-of-phase stretching absorption band was also observed to shift, from 1419 to 1371 cm^{-1}. In aragonite and strontium carbonate the lower crystalline symmetry present brings about a splitting of the degenerate ν_3 and ν_4 vibrations of the CO_3^{2-} ion into two branches for each vibration. While the ν_4 bands are practically unaffected by the ^{14}C-labelling, the various ν_3 branches are shifted by 90–121 cm^{-1} as compared with the 86 cm^{-1} value calculated for the ν_3 fundamental of aragonite and the 85 cm^{-1} shift expected for the ν_3 vibration of $Sr^{13}CO_3$.

$Ba^{14}CO_3$ was found to exhibit its single ν_3 band at[29] 1346 cm^{-1} (i.e. shifted by 107 cm^{-1}).

5.8 ^{13}C-Methane

The absorption of a 92 atom % sample of ^{13}C-methane in the 2870–3130 cm^{-1} region, measured under a resolution of 0·3 cm^{-1}, exhibited[30] the centre of its ν_3, out-of-phase C—H stretching, band at 3009·53 ± 0·03 cm^{-1}, as compared with the respective ^{12}C—H frequency of 3019·48 cm^{-1}.

The ω_3 vibration of $^{13}CH_4$ was calculated to be at 3142·9 cm^{-1} while its infrared inactive ν_1 and ν_2 and its infrared active ν_4 fundamentals were calculated to be at 2916·6, 1533·9 and 1296·0 cm^{-1}, respectively.[30]

A number of other vibration–rotation constants of $^{13}CH_4$ (such as B_1 = 5,2006 cm^{-1}, etc.) have also been determined.[30]

5.9 Labelled Substituted Methanes

The ν_3 and ν_4 fundamentals of $^{13}CF_4$ appear[31] at 1241·2 and 628·8 cm^{-1}, respectively, as compared with the corresponding 1282·6 and 631·3 cm^{-1} values observed for $^{12}CF_4$. Other bands observed for $^{13}CF_4$ were found at:[31] 1259·7 ($2\nu_4$), 1537·1 ($\nu_1 + \nu_4$), 1867·1 ($\nu_3 + \nu_4$, 1888·4 ($3\nu_4$), 2145·1 ($\nu_1 + \nu_3$) and 2477·8 cm^{-1} ($2\nu_3$).

The 2·5 cm^{-1} ^{13}C-isotopic shift found thus for the ν_4 fundamental of $^{13}CF_4$ was used to fix effectively the force constants of the CF_4 molecule.[31]

The values of the Coriolis constants for the ν_3 and ν_4 bands of $^{13}CF_4$ have been calculated.[32]

The highly resolved absorption spectrum of a *ca.* 57 atom % sample of $^{13}CH_3Cl$ yielded the location of the centres of the ν_2–ν_6 vibration–rotation fundamental bands of $^{13}CH_3{}^{35}Cl$ as well as its main vibrational–rotational constants.[33] Its six normal vibrations appear at: 2964·38 (ν_1; symmetrical C—H stretching), 1349·42 (ν_2; symmetrical C—H bending), 714·885 (ν_3; C—Cl stretching), 3030·42 (ν_4; asymmetric C—H stretching), 1449·54 (ν_5; asymmetric C—H bending) and 1011·42 cm^{-1} (ν_6; asymmetric deformation). Its rotational ground state constants A_0 and B_0 have the values of 5·124 and 0·42684 cm^{-1}, respectively.[33]

57 Atom % $^{13}CH_3I$ exhibits its fundamentals[34] at: 2967·96 (ν_1), 1245·09 (ν_2), 517·04 (ν_3), 3049·37 (ν_4), 1431·80 (ν_5) and 878·16 (ν_6) cm^{-1}. Its A_0 and B_0 constants have the values of 5·158 and 0·23747 cm^{-1}, respectively.[34]

5.10 Other Non-Aromatic Labelled Hydrocarbons

The rotational structure of the $\nu_2 + \nu_6$ combination band of a sample of 60% $^{12}CH_3{}^{13}CH_3$ (centred at about 2750 cm^{-1}) was analysed to yield its B_0 rotational constant[35] which is equal to 0·64865 cm^{-1}.

The absorption spectrum of 3-^{13}C-labelled 2,4-dimethylpentane was used to show that the 867, 982 and 1164 cm^{-1} bands of 2,4-dimethylpentane belong to skeletal modes of vibration since they appear in the case of the labelled compound at 862, 972 and 1156 cm^{-1}, respectively.[36]

The 400–3650 cm^{-1} spectrum of 55 atom % 1-^{13}C-propene[37] shows two isotopic shifts, from 1652 to 1642 cm^{-1} (C=C stretching band) and from 912·3 to 903·6 cm^{-1} (out-of-plane C—H bending). In the case of 1-^{13}C-2,2-diisopropylethylene[36] the labelled C=C group stretching fundamental was observed to move from 1639 cm^{-1} to 1608 cm^{-1}, besides another isotopic shift observed for the 1101 cm^{-1} band which goes over to 1097 cm^{-1}.

The 2600–2800, 3100–3400 and 6400–6600 cm^{-1} absorption of a ^{13}C-labelled acetylene sample which contained *ca.* 16% of $^{12}C_2H_2$, 48% of $^{12}C^{13}CH_2$ and 36% of $^{13}C_2H_2$ was investigated under high resolution.[38] The analysis of the spectra obtained thus showed the ν_1, symmetric C—H stretching, frequency of $^{12}C^{13}CH_2$ (which has become infrared active as a result of the isotopic substitution) to have the value of 3361·59 cm^{-1} (in vacuum) as compared with the calculated ν_1 value of 3362 cm^{-1}. Its ν_2, C≡C stretching, fundamental was observed at 1941 cm^{-1} while its ν_3, antisymmetric C—H stretching, was found at 3284·20 cm^{-1} (the calculated value being 3283 cm^{-1}). The ν_4^1 and ν_5^1 bending vibrations of this acetylene were calculated to have the 607·21 and 728·23 cm^{-1} values, respectively. The strong $2\nu_3$ 6493 cm^{-1} overtone and $\nu_1 + \nu_3$ 6542 cm^{-1} combination bands, the ternary $\nu_2 + \nu_4^1 + \nu_5^1$ 3250·50 cm^{-1} absorption, as well as a number of other $^{12}C^{13}CH_2$ binary and multiple bands, have also been studied.

The B_0 and D_0 rotational constants of $^{12}C^{13}CH_2$ have the values of

$1{\cdot}14840 \pm 0{\cdot}00002$ cm^{-1} and $1{\cdot}56 \pm 0{\cdot}02 \times 10^{-6}$ cm^{-1}, correspondingly.[38]

In the case of the $^{13}C_2H_2$ isotopomer the infrared inactive ν_1 normal vibration was calculated to have a value of 3349·51 cm^{-1}, as compared with the respective 3373 cm^{-1} frequency of the normal acetylene.[38] The $^{13}C_2H_2$ ν_3 fundamental was observed at 3279·494 cm^{-1}, i.e. about 9·5 cm^{-1} below the parallel frequency of $^{12}C_2H_2$. The observed isotopic frequency ν_3^i/ν_3 ratio of 0·997018 is in an excellent agreement with the theoretically calculated ν_3^i/ν_3 ratio of 0·997006.

The ν_4 normal vibration of $^{13}C_2H_2$ was calculated to have a frequency of 602·78 cm^{-1}.

Its $\nu_1 + \nu_3$ and $\nu_2 + \nu_4^1 + \nu_5^1$ combination bands were observed at 6521·94 and 3213·40 cm^{-1}, respectively, while its B_0 and D_0 rotational constants were found to have the values of 1·11949 and $1{\cdot}47 \times 10^{-6}$ cm^{-1}, correspondingly.[38]

5.11 Aromatic ^{13}C-Labelled Compounds

The 250–3800 cm^{-1} absorption of a mixture of equal parts of $^{12}C_6H_6$ and $^{12}C_5{}^{13}CH_6$ was measured both for the liquid and the gas phases.[39]

The ^{13}C-sensitive fundamentals ($\Delta\nu > 3$ cm^{-1}) of liquid $^{12}C_5{}^{13}CH_6$ were observed at 977 ($\Delta\nu = 15$), 984 ($\Delta\nu = 8$; ν_1), 1303 ($\Delta\nu = 6$; ν_{14}), 1473 ($\nu = 6$; ν_{19}), 1577 ($\Delta\nu = 9$; ν_8) and 3032 cm^{-1} ($\Delta\nu = 5$; ν_{20}).

The ν_6 normal vibration of gaseous $^{12}C_5{}^{13}CH_6$ was observed[40] to absorb very weakly at 607 cm^{-1}.

A Nujol mull of a mixture of 65 atom % $(C_6H_5)_3{}^{13}CCl$ and 35 atom % $(C_6H_5)_3{}^{12}CCl$ showed[41] the ^{13}C-sensitive fundamentals of the labelled trityl chloride at 617 (*m*), 730 (*s*), 741 (*s*; both C—Cl stretchings), 803 (*sh*), 808 (*m*), 883 (*s*; C—H bending) and 1186 (*m*; C_6H_5—C stretching) cm^{-1}, as compared with the respective normal trityl chloride frequencies of 637, 735, 746, 813, 818, 892 and 1211 cm^{-1}.

A similar mixture containing $(C_6H_5)_3{}^{13}C^+BF_4^-$ exhibits characteristic bands at[41] 641 (*w*), 652 (*m*), 696 (*s*; C—C—C bending), 802 (*s*; C—H bending), 906 (*m*; C—H bending), 1294 cm^{-1} (*s*; C—H bending) and 1342 (*s*; C_6H_5—C stretching), the corresponding normal trityl fluoborate bands appearing at 647, 658, 701, 809, 915, 1299 and 1359 cm^{-1}, respectively.

5.12 ^{13}C-Labelled Hydroxy- and Oxo-Compounds

3-^{13}C-2,4-dimethylpentan-3-ol solutions show the normal 848, 977, 993, 1096 and 1152 cm^{-1} skeletal bands at 844, 966, 982, 1094 and 1147 cm^{-1}, respectively.[36]

Nujol mulls of 65 atom % $(C_6H_5)_3{}^{13}COH$ exhibit their characteristic absorption at about 871 (*m*), 888 (*sh*), 894 (*m*; all three due to CH bendings),

994 (*s*; C—O stretching), 1193 (*w*; C_6H_5—C stretching) and 1321 cm^{-1} (*s*; O—H bending), as compared with the corresponding normal trityl carbinol values of 889, 906, 912, 1010, 1202 and 1330 cm^{-1}, respectively.[41]

1-^{13}C-2,4,6-triphenylphenol[42] shows the 1226 cm^{-1} band of the normal phenol (probably due to its C—O stretch) shifted to 1214 cm^{-1}.

The infrared absorption spectrum of 62 atom % 3-^{13}C-2,4-dimethyl-3-pentanone solutions (in CS_2) exhibits ^{13}C-sensitive bands at: 853, 972, 1004, 1120, 1190 and 1675 cm^{-1}, corresponding to the normal diisopropyl ketone frequencies at 858, 983, 1024, 1129, 1203 and 1712 cm^{-1}, respectively.[43] While the 853, 972 and 1120 cm^{-1} bands belong to skeletal vibrations the 1004 cm^{-1} fundamental must be due to an uncoupled C—^{13}C—C stretching mode of vibration and the absorption at 1190 cm^{-1} is probably connected with a stretching of the C—^{13}C bond adjacent to the carbonyl group, which affects also the CO distance. The 1675 cm^{-1} band is the result of the ^{13}C=O stretching vibration and its 37 cm^{-1} shift from the corresponding 1712 cm^{-1} band of the normal ketone is in good agreement with the 38 cm^{-1} ^{13}C-shift calculated for the diatomic ^{13}C=O oscillator (having a frequency of 1712 cm^{-1}).[43] The location of the 1004, 1190 and 1675 cm^{-1} bands is retained also in the case of the 2,4-D_2-isotopomer of the 3-^{13}C-diisopropyl ketone.[43]

The infrared absorption of 1,4-^{13}C-labelled cyclohexane-1,4-dione, containing 33% of the 1,4-$^{13}C_2$-modification and 49% of the mono-^{13}C-ketone, was measured in CCl_4 or CS_2 solutions, in the vapour phase and in potassium bromide pellets.[44] The location of the ^{13}C-sensitive ($\Delta\nu > 3$ cm^{-1}) bands of solutions of the $^{13}C_2$-compound is given in Table 5.4.

TABLE 5.4 The Location (cm^{-1}) of the ^{13}C-sensitive Bands of 1,4-$^{13}C_2$-cyclohexane-1,4-dione in Solution[a]

$^{13}C_2$-cyclohexane-1,4-dione	^{12}C-ketone[b]	Approximate description
797[c]	802 (0·4)[c]	C—H bending
1049	1055 (low)	C—C stretching
1123	1130 (1·2)	C—H bending
1191	1200 (low)	C—C stretching
1246	1260 (0·8)	C—H bending
1252	1260 (0·8)	C—H bending
1289	1294 (low)	C—C stretching
1686	1728 (1·0)	C=O stretching
3355	3436 (low)	overtone

[a]Unless otherwise stated, in CCl_4.
[d]The optical density of each band is given in parentheses following its frequency.
[c]In CS_2.

The dimer of the 2,4,6-triphenylphenol radical, which has the structure given in Fig. 5.3, exhibits its normal C=O stretching band at 1665 cm^{-1}.

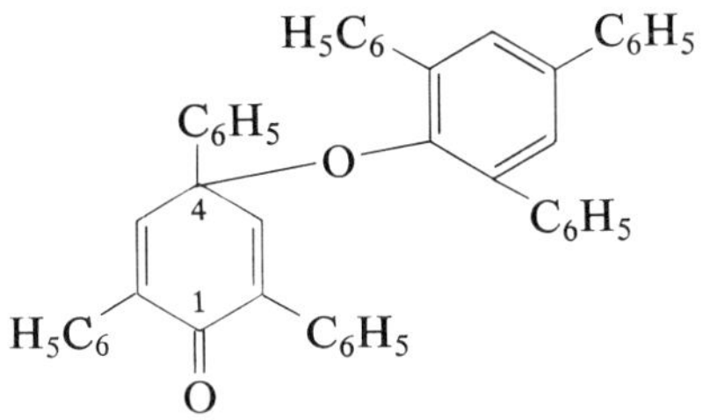

Fig. 5.3 The structure of the dimer of the 2,4,6-triphenylphenol radical.

Labelling this dimer at the 1-position with a ^{13}C atom shifts this absorption to 1616 cm^{-1} while labelling it at the 4 position lowers its C—O—C stretching bands from 958 and 1201 cm^{-1} to 948 and 1172 cm^{-1}, respectively.[42]

The infrared absorption of 1,4-^{13}C-labelled *p*-benzoquinone which contained about 49% of $^{13}C_2$-quinone and 33% of the mono-labelled species was investigated in the vapour phase, in solution in carbon tetrachloride and in a potassium bromide pellet.[44] Its ^{13}C-sensitive bands were observed at: 874 ($\Delta\nu = 9$; out-of-plane C—H bending of the $^{13}C_2$-compound), 879 ($\Delta\nu = 4$; the respective vibration of the mono-labelled species), 1059 ($\Delta\nu = 7$; C—H bending of the $^{13}C_2$-quinone), 1062 ($\Delta\nu = 4$; same vibration of the ^{13}C-compound), 1275 ($\Delta\nu = 27$; C—C stretching of the $^{13}C_2$-compound), 1290 ($\Delta\nu = 12$; C—C stretching of the ^{13}C-quinone), 1628 ($\Delta\nu = 30$), 1638 ($\Delta\nu = 32$; both due to a resonance between a ^{13}C=O stretching and a combination vibration) and 3059 cm^{-1} ($\Delta\nu = 5$; ^{13}C—H stretching).

5.13 Labelled Acids and Their Derivatives

The ^{13}C=O stretching band of 1-^{13}C-2-methylpropionic and 1-^{13}C-2,2-dimethylpropionic acids is shifted by 37 cm^{-1} from the location of the ^{12}C=O stretching vibration in the case of the corresponding normal acids[43] and appears in the spectrum of $(CH_3)_3C^{13}COOH$, e.g. at 1661^{-1}.

Highly enriched 1-^{14}C-palmitic acid in CCl_4 solution was observed to exhibit its ^{14}C=O stretching bands at 1695 (*s*) and 1637 cm^{-1} (*vs*) besides the 1760 (*w*) and 1715 (*s*) cm^{-1} ^{12}C=O stretching bands of (monomeric and dimeric) normal palmitic acid.[45] 1-^{14}C-stearic acid showed these bands at exactly the same frequencies.[45] The ^{14}C-shift of the C=O stretching fundamental of the (dimeric) aliphatic acids can thus be taken to be equal to about 1715–1637 = 78 cm^{-1}, i.e. almost exactly twice the isotopic shift (= 37 cm^{-1}) reported above for the ^{13}C=O stretching band of such acids, as can be expected.

These isotopic shifts are in good agreement with the value of 38 cm^{-1} calculated[46] for the ^{13}C-shift of a diatomic CO oscillator the stretching frequency of which is 1715 cm^{-1}.

The ^{13}C=O stretching band of dilute chloroform solutions of (monomeric)

$C_6H_5{}^{13}COOH$ appears at 1680 cm^{-1} as compared with the corresponding 1715 cm^{-1} value observed for the normal benzoic acid.[46]

The only case of C-labelled esters the infrared absorption of which has been studied so far, seems to be that of methyl 1-^{14}C-stearate which shows its $^{14}C{=}O$ stretching band at 1661 cm^{-1} besides the 1742 cm^{-1} CO band of the normal stearate.[45]

The infrared spectrum of 54 atom % ^{13}C-labelled sodium formate was investigated in aqueous solutions, in mulls and in KBr pellets.[47] Its mull showed the $H^{13}COONa$ A_1 fundamentals at: 2818 (ν_1), 1336 (ν_2) and 766 (ν_3) cm^{-1} while one of its two B_1 normal vibrations was observed at 1559 (ν_4) cm^{-1} and its only B_2 fundamental was found at 1054 cm^{-1}. The corresponding $H^{12}COONa$ frequencies appear at:[47] 2830 (C—H stretching), 1363 (C—O stretching), 774 (COO^- bending), 1602 (out-of-phase C—O stretching) and 1071 cm^{-1} (out-of-plane deformation).

All the 700–1700 cm^{-1} absorption peaks of a KBr pellet of a ^{14}C-enriched sodium formate sample have been observed to split as a result of the labelling which has also been found to give rise to a weak shoulder at *ca*. 3260 cm^{-1}.

5.14 Nitrogen-Containing ^{13}C-Labelled Compounds

3-^{13}C-labelled 2,4-dimethyl-3-pentylamine shows its $^{13}C{-}N$ stretching band at 1095 cm^{-1} as compared with the 1117 cm^{-1} frequency observed in the case of the normal amine.[36]

The structure of the ν_3 (C—H stretching) vibrational–rotational parallel band of gaseous $H^{13}CN$ was resolved and its centre was located at[48] 3293·51 cm^{-1}.

In the case of $D^{13}CN$ both the ν_1 (CN stretching) and the ν_3 fundamentals were studied and their analysis yielded the values of 1911·84 and 2590·07 cm^{-1}, respectively, for their centres.[48]

The 625–5000 cm^{-1} absorption of gaseous $Cl^{13}CN$ exhibits bands at 708 (C—Cl stretching), 765 (bending overtone) and 2157 cm^{-1} ($C{\equiv}N$ stretching) which correspond to the 714, 783 and 2213 cm^{-1} frequencies, respectively, of the normal cyanogen chloride molecule.[49]

Mono-^{13}C-cyanogen shows[50] its in-phase $C{\equiv}N$ stretching band at 2298 cm^{-1} and its nearly-symmetric (but for the isotopic difference) deformation fundamental at 495 cm^{-1}. These bands are infrared inactive in the case of $^{12}C_2N_2$.

A 10 atom % ^{13}C-enriched sample of silyl cyanide shows its $^{13}C{\equiv}N$ stretching band shifted from 2212 cm^{-1} to 2161 cm^{-1}, as expected for its cyanide, rather than isocyanide, structure.[51] The 2095 cm^{-1} band of trimethylsilyl isocyanide is shifted, however, only by -43 cm^{-1} as a result of labelling its isocyanide group with ^{13}C. This shift proves the existence of this compound in this structure since the calculated shift in this case amounts to

40 cm^{-1} as compared with the value of 51 cm^{-1} calculated for (and observed in the case of) the 2198 cm^{-1} C≡N stretching band of $(CH_3)_3Si^{13}CN$.[52]

3-^{13}C-2,4-dimethylpentan-3-one oxime displays its ^{13}C=N stretching band shifted from 1656 cm^{-1} to 1629 cm^{-1}. Its other ^{13}C-sensitive bands appear at 1015, 1050, 1151, 1214 and 1309 cm^{-1}, corresponding to the normal oxime bands at 1023, 1062, 1163, 1221 and 1314 cm^{-1}, respectively.[36]

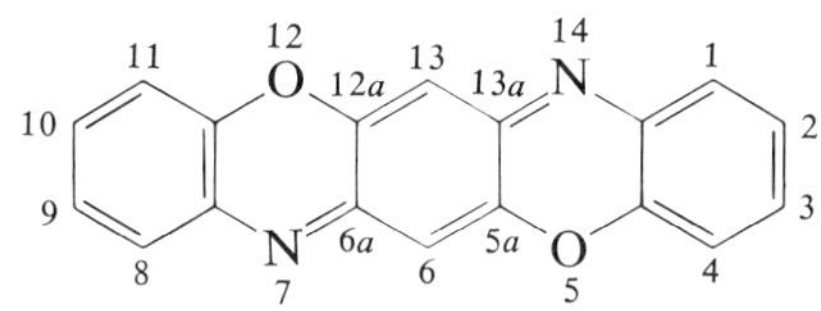

Fig. 5.4 The structure of triphendioxazine.

A potassium bromide pellet of 5*a*,12*a*-$^{13}C_2$-triphendioxazine (see Fig. 5.4) exhibits its C_{ar}–C_{ar} (ar = aromatic) stretching bands shifted from 1573 and 1378 cm^{-1} to 1549 and 1369 cm^{-1}, respectively, its C—O stretching fundamental moved from 1171 cm^{-1} to 1158 cm^{-1} and another ^{13}C-sensitive absorption lowered from 905 cm^{-1} to 900 cm^{-1}. In the case of its 6*a*,13*a*-$^{13}C_2$-isomer the C_{ar}–C_{ar} stretching bands observed normally at 1527 and 1378 cm^{-1} give rise to absorption peaks at 1513, and 1362 and 1367 cm^{-1}, respectively.[53]

5.15 Various ^{13}C-Labelled Compounds

A KBr pellet of *p*-$CH_3C_6H_4SO_2N(^{13}CH_3)_2$ shows its ^{13}C-sensitive ($\Delta\nu > 4$ cm^{-1}) bands at[54] 952 (*vs*; $\Delta\nu = 5$; S—$N(CH_3)_2$ stretching), 1183 (*vs*; $\Delta\nu = 5$; N—CH_3 rocking) and 2865 cm^{-1} (*w*; $\Delta\nu = 5$ cm^{-1}; ^{13}C—H stretching).

The infrared spectra of copper (II) 1-^{13}C-, 3-^{13}C-, 1,5-$^{13}C_2$-, 2,4-$^{13}C_2$- and 1,3,5-$^{13}C_3$-acetylacetonate, as well as those of beryllium and palladium 2,4-$^{13}C_2$- and 1,3,5-$^{13}C_3$-acetylacetonates, lithium, sodium and magnesium 2,4-$^{13}C_2$-acetylacetonates and potassium 1,3,5-$^{13}C_3$- acetylacetonate, were all investigated in connection with the assignment of the acetylacetonates infrared absorption bands.[55, 56] The main isotopic ^{13}C-shifts observed for the copper 1-^{13}C-complex were found for the 682 (3 cm^{-1}; ring deformation) and 1017 cm^{-1} (3 cm^{-1}; CH_3 rocking) bands while in the case of the 3-^{13}C-isotopomer these shifts were seen in the peaks at 648 (4 cm^{-1}; C—C—C deformation), 920 (17 cm^{-1}), 1394 (3 cm^{-1}), 1505 (22 cm^{-1}; C_{ar}—C_{ar} stretching) and 1575 cm^{-1} (3 cm^{-1}; C=O stretching). The main ^{13}C-shifts observed for the 2,4-$^{13}C_2$-acetylacetonate are for its bands at 642 (10 cm^{-1}; C—C—C deformation), 1184 (5 cm^{-1}), 1242 (33; symmetric C—C—C stretching), 1378 (19 cm^{-1}), 1520 (7 cm^{-1}; C_{ar}—C_{ar} stretching) and 1548 cm^{-1} (30 cm^{-1}; C=O stretching).

The infrared spectrum of the 2,4-$^{13}C_2$-complex also showed absorption bands due to the presence of a certain amount of the mono-labelled acetylacetonate in the studied sample.[55, 56]

The chelated copper complex of ethyl-1,3-$^{13}C_2$-acetoacetate (in CCl_4) shows its C—C stretching band at 1251 cm^{-1} near the 1265 cm^{-1} band of the mono-^{13}C-complex and the 1284 cm^{-1} band of the respective normal compound.[56]

References

1. Mills, I. M. and Thompson, H. W. (1953). *Trans. Faraday Soc.* **49**, 224.
2. Plyler, E. K., Benedict, W. S. and Silverman, Sh. (1952). *J. Chem. Phys.* **20**, 175.
3. Migeotte, M., Neven, L. and Swensson, J. (1957). *Mem. soc. roy. sci. Liège No. 2.*
4. Alben, J. O. and Caughey, W. S. (1968). *Biochemistry, N.Y.* **7**, 175; (1969). **8**, 59.
5. Nielsen, A. H. (1938). *Phys. Rev.* **53**, 983.
6. Pariseau, M. A., Suzuki, I. and Overend, J. (1965). *J. Chem. Phys.* **42**, 2335.
7. Courtoy, C. P. and Triaille, E. A. (1962). *J. Mol. Spectroscopy* **9**, 339.
8. Gordon, H. R. (1966). *Diss. Abs.* **26**, 6781.
9. Andrade, M. H., Silva, E. and Amat, G. (1962). *J. Mol. Spectroscopy* **9**, 354.
10. Andrade, M. H. and Silva, E. (1967). *Ann. Physique* **2**, 217.
11. Nielsen, A. H. and Lagemann, R. T. (1954). *J. Chem. Phys.* **22**, 36.
12. Maki, A. G., Plyler, E. K. and Tidwell, E. D. (1962). *J. Res. Nat. Bur. Stand.* **A66**, 163.
13. Verderame, F. D. and Nixon, E. R. (1966). *J. Chem. Phys.* **44**, 43.
14. Triaille, E. A. (1965). *Ann. Soc. sci. Bruxelles*, Ser. I, **79**, 193.
15. Jones, L. H., McDowell, R. S. and Goldblatt, M. (1968). *J. Chem. Phys.* **48**, 2663.
16. Bouquet, G. and Bigorgne, M. (1965). *Compt. rend.* **261**, 2865.
17. Haas, H. and Sheline, R. K. (1967). *J. Chem. Phys.* **47**, 2996.
18. Bor, G. (1967). *J. organometall. Chem.* **10**, 343.
19. Harrill, R. W. and Kaesz, H. D. (1968). *J. Amer. Chem. Soc.* **90**, 1449.
20. Fellmann, W., Harrill, R. W. and Kaesz, H. D. (1966). *Inorg. nucl. Chem. Lett.* **2**, 69.
21. Braterman, P. S., Harrill, R. W. and Kaesz, H. D. (1967). *J. Amer. Chem. Soc.* **89**, 2851.
22. Kaesz, H. D., Bau, R., Hendrickson, D. and Smith, M. (1967). *J. Amer. Chem. Soc.* **89**, 2844.
23. Poletti, A., Cataliotti, R. and Foffani, A. (1968). *Inorg. Chim. Acta* **2**, 157.
24. Noack, K. (1968). *J. organometall. Chem.* **12**, 181.
25. Lewis, J., Manning, A. R. and Miller, J. R. (1966). *J. Chem. Soc.* (*A*), 845.
26. Braterman, P. S. (1968). *J. Chem. Soc.* (*A*) 2907.
27. Maki, A. and Decius, J. C. (1959). *J. Chem. Phys.* **31**, 772.
28. Jones, L. H. (1956). *J. Chem. Phys.* **25**, 1069.
29. Sterzel, W. and Chorinsky, E. (1968). *Spectrochim. Acta* **24A**, 353.
30. McDowell, R. S. (1966). *J. Mol. Spectroscopy* **21**, 280.
31. Chalmers, A. A. and McKean, D. C. (1966). *Spectrochim. Acta* **22**, 251.
32. Ruoff, A. (1967). *Spectrochim. Acta* **23A**, 2421.
33. Duncan, J. L. and Allan, A. (1968). *J. Mol. Spectroscopy* **25**, 224.
34. Duncan, J. L. and Allan, A. (1969). *Spectrochim. Acta* **25A**, 901.

35. Lafferty, W. J. and Plyler, E. K. (1963). *J. Res. Nat. Bur. Stand.* **A67**, 225.
36. Karabatsos, G. J. (1960). *J. Org. Chem.* **25**, 1409.
37. George, W. O. (1962). *Nature* **194**, 672.
38. Lafferty, W. J. and Thibault, R. J. (1964). *J. Mol. Spectroscopy* **14**, 79.
39. Brodersen, S., Christoffersen, J., Bak, B. and Nielsen, J. T. (1965). *Spectrochim. Acta* **21**, 2077.
40. Miller, F. A. (1956). *J. Chem. Phys.* **24**, 996.
41. Weston, R. E. Jr., Tsukamoto, A. and Lichtin, N. N. (1966). *Spectrochim. Acta* **22**, 433.
42. Dimroth, K., Berndt, A. and Volland, R. (1966). *Chem. Ber.* **99**, 3040.
43. Karabatsos, G. J. (1960). *J. Org. Chem.* **25**, 315.
44. Doepp, D. and Musso, H. (1966). *Spectrochim. Acta* **22**, 1813.
45. Baxter, B. H. and Horsler, A. F. C. (1964). *Nature* **204**, 675.
46. Pinchas, S. (1965). *J. Phys. Chem.* **69**, 2256.
47. Hammacker, R. M. and Walters, J. P. (1964). *Spectrochim. Acta* **20**, 1311.
48. Maki, A. G., Plyler, E. K. and Thibault, R. (1964). *J. Opt. Soc. Amer.* **54**, 869.
49. Richardson, W. S. and Wilson, E. B. Jr. (1950). *J. Chem. Phys.* **18**, 155.
50. Schultz, J. W. and Eggers, D. F. Jr. (1958). *J. Mol. Spectroscopy* **2**, 113.
51. Linton, H. R. and Nixon, E. R. (1958). *Spectrochim. Acta* **10**, 299.
52. Booth, M. R. and Frankiss, S. G. (1968). *Chem. Commun.* **21**, 1347.
53. Musso, H., Doepp, D. and Kuhls, J. (1965). *Chem. Ber.* **98**, 3937.
54. Rademacher, P., Wiegraebe, W. and Luettke, W. (1967). *Chem. Ber.* **100**, 1213.
55. Junge, H. and Musso, H. (1968). *Spectrochim. Acta* **24A**, 1219.
56. Musso, H. and Junge, H. (1968). *Chem. Ber.* **101**, 801.

CHAPTER 6

Nitrogen Labelled Compounds

6.1 INTRODUCTION

The only nitrogen isotope which can practically be used for labelling purposes is the stable isotope nitrogen-15, the natural abundance of which is 0·38 atom %, since the half-life of the other existing nitrogen isotopes, nitrogen-13, nitrogen-16 and nitrogen-17, is less than 11 minutes.

The most important ^{15}N-induced isotopic shifts are observed in the cases of the various ^{15}N—H, C=^{15}N, ^{15}N=O and C≡^{15}N stretching bands, although the shifts found for many ^{15}N—H bending and C—^{15}N, ^{15}N—O, ^{15}N=N, P=^{15}N and S—^{15}N stretching fundamentals, as well as for other bands, are often also very significant.

Applying eqn. 2.17 to the stretching frequencies of these groups gives the stretching frequencies expected for the corresponding ^{15}N-labelled diatomic oscillators, (which are a good approximation for the respective ^{15}N-labelled group frequencies—see Section 3.2) and these are reported in Table 6.1.

TABLE 6.1 The Expected Stretching Frequencies (cm^{-1}) of some ^{15}N-labelled Groups

Group	Normal frequency[a]	^{15}N-labelled group frequency	Shift
N—H	3400	3392·5	7·5
C=N	1670	1644	26
N=O	1550	1522	28
C≡N	2250	2215	35

[a] An approximate mean value is always used.

6.2 LABELLED MOLECULAR NITROGEN

Although mono-labelled molecular nitrogen can, as a result of its isotopic asymmetry, be expected to show a weakly infrared active stretching fundamental such a band does not seem to have been reported in the literature so far. In case, however, of the complexes it forms with pentammineruthenium(II) or osmium(II) salts of the general structure $[M(NH_3)_5N_2]X_2$ (X = Br, I, BF_4) the absorption bands due to its $^{14}N{\equiv}^{15}N$ stretching vibrations have recently been studied.[1] For the osmium salts they appear at 2050, 2100 and 2158 cm^{-1} (the last band being the strongest) as compared with the 2035 and 2085 (stronger) cm^{-1} values observed for the respective two bands of the corresponding $^{15}N_2$ complexes.

6.3 ^{15}N-Nitric Oxide

Nitric oxide possesses an odd number of electrons hence an electronic angular momentum due to the spin of its odd electron. Its ground state is thus split into the $^2\Pi_{1/2}$ and $^2\Pi_{3/2}$ electronic sub-levels as a result of a coupling between this momentum and its orbital angular momentum. When this molecule is excited vibrationally, it usually undergoes either a $^2\Pi_{1/2} \rightarrow {}^2\Pi_{1/2}$ fundamental vibrational transition or a $^2\Pi_{3/2} \rightarrow {}^2\Pi_{3/2}$ transition, but in a number of cases it is excited from $^2\Pi_{1/2}$ to an excited $^2\Pi_{3/2}$ level or from $^2\Pi_{3/2}$ to an excited $^2\Pi_{1/2}$ level. Its infrared absorption spectrum therefore shows two fundamental sub-bands due to the $^2\Pi_{1/2} \rightarrow {}^2\Pi_{1/2}$ and $^2\Pi_{3/2} \rightarrow {}^2\Pi_{3/2}$ transitions besides satellite bands, separated by about 120 cm^{-1}, due to the $0 \rightarrow 1\,{}^2\Pi_{1/2} \rightarrow {}^2\Pi_{3/2}$ and $^2\Pi_{3/2} \rightarrow {}^2\Pi_{1/2}$ excitations. These bands in the case of ^{14}NO and ^{15}NO were resolved under a resolution of 0·07 cm^{-1} and analysed.[2] The main ^{14}NO fundamentals were centred at 1876 cm^{-1} (with a difference of about 0·3 cm^{-1} between the two sub-bands) while the corresponding ^{15}NO bands were centred at 1843 cm^{-1}. The ω_e frequencies of ^{15}NO were calculated to be at 1870 cm^{-1} and its B_e and D_e rotational constants were determined as $1{\cdot}64449 \pm 0{\cdot}00008$ and $(5{\cdot}1 \pm 0{\cdot}1) \times 10^{-6}$ cm^{-1}, respectively.[2, 3]

The $K[Ir(^{15}NO)Br_5]$ complex of ^{15}NO and potassium iridium bromide exhibits its ^{15}NO stretching band at 1899 (*s*) cm^{-1} and its $Ir{-}^{15}NO$ vibrations at 525 (*w*), 535 (*w*) and 552 cm^{-1} (*vw*) as compared with the corresponding ^{14}NO complex values of 1953, 536, 551 and 562 cm^{-1}, respectively.[4]

Various ^{15}NO complexes of the general structure $M_2[Ru(NO)X_5]$ (M = K, Cs; X = Cl, Br, I) have also been studied.[4, 5, 6] In the case of M = K, X = Br, the ^{15}NO stretching bands are shifted from 1880 and 1888 cm^{-1} to 1843 and 1850 cm^{-1}, while the M—NO stretching fundamental moves from 606 to 605 cm^{-1}. These results agree well with the 40 and 5 cm^{-1} isotopic shifts calculated for the stretching frequencies of a triatomic $Ru{-}^{15}N{-}O$ oscillator but are at variance with the 27 and 10 cm^{-1} ^{15}N-shifts calculated for a Ru—O—N model, thus proving the Ru—NO structure, rather than the Ru—O—N form, of this complex.[5] Its $Ru{-}^{15}N{-}O$ bending vibration was found to give rise to a band at 557 cm^{-1}, as compared with the respective $Ru{-}^{14}N{-}O$ value of 573 cm^{-1}.

The $K_3[Cr(^{15}NO)(CN)_5]\ H_2O$ complex[5] shows its ^{15}N-sensitive bands at 600 (*w*; Cr—N—O bending), 617 (*s*; Cr—N stretching), 1590 (*sh*) and 1610 cm^{-1} (*vs*; N=O stretching), corresponding to the normal complex bands at 610, 620, 1635 and 1643 cm^{-1}, respectively. The 200–3300 cm^{-1} absorption spectrum of the ^{15}N-nitrosyl $[Cr(NO)(NH_3)_5]Cl_2$ complex[4,5] shows its ^{15}NO stretching band at 1642 (*vs*) cm^{-1} and its Cr—N stretching fundamental at 570 (*m*) cm^{-1} while the Cr—N—O band appears at 527 (*m*) cm^{-1}. When Cl^- is exchanged in this compound for ClO_4^- the $^{15}N{=}O$ fundamental is found[5] to increase by 43 cm^{-1}.

The fact that the NO band of black $[Co(^{15}NO)(NH_3)_5]Cl_2$ exhibits a ^{15}N-shift of 25 cm^{-1} only (from 1614 to 1589 cm^{-1}) while its $Co(^{15}NO)$ bending vibration moves by 564–558 = 6 cm^{-1} only, shows that in this case the compound has a Co—ON structure, which can be calculated to bring about ^{15}N-shifts equal to 23 and 4 cm^{-1}, respectively, as compared with the values of 35 and 15 cm^{-1} estimated for the case of a Co—NO structure.[5] The ^{15}NO frequency of the corresponding deuterated complex appears also at 1589 cm^{-1} but the respective Co—O—N bending vibration in this case is displayed at[5] 561 cm^{-1}.

6.4 Labelled Nitrous Oxide

The infrared absorption of gas samples which contained 99 and 97 atom % of $^{15}N_2O$ and $^{15}N^{14}NO$, respectively, was measured under high resolution and analysed to yield their main vibrational–rotational constants.[7,8] The more important results of this study are given in Table 6.2.

Table 6.2 The Main Infrared Absorption Bands (cm^{-1}) of ^{15}N-labelled Nitrous Oxides[7,8]

Band	Description	$^{14}N_2O$	$^{15}N_2O$	$^{15}N^{14}NO$
ν_1	$N^+{-}O^-$ stretching	1284·907	1265·332	1269·894
ν_2^1	Bending	588·767	571·887	585·32
$2\nu_2^2$	Overtone	1177·75	1143·95	1170·85
ν_3	N≡N stretching	2223·756	2154·73	2201·605
$2\nu_3$	Overtone	4417·38	4281·34	4373·61
$\nu_1 + \nu_3$	Combination	3480·856	3394·185	3443·659
$\nu_2^1 + \nu_3$	Combination	2798·288	2713·143	2772·71

The values of the B_0 and D_0 rotational constants of $^{15}N_2O$ and $^{15}N^{14}NO$ have been found to be 0·404859, 0·166 and 0·404856, 0·165 cm^{-1}, respectively.[8]

The 420–3600 cm^{-1} absorption bands of the atmospheric $^{15}N^{14}NO$ have also been recorded.[9]

6.5 LABELLED NITROGEN DIOXIDE

The main infrared absorption bands of gaseous $^{15}NO_2$ appear at 740·15 (ν_2), 1306 (ν_1), 1580·32 (ν_3) and 2858·707 cm^{-1} ($\nu_1 + \nu_3$), corresponding to the $^{14}NO_2$ bands at 749·8, 1318, 1617·75 and 2907 cm^{-1}, respectively.[10, 11] The equilibrium-point fundamental frequencies of $^{15}NO_2$ have the values of 1342·5 (ω_1), 747·1 (ω_2) and 1628·0 (ω_3) cm^{-1}.

An analysis of the fine structure of the $^{15}NO_2$ vibration–rotation bands in the 2200–3800 cm^{-1} region yielded the values of its A_0, B_0, C_0 constants (= 7·63062, 0·433717, 0·409492 cm^{-1}, respectively) as well as many other of its rotational constants.[11]

The shifts brought about in the harmonic ω_1 and ω_2 frequencies of the NO_2 molecule as a result of labelling it at its bridge position were used to define a more self-consistent force field[12] for this molecule than the field calculated on the basis of other spectral data.[10, 13]

The anharmonicity constants of $^{15}NO_2$ have also been estimated satisfactorily.[10, 14]

The $K_2[Pt(^{15}NO_2)Cl_3]$ complex shows its ^{15}N-sensitive bands at 297 (*sh*; in-plane ···NO_2 rock), 346 (*m*; Pt—NO_2 stretching), 597 (*m*; out-of-plane ···NO_2 wag), 837 (*s*; NO_2 bending), 1301 (*vs*; in-phase NO_2 stretching) and 1370 cm^{-1} (*vs*; out-of-phase NO_2 stretching), as compared with the corresponding normal complex bands at 304, 350, 614, 844, 1325 and 1401 cm^{-1}, respectively.[4] In the case of $Cs_3[Ir(^{15}NO_2)_2Br_4]$, $K_2[OsO_2(^{15}NO_2)_2(OH)_2]$ and $[Rh(^{15}NO_2)(NH_3)_5]Cl_2$ the $^{15}NO_2$ stretching bands appear at 1301, 1345; 1309, 1329, 1364; 1333 and 1388 cm^{-1}, respectively.[4]

The similar nitrito complex $[Rh(O^{15}NO)(NH_3)_5]Cl_2$ exhibits its $O^{15}NO$ bending fundamental at 825 (*sh*) cm^{-1}, its in-phase O—N—O stretching band at 1049 cm^{-1} and its out-of-phase O—N—O stretching absorption at 1382 (*sh*) and 1412 cm^{-1}, respectively.[4] The corresponding normal complex bands are shown at 830, 1063, 1445 and 1461 cm^{-1}.

6.6 ^{15}N-NITROSYL HALIDES

The infrared absorption spectra of gaseous ^{15}N-labelled nitrosyl fluoride[15] and nitrosyl chloride[16] were studied under a resolution of 0·3–0·5 cm^{-1} in connection with the determination of the force fields of these molecules. Their main absorption vibrations are reported in Table 6.3.

6.7 LABELLED AMMONIA

The pyramidal NH_3 molecule, which belongs to the $\boldsymbol{C_{3v}}$ point group, can be expected to show six main bands in its infrared absorption spectrum as a

TABLE 6.3 The Main Vibrational Frequencies (cm^{-1}) of $O^{15}NF$ and $O^{15}NCl$[a]

Vibration	Description	$O^{14}NF$	$O^{15}NF$	$O^{14}N^{35}Cl$	$O^{15}N^{35}Cl$
ν_1	N=O stretching	1843·5	1811·8	1799·7	1768·9
ν_2	Mainly bending	765·8	748·7	595·8	581·7
ν_3	Mainly N—F (Cl) stretching	519·9	517·4	331·9	329·9
ω_1	Harmonic NO stretching	1876·8	1843·9	1835·6	1803·6
ω_2	Harmonic bending	775·5	757·9	603·2	588·8
ω_3	Harmonic N—F (Cl) stretching	522·9	520·4	336·4	334·3

[a] The fundamental frequencies of $O^{15}NBr$ have very recently been found at 1768·3 (ν_1), 527·8 (ν_2) and 264 (ν_3) cm^{-1} (Laane, T., Jones, L. H., Ryan, R. R. and Asprey, L. B. (1969). *J. Mol. Spectroscopy* **30**, 485.

result of a splitting of its ν_1 and ν_2 fundamentals due to its capability to undergo an inversion, a fact which brings about a (resonance) splitting of its vibrational energy levels. In case of the $^{14}NH_3$ species, however, only four such bands were observed: at 931·0 and 966·6 cm^{-1} (the two ν_2 symmetrical bending branches), at 1627·5 cm^{-1} (asymmetrical ν_4 deformation) and at 3337 cm^{-1} (unresolved ν_1 symmetrical stretching doublet). The corresponding bands in the case of the $^{15}NH_3$ isotopomer were found at 926·3, 961·0, 1625·0 and 3335 cm^{-1}, respectively, in very good agreement with the ^{15}N-shifts of 4·7, 2·7 and 3·4 cm^{-1}, calculated theoretically for the corresponding ν_2, ν_4 and ν_1 vibrations of $^{15}NH_3$.[17]

A film of $^{15}NH_4Cl$ exhibits its $^{15}NH_4^+$ ν_3 (asymmetric N—H stretching) and ν_4 (degenerate N—H bending) fundamentals at 3137 and 1399 cm^{-1}, respectively, as compared with the parallel 3146 and 1404 cm^{-1} frequencies observed for the normal ammonium chloride.[17]

The calculated isotopic shifts in this case are 10 and 6 cm^{-1}, respectively.

The $^{15}NH_4Cl$ $\nu_2 + \nu_4$ combination band was observed at 3045 cm^{-1}, being shifted by 8 cm^{-1} from its position in the normal salt absorption spectrum.

6.8 ^{15}N-LABELLED HYDRAZOIC ACIDS

The infrared absorption of $H^{15}N^{14}N_2$ and $H^{14}N_2{}^{15}N$ was studied both in the gaseous phase and in a matrix of solid nitrogen.[18] These planar molecules belong to the C_s point group and exhibit 6 fundamental bands (the N_3 group being linear). In the case of gaseous $H^{15}N^{14}N_2$ the ^{15}N-labelling was observed to shift the ν_1 (N—H stretching) fundamental by 6·2 cm^{-1}, the ν_2 (N≡N stretching) band from 2139·8 to 2133·7 cm^{-1} and the ν_4 (N—H bending) absorption peak by *ca.* 30 cm^{-1}. For gaseous $H^{14}N_2{}^{15}N$ the ^{15}N-shifts observed were equal to 23·3, 8, 4·7 and 3·3 cm^{-1} for its ν_2, ν_3, ν_4 and ν_5 frequencies, respectively.[18]

The ν_3 fundamental of gaseous $D^{15}N^{14}N_2$ appears at 1164·0 cm^{-1}, as compared with the 1184·7 cm^{-1} value for the ν_3 vibration of $D^{14}N_3$ and the 1172·1 cm^{-1} ν_3 frequency of $D^{14}N_2{}^{15}N$.

The ^{15}N-induced shifts in the location of the ν_4 band were 7·1 cm^{-1} in the case of $D^{15}N^{14}N_2$ and 0·3 cm^{-1} for $D^{14}N_2{}^{15}N$.

6.9 Labelled Nitrous Acids

Nitrous acid usually exists in an equilibrium between its *cis* and *trans* (hydrogen atom in respect to the nitrosyl oxygen atom) forms, each of which showing a discrete set of infrared absorption bands. Their frequency in the case of gaseous ^{15}N-labelled HONO and DONO[19] is given in Table 6.4.

Table 6.4 The $HO^{15}NO$ and $DO^{15}NO$[19] Fundamentals (cm^{-1})

	Band	Approximate description	$HO^{14}NO$	$HO^{15}NO$	$DO^{14}NO$	$DO^{15}NO$
trans	ν_1	O—H(D) stretching	3588	3588	2648	2646
	ν_2	N=O stretching	1699	1669	1692	1661
	ν_3	O—H(D) bending	1265	1261	1014	1008
	ν_4	N—OH(D) stretching	791	774	737	723
	ν_5	O—N—O bending	593	589	586	583
	ν_6	Out-of-plane deformation	540	538	416	413
cis	ν_1	O—H(D) stretching	3424	—	2525	—
	ν_2	N=O stretching	1640	1611	1625	1599
	ν_3	O—H(D) bending	1261	1258	1008	1002
	ν_4	N—OH(D) stretching	853	836	814	802
	ν_5	O—N—O bending	608	604	601	596
	ν_6	Out-of-plane deformation	638	636	504	502

The observed values reported in Table 6.4 agree reasonably well with those calculated for them.[19] From the isotopic frequencies of the out-of-plane deformation (torsion) fundamental and the temperature dependence of the relative intensity of pairs of *cis* and *trans* fundamentals the potential-energy minimum of the *cis* form, relative to that of the *trans* modification, and the height of the torsion barrier above the *trans* nitrous acid were both evaluated.[19]

6.10 Other ^{15}N-Labelled Tetra-Atomic Compounds

^{15}N-labelled nitryl fluoride and nitryl chloride were studied both in the gas phase and as solids at -190°C. The six fundamental bands of gaseous $F^{15}NO_2$ were observed at 558·9 (*s*; ν_5), 567·1 (*s*; ν_3), 722·0 (*m*; ν_6), 808·5 (*vs*; ν_2), 1298·0 (*vs*; ν_1) and 1750·5 cm^{-1} (*vs*; ν_4), as compared with the corresponding $F^{14}NO_2$ values of 559·6, 567·8, 742·0, 822·4, 1309·6 and 1791·5 cm^{-1}, respectively.[20]

In the case of gaseous $Cl^{15}NO_2$[20] the ν_5 frequency was not observed directly but was rather estimated from an overtone, at 408 cm^{-1}. Its other fundamentals were found at 367·4 (*vs*; ν_3), 635·6 (*m*; ν_6), 782·0 (*vs*; ν_2), 1292·5 (*vs*; ν_1 in Fermi resonance with the $2\nu_6$ vibration) and 1646·8 cm^{-1} (*vs*; ν_4).

The force constants of these molecules have also been evaluated.[20, 21]

^{15}N-nitrogen trichloride solutions in carbon tetrachloride or *n*-hexane show their infrared absorption bands at 258 (*m*; ν_4), 349 (*w*; ν_2), 371 (*w*; ν_3–ν_4), *ca.* 510–530 (*w*; ν_1) and 627 cm^{-1} (*vs*; ν_3). The corresponding $^{14}NCl_3$ bands appear at 258, 349, 385, 520–540 and 643 cm^{-1}, respectively.[22]

6.11 Labelled Penta-atomic Molecules

The infrared absorption of ^{15}N-labelled nitric acids was investigated both in the monomeric gaseous phase and in the hydrogen bonded solid phase[23] (at −190°C). The fundamental bands of gaseous $H^{14}NO_3$, $H^{15}NO_3$, $D^{14}NO_3$ and $D^{15}NO_3$ are reported in Table 6.5.

TABLE 6.5 The Fundamental Bands (cm^{-1}) of Isotopic Nitric Acids[23, 24]

Band	Approximate description	$H^{14}NO_3$	$H^{15}NO_3$	$D^{14}NO_3$	$D^{15}NO_3$
ν_1	O—H (D) stretching	3550·0	3550·0 (*m*)	2621·5	2621·9
ν_2	Asymmetric NO stretching	1706; 1712	1669; 1675 (*vs*)	1683; 1691	1654; 1658
ν_3	N—O—H (D) bending	1330·7	1327·0 (*s*)	1013·6	1011·5
ν_4	Symmetric NO stretching	1324·9	1320·6 (*s*)	1308·4	1290·7 (*s*)
ν_5	NO_2 bending	878·6	870·8 (*s*)	888·0	876·3 (*s*)
ν_6	N—OH (D) stretching	646·6	646·6 (*w*)	641·0	640·9
ν_7	(D) HO—N=O bending	579·0	578·0 (*w*)	541·0	541·0
ν_8	Out-of-plane NO_2 bending	762·2	743·6 (*s*)	762·8	743·4
ν_9	Torsion	455·8	455·9 (*m*)	342·0	—

The values calculated for the various normal vibrations of these isotopic nitric acids (on basis of their force constants[23, 24]) agree well with the frequencies reported in Table 6.5.

The planar $^{15}N_2O_3$ molecules showed their fundamental bands in the solid phase at 308 (ν_6), 397 (ν_5), 771 (ν_4), 1277 and 1299 (both due to branches of ν_3), 1546 (ν_2) and 1830 (ν_1) cm^{-1}, as compared with the corresponding $^{14}N_2O_3$ bands at 313, 407, 783, 1297, 1589 and 1863 cm^{-1}, respectively.[25, 26]

6.12 ^{15}N-Labelled Nitrogen Tetroxide

The 340–5500 cm^{-1} absorption spectra of gaseous and liquid $^{15}N_2O_4$ have yielded the frequencies of all its 11 infrared active fundamentals which obey the product rule.[27] Gaseous $^{15}N_2O_4$ exhibits its main absorption bands at *ca.* 377 (m; ν_{10}), *ca.* 420 (s; ν_7), *ca.* 685 (m; $\nu_3 + \nu_7$), 739 (vs; ν_{12}), 1251 (s; ν_{11}), 2600 (m; $\nu_1 + \nu_{11}$), 2921 (m; $\nu_5 + \nu_{11}$) and 3058 cm^{-1} (m; $\nu_1 + \nu_9$) corresponding to the normal tetroxide bands at 381, 429, 695, 750, 1262, 1748, 2630, 2971 and 3111 cm^{-1}, respectively.[27]

6.13 ^{15}N-Mononitrosylcobalt Carbonyl

^{15}N-labelling $Co(NO)(CO)_3$ shifts its NO stretching band from 1822 to 1786 cm^{-1}, i.e. by 36 cm^{-1}, as compared with the ^{15}N-shift of only 33 cm^{-1} observed for the free nitric oxide itself (Section 6.3). Its Co—NO stretching fundamental shows an ^{15}N-isotopic shift from 594 to 591 cm^{-1} while the bending of the Co—^{15}N=O group gives rise to an absorption at 555 cm^{-1} corresponding to the Co—^{14}N=O deformation band at[28] 565 cm^{-1}.

6.14 Labelled Cyanate and Fulminate Ions

The 600–5000 cm^{-1} absorption spectra of dilute solid solutions of potassium cyanate in various MX (M = Na, K; X = Cl, Br, I) single crystals at a low temperature showed bands at 1189, 1283 and 2152 cm^{-1}, due to its natural abundance $^{15}NCO^-$ content.[29] The first two bands are the result of a Fermi resonance between the ν_1 (C—O stretching) and $2\nu_2$ (bending) vibrations of this ion while the 2152 cm^{-1} peak belongs to its ν_3 (N≡C stretching) frequency. The corresponding normal vibrations appear at 1206, 1293 and 2170 cm^{-1}, respectively. In the case of the $^{15}N^{13}CO$ species[29] the ν_3 band was observed at 2095 cm^{-1}.

The infrared spectrum of a Nujol mull of mercury ^{15}N-fulminate exhibited its C≡^{15}N—O$^-$ fundamental bands at 477 (ν_2; bending), 1220 (ν_1; mainly N—O stretching) and 2165 cm^{-1} (ν_3; mainly C≡N stretching) as compared with the corresponding C≡^{14}N—O$^-$ vibrations at 480, 1220 and 2202 cm^{-1}, respectively.[30]

6.15 Other ^{15}N-Labelled Triatomic Ions

Of the three fundamental frequencies of the azide ion, its symmetrical stretching ν_1 vibration is only infrared active in the case of the end-labelled species $^{14}N_2{}^{15}N^-$, while its ν_2 bending and ν_3 asymmetrical stretching fundamentals are infrared active for all of its isotopomers.[31] The frequency of their infrared absorption bands in potassium bromide pellets is given in Table 6.6.

When a single crystal of potassium azide is irradiated at 77°K with UV light, a single sharp band appears in its absorption spectrum at 1638 cm^{-1}.[32] This strong band is due to the formation of cyclic nitrogen (N_3) and, in the

TABLE 6.6 The Infrared Bands (cm^{-1}) of the Isotopic Azide Ions[31]

Band	$^{14}N_3{}^-$	$^{15}N_3{}^-$	$^{14}N_2{}^{15}N^-$	$^{14}N^{15}N^{14}N^-$
ν_1	—	—	1352·0[a]	—
ν_2	640·1	618·3	636·5	625·7
ν_3	2037·1	1968·3	2025·0	1991·0

[a] In Fermi resonance with the $2\nu_2$ overtone band at 1244·5 cm^{-1}.

case of $K^{15}N_3$, it is observed at 1583 cm^{-1}, while for the two mono-labelled azide modifications it is split into three bands, at 1611 (*w*), 1623 (*s*) and 1635 (*w*) cm^{-1}.

Potassium thiocyanate shows an ^{15}N-isotopic shift of *ca.* 25 cm^{-1} (from *ca.* 2051 to 2026 cm^{-1}) for the ν_1 fundamental of its NCS^- ion.[33]

A potassium bromide pellet of sodium ^{15}N-labelled nitrite was observed to show its $O^{15}NO^-$ absorption bands at 824 (*m*; ν_2; bending), 1243 (*vs*; ν_3; out-of-phase stretching), 1306 (*w*; ν_1; in-phase stretching) and 2485 cm^{-1} (*w*; $2\nu_3$; overtone), corresponding to the analogous $O^{14}NO^-$ bands at 829, 1270, 1325 and 2550 cm^{-1}, respectively.[34]

6.16 ^{15}N-LABELLED TETRA-ATOMIC IONS

A ^{15}N-labelled sodium hyponitrite sample, which contained about 55% of the ^{15}N–^{15}N species and 42 atom % of the ^{14}N–^{15}N modification, showed its $N_2O_2{}^{2-}$ absorption bands[35] (in a KBr pellet) at 362 (*m*; ν_6 of $^{15}N_2O_2{}^{2-}$), 366 (*m*; ν_6 of $^{14}N^{15}NO_2{}^{2-}$), 482 (*m*; ν_4 of $^{15}N_2O_2{}^{2-}$), 487 (*m*; ν_4 of $^{14}N^{15}NO_2{}^{2-}$), 696 (*vvw*; ν_3 of $^{14}N^{15}NO_2{}^{2-}$), 1015 (*s*; ν_5 of $^{15}N_2O_2{}^{2-}$), 1022 (*s*; ν_5 of $^{14}N^{15}NO_2{}^{2-}$), 1113 (*vw*; ν_2 of $^{14}N^{15}NO_2{}^{2-}$), 1398 (*w*; ν_1 of $^{14}N^{15}NO_2{}^{2-}$), 1822, 1845 and 2366 cm^{-1} (all three: *vw*; combinations). The observed $^{14}N_2O_2{}^{2-}$ fundamentals appear at 371 (ν_6), 492 (ν_4) and 1031 (ν_5) cm^{-1}.

The infrared absorption of the unstable $^{15}N_2$-hyponitrous acid, obtained from this salt, exhibits its bands at 436 (*w*; out-of-plane NO bending), 980 and 991 (both: *s*; NO stretching), 1385 (*vw*; HON bending), *ca.* 2960 and *ca.* 3040 cm^{-1} (both: *m*; OH stretching), as compared with the corresponding ^{14}N-^{15}N-species fundamentals at 439, 990, 999, 1385, 2960 and 3040 cm^{-1}, respectively.[35]

A 50% $Na^{14}NO_3$–50% $Na^{15}NO_3$ crystal was found to show its $^{14}NO_3^-$ ν_2 bending fundamental at 836·1 cm^{-1} and its $^{15}NO_3^-$ ν_2 band at 814·4 cm^{-1}, with similar absorption intensities.[36] The observed Teller–Redlich product (Section 3.4) in this case is thus equal to 0·9740, being in very good agreement with the respective calculated value of 0·9739. In the case of potassium nitrate, which exists in a number of polymorphic phases, however, a similar situation exists only above 130°C, when the nitrate ions are already randomly oriented, but not at room temperature when they are coupled to each other as a result of dipole–dipole interaction between the nearest NO_3^- neighbours[37] which are here only 3·2 Å apart. This coupling brings about a splitting of both the $^{14}NO_3^-$ and the $^{15}NO_3^-$ ν_2 fundamentals of a 50% $K^{14}NO_3$–50% $K^{15}NO_3$ crystal into bands at 807·3, 810·4 ($^{15}NO_3^-$), and 828, 830·6 and 836·5 cm^{-1} (all three: $^{14}NO_3^-$ ν_2 branches), the intensity of the $^{15}NO_3^-$ bands being stronger.[38]

The main ^{15}N-sensitive bands of labelled gaseous fluorine nitrate appear at 690 (*m*; ν_8), 792 (*s*; ν_4), 1290 (*vs*; ν_2), 1720 (*vs*; ν_1) and 2993 cm^{-1} (*m*; $\nu_1 + \nu_2$), as compared with the corresponding $FO^{14}NO_2$ bands at 708, 804, 1301, 1759 and 3042 cm^{-1}, respectively.[39] In the case of the ^{15}N-labelled chlorine nitrate the main characteristic absorption bands were observed at 773 (*s*; ν_4), 805 (*m–s*; ν_3), 820 (*m*; $\nu_5 + \nu_7$), 1280 (*vs*; ν_2), 1391 (*m*; $2\nu_8$) and 1694 cm^{-1} (*vs*; ν_1), corresponding to the normal compound absorption peaks at 780, 809, 835, 1292, 1424 and 1735 cm^{-1}, respectively.[39]

6.17 Other ^{15}N-Labelled Ions

The 400–1600 cm^{-1} absorption spectrum of Nujol mulls of $K_2{}^{15}NClO_3$ shows the normal vibrations of its $NClO_3^{2-}$ ions (C_{3v} point group) at[40] 597 (ν_3 and ν_6), 625 (ν_5), 829 (ν_2), 893 (ν_4) and 1267 (ν_1) cm^{-1}, the respective $^{14}NClO_3^{2-}$ bands appearing at 592, 622, 829, 893 and 1238 cm^{-1}.

The Os—N—Os asymmetric stretching band of the $[Os_2{}^{15}N(NH_3)_8Br_2]Br_3$ complex appears at 1062 cm^{-1} as compared with the 1098 cm^{-1} value observed for the respective ^{14}N-complex.[41] The fact that the frequency calculated for the asymmetric stretching of a linear Os—^{15}N—Os triatomic model is very near (1061 cm^{-1}) to the value observed for the labelled complex indicates that such complexes possess a linear Os—N—Os structure.[41]

6.18 ^{15}N-Labelled Amines

The infrared absorption of gaseous, *ca.* 90 atom % ^{15}N-labelled dimethylamine[42] exhibits its ^{15}N-characteristic bands below 1700 cm^{-1} at 380·3 (C—N—C bending; shifted by *ca.* 3 cm^{-1}), *ca.* 732 (N—H bending; $\Delta\nu = 3$ cm^{-1}), 925 (C—N stretching; $\Delta\nu = 3·5$ cm^{-1}), 1130, 1138 and 1147 cm^{-1}. Fig. 6.1 shows its absorption in the 1100–1200 cm^{-1} region.

^{15}N-labelling aniline brings about an isotopic shift of its C—N stretching band[43] from 1276 to 1266 cm^{-1}. This labelling also shifts its symmetric and antisymmetric N—H stretching fundamentals (in solution in CCl_4) from 3395 and 3481 cm^{-1} to 3390 and 3472 cm^{-1}, respectively.[44] Other characteristic $C_6H_5{}^{15}NH_2$ bands appear at 1112 and 1613 (mainly due to the NH_2

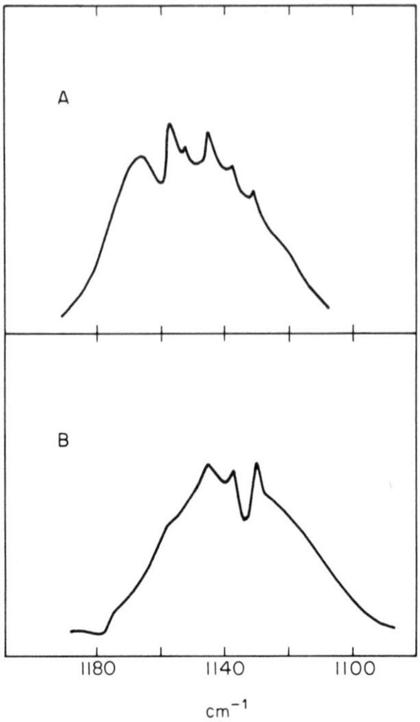

Fig. 6.1 A. $(CH_3)_2{}^{14}NH$, gas phase, grating resolution, 1100–1200 cm^{-1};
B. $(CH_3)_2{}^{15}NH$, gas phase, prism resolution, 1100–1200 cm^{-1}.
Cell length 10 cm, pressure *ca.* 25 mm.

bending vibration) cm^{-1}, as compared with the corresponding normal aniline absorption peaks at 1115 and 1619 cm^{-1}, respectively. The comparison of the ^{15}N-shifts observed for the N—H stretching frequencies with the values calculated for them on basis of various structural models indicates that the structure of the aniline NH_2 group is close to planar.[44]

The C—^{15}N stretching frequency of labelled trimethylsilyl-, trimethylgermyl-, and trimethylstannylaniline was observed at 1287, 1287 and 1284 cm^{-1}, i.e. shifted by 7, 8 and 8 cm^{-1}, respectively, from the corresponding C—^{14}N stretching vibration. The parallel Si—^{15}N, Ge—^{15}N and Sn—^{15}N fundamentals were found at 890 ($\Delta\nu = 9$ cm^{-1}), 845 ($\Delta\nu = 10$ cm^{-1}) and 843 cm^{-1} ($\Delta\nu = 8$ cm^{-1}), respectively.[45]

The 250–4000 cm^{-1} absorption spectra of gaseous and liquid ^{15}N-ethyleneimine were measured in connection with the assignment of the C_2H_5N (point group: C_s) fundamentals.[46] The frequency of the observed gaseous isotopic ethyleneimine normal vibrations is given in Table 6.7.

N-alkyltrimethylammonium halides of the general form $RN^+(CH_3)_3X^-$ (R = CH_3, n-$C_{14}H_{29}$, n-$C_{16}H_{33}$; X = Br, I) show three strong C—N stretching bands, at 913 ± 3, 944 ± 7 and 958 ± 7 cm^{-1}. ^{15}N-labelling shifts

TABLE 6.7 The Observed Fundamental Bands (cm^{-1}) of $C_2H_4{}^{14}NH$, $C_2H_4{}^{15}NH$ and $C_2H_4{}^{14}ND$ [46]

Species	Approximate description	$C_2H_4{}^{14}NH$	$C_2H_4{}^{15}NH$	$C_2H_4{}^{14}ND$
A′	N—H stretching	3346	3337	2482
	CH_2 stretching	3079[a]	3077[a]	3079[a]
	CH_2 stretching	3015	3012	3015
	CH_3 deformation	1483	1482	1487
	Ring breathing	1210	1204	1210
	N—H deformation	1096	1095	864
	CH_2 twisting	1090	1088	1090
	CH_2 wagging	998	997	997
	Ring deformation	856	845	856
	CH_2 rocking	772	769	724
A″	CH_2 stretching	3003	2998	3003
	CH_2 deformation	1462	1462	1462
	Ring deformation (antisymmetrical)	1268	1262	1223
	CH_2 twisting	1237	1236	1237
	CH_2 wagging	1131	1131	1082
	N—H deformation	904	903	728

[a] This band is also shared by the respective A″ mode of vibration.

these bands to 907 ± 5, 939 ± 6 cm^{-1} and 949 ± 8 cm^{-1}, respectively, as compared with the 10 cm^{-1} ^{15}N-shift calculated for such vibrations.[47]

6.19 LABELLED HYDROGEN CYANIDES

The natural abundance of $HC^{15}N$ in a gaseous HCN sample was observed to exhibit its ν_3 (C—H stretching) vibration–rotation band[48] centred at 3310·09 cm^{-1} (in vacuum), corresponding to the $HC^{14}N$ ν_3 frequency of 3312 cm^{-1}. The analysis of its fine structure yielded the values for the main rotational constants of $HC^{15}N$ as $B_0 = 1{\cdot}4353$ cm^{-1} and $D_0 = 2{\cdot}5 \times 10^{-6}$ cm^{-1}.

In the case of $DC^{15}N$ the centre of the ν_3 vibration–rotation band was found at 2621·19 cm^{-1} (compared with the $DC^{14}N$ ν_3 value of 2630·30 cm^{-1}) and the B_0 and D_0 constants were calculated to have the values of 1·1730 cm^{-1} and $1{\cdot}3 \times 10^{-6}$ cm^{-1}, respectively.[48]

6.20 ^{15}N-LABELLED COMPOUNDS CONTAINING MORE THAN ONE NITROGEN ATOM

A methylene chloride solution of 93 atom % 1,2-$^{15}N_2$—N^1, N^1-bis (β-chloroethyl)-N^2-*p*-nitrobenzalhydrazine shows its $C{=}^{15}N$ stretching band at 1552 cm^{-1} as compared with the respective 1567 cm^{-1} $C{=}^{14}N$ stretching frequency.[49] A potassium bromide pellet of a 50 atom % ^{15}N-labelled sample of this hydrazone also displayed 3 and 5 cm^{-1} ^{15}N-induced shifts for the 1110 and 1130 cm^{-1} $ClCH_2CH_2$—^{14}N stretching bands, respectively, of this compound.[50]

The ^{14}N=^{15}N stretching band of ^{15}N-labelled *cis*-azobenzene appears at 1486 cm^{-1}, as compared with the respective 1511 cm^{-1} (*m–s*) ^{14}N=^{14}N stretching fundamental. The various ^{15}N-labelled *p*-hydroxy-, *p*-amino- and *p*-N,N-dimethylaminoazobenzenes exhibit their labelled N=N stretching band at *ca.* 1400 cm^{-1} (*m–s*), shifted by 15–35 cm^{-1} relative to the respective ^{14}N=^{14}N frequency.[51] In the case of $C_6H_5{}^{14}N$=$^{15}NC_6H_4{}^{14}NH_2$, e.g. this band appears at 1402 cm^{-1} while the corresponding normal compound fundamental is shown at 1418 cm^{-1} and that of the $^{15}N_3$-isotopomer appears at 1383 cm^{-1}. The C_6H_5—N bands of the ^{14}N=^{15}N compound were observed at 521, 785 and 1248 cm^{-1}, as compared with the 526, 790 and 1256 cm^{-1} frequencies, respectively, found for the normal azobenzene and the 513, 716, 1260 cm^{-1} and 513, 706 and 1256 cm^{-1} values determined for *p*-$C_6H_4{}^{14}NH_2{}^{15}NH_2$ and *p*-$C_6H_4({}^{15}NH_2)_2$, correspondingly.[52]

The 500–4000 cm^{-1} absorption spectrum of gaseous $^{15}N_2$-cyanogen exhibits bands at 729·9 (*s*; $\nu_4 + \nu_5$), 2077·8 (*w*; $\nu_1 - \nu_5$), 2122·1 (*s*; ν_3), 2532·8 (*m*; $\nu_1 + \nu_5$) and 2621·3 cm^{-1} (*m*; $\nu_3 + \nu_4$), corresponding to the $C_2{}^{14}N_2$ bands at 732, 2092, 2157·6, 2563 and 2663 cm^{-1}, respectively.[53] An analysis of the fine structure of three vibration–rotation absorption bands of $C_2{}^{15}N_2$ yielded the value of *ca.* 498 cm^{-1} for its ν_4 fundamental as well as its B_0 and D_0 rotational constants.[54]

The ^{15}N≡^{15}N stretching band of $^{15}N_2$-benzenediazonium chloride (in a KBr pellet) was observed at 2227·5 cm^{-1}, i.e. shifted by 74·5 cm^{-1} from the respective ^{14}N≡^{14}N frequency. Its C_6H_5—^{15}N stretching fundamental was found at 1478 cm^{-1} and the absorption peak due to the in-plane bending of the C—^{15}N≡^{15}N moiety was discerned at 511 cm^{-1} while the corresponding normal compound bands appear at 1500 and 518 cm^{-1}, respectively.[55]

Gaseous ^{15}N-labelled diazomethane shows its nine fundamentals at 3077 (ν_1), 2060 (ν_2), 1414·5 (ν_3), 1167 (ν_4), 3192 (ν_5), 1104·1 (ν_6), 414·1 (ν_7), 553·4 (ν_8) and 404·3 (ν_9) cm^{-1}. Its infrared spectrum, however, displays only five bands which are shifted by more than 1% as a result of its labelling. These bands are due to its ν_2, ν_7 and ν_8 normal vibrations and its 958 ($\nu_8 + \nu_9$) and 3211 cm^{-1} ($\nu_2 + \nu_4$) combination frequencies, which are shifted by 42·4, 7·1, 10·6, 14 and 43 cm^{-1}, respectively.[56]

6.21 Labelled Amides and Lactams

^{15}N-labelled dimethylformamide exhibits its CH_3—N and OC—N stretching bands at 1233 and 1370 cm^{-1}, as compared with the corresponding 1257 and 1383 cm^{-1} $HCO^{14}N(CH_3)_2$ frequencies, respectively.[57]

Complexes of this amide with Lewis acids also showed such ^{15}N-shifts for their CH_3—N and OC—N stretching bands.[57]

The 400–10,000 cm^{-1} absorption of ^{15}N-benzamide was measured for its crystalline and melted phases and in a potassium bromide pellet.[58] The main ^{15}N-sensitive ($\Delta\nu > 3$ cm^{-1}) fundamentals observed in the latter case appeared at 1118 (NH_2 twisting), 1140 (bending), 1398 (mainly C—N stretching), 1621 (mainly NH_2 scissoring), 1656 (CO stretching), 3165 (symmetric hydrogen bonded NH_2 stretching) and 3356 cm^{-1} (antisymmetric NH_2 stretching), as compared with the corresponding $C_6H_5CO^{14}NH_2$ bands at 1124, 1144, 1404, 1626, 1661, 3175 and 3372 cm^{-1}, respectively.[58]

The solution infrared absorption spectra of over 95 atom % ^{15}N-labelled 2(lH)-, 2(lD)- and 2(l-methyl)-pyridones were reported.[59] The frequency of their main bands is given in Table 6.8.

TABLE 6.8 The Main Bands (cm^{-1}) of Various Isotopic 2-Pyridones[59]

$C_5H_5{}^{14}NO$	$C_5H_5{}^{15}NO$	$C_5H_4D^{14}NO$	$C_5H_4D^{15}NO$	$C_6H_7{}^{15}NO$
991[a]	984	980	973	873[b]
1097	1091			1149[b]
1251	1241	1310	1295	1302[b]
1477	1472			
1544	1542	1539		
1620	1617	1610		
1658	1653	1586, 1575 (*sh*)	1584, 1572	1583[b]
1682 (*sh*)	1675	1667		1657[b]

[a] Unless stated otherwise, in CCl_4.
[b] In $CHCl_3$.

In the case of the ^{15}N-labelled 4-pyridone the only significant isotopic shifts observed in the 800–2000 cm^{-1} spectrum were for the 988 (-13 cm^{-1}) and 1632 cm^{-1} (-3 cm^{-1}) bands of the normal lactam.[60] ^{15}N-labelling of the 1-deuterated-4-pyridone shifts its C—N stretching band[60] from 1352 cm^{-1} to 1338 cm^{-1}.

6.22 OTHER OXYGEN CONTAINING COMPOUNDS

^{15}N-labelled 3-(N-methylamino)-5,5-dimethyl-2-cyclohexen-1-one shows its N—H deformation band shifted from 1519 cm^{-1} to 1505 cm^{-1} while its C—N stretching fundamental appears at 1263 cm^{-1} as compared with the 1266 cm^{-1} value in the case of the respective ^{14}N-ketone.[61]

The strongly intramolecularly hydrogen bonded 3-(N-methylamino)-crotonophenone exhibits only small ^{15}N-induced isotopic shifts, this being the result of an appreciable coupling of its nitrogen atom centred vibrations with other modes of vibration, due to the powerful hydrogen bonds existing in it. Its largest ^{15}N-shift was observed for its 1281 cm^{-1} C—N stretching band which moves to 1277 cm^{-1} upon its labelling.[61]

^{15}N-3-(N-methylacetimidoyl)-2-naphthol displays its C=^{15}N stretching absorption peak at 1612 cm^{-1} and another characteristic band at 1059 cm^{-1}, corresponding to the normal compound bands at 1630 and 1065 cm^{-1}, respectively.[61]

The 400–5000 cm^{-1} absorption of the *syn* and *anti* isomers of $C_6H_5{}^{15}N{=}^{14}NOK$, $C_6H_5{}^{14}N{=}^{15}NOK$, *p*-$ClC_6H_4$—N=^{15}NOK and *p*-$CH_3OC_6H_4N{=}^{15}NOK$ as well as the *anti* modifications of the silver outer-^{15}N-benzenediazotate and the *p*-NO_2-$C_6H_4N{=}^{15}NONa$ diazotate was investigated in order to elucidate the structure of the *syn* and *anti* diazotates.[62] Table 6.9 reports the ^{15}N-sensitive absorption bands of some of these compounds in the solid phase.

TABLE 6.9 The Characteristic Absorption Bands (cm^{-1}) of ^{15}N-labelled Diazotates

Approximate description	*syn* diazotates			*anti* diazotates		
	$R^{14}N_2OK$	$R^{15}N^{14}NR_1$	$R^{14}N^{15}NR_1$	$R^{14}N_2OK$	$R^{15}N^{14}NR_1$	$R^{14}N^{15}NR_1$
			R = C_6H_5			
NNO bending	585	578	581	416		412
Out-of-plane C_6H_5—N deformation				512		508
	885	879	885			
C_6H_5—N stretching	893	883	893	912	899	910
	943	935	942			
N—O stretching	1139	1138	1124	1165	1165	1145
				1179	1179	1156
C_6H_5—N stretching				1222	1215	1218
C_6H_5 vibration				1306	1301	1304
N=N stretching	1344	1326	1330	1379	1355	1362
			R = *p*-Cl-C_6H_4			
NNO bending	503		499			
Out-of-plane deformation				548		544
N—O stretching	1125		1116	1181		1161
	1168		1179			
N=N stretching	1351		1335	1370		1348
C_6H_5 vibration				1414		1410

^{15}N-labelled *n*-hexyl nitrite which exists in an equilibrium between its *cis* and *trans* forms exhibits its ^{15}N-characteristic *cis* modification bands at 671 (in-phase C—O—N stretching), 678 (O—N—O bending), 1220·8 (out-of-phase C—O—N stretching) and 1583 cm^{-1} (N=O stretching). Its *trans* form bands appear at 774 (ONO bending), 918 (C—O—N stretching) and 1622 cm^{-1} (N=O stretching) while both species share a C—O—N bending absorption peak at 595 cm^{-1}. The corresponding normal *cis* and *trans n*-hexyl nitrite vibrations absorb at 675, 686, 1236, 1607, 784, 931, 1652 and 599 cm^{-1}, respectively.[55]

The main ^{15}N-sensitive bands of liquid ^{15}N-nitrobenzene are displayed at 694, 786 (both: out-of-plane ring deformations), 845 (NO_2 bending[63]), 1323 (symmetrical NO_2 stretching) and 1497 cm^{-1} (antisymmetrical NO_2 stretching), corresponding to the $C_6H_5{}^{14}NO_2$ bands at 704, 794, 852, 1348 and 1522 cm^{-1}, respectively.[62]

The normal nitrobenzene combination bands at about 2451, 2863 and 2935 cm^{-1} were also observed to shift, as a result of the ^{15}N-labelling, to 2427, 2812 and 2914 cm^{-1}, respectively.[64]

N-methyl-*p*-nitrophenylnitrosamine ^{15}N-labelled in the amino group exhibits its ^{15}N-characteristic absorption bands at 936 (N—N stretching), 1073 (N—CH_3 stretching), 1178 ($C_{aromatic}$—H bending) and 1189 cm^{-1} (C_6H_5—N stretching), corresponding to the normal compound bands at 942, 1079, 1182 and 1195 cm^{-1}, respectively.[62] When this compound is ^{15}N-labelled in its nitroso group the normal nitrosamine bands at 815, 942, 1401 and 1466 cm^{-1} are observed to shift to 808 (in-phase C—N—N stretching), 936 (out-of-phase C—N—N stretching), 1395 (ring vibration) and 1449 cm^{-1} (N=O stretching), respectively.[62]

The main ^{15}N-sensitive absorption bands of *p*-nitrophenyldiazo-methyl ether labelled at the inner position appear at 559 (C—N=N—O—C bending), 820 (C_6H_5—N stretching), 993 (N—O stretching), 1212 (C_6H_5—N stretching) and 1490 cm^{-1} (N=N stretching; *sh*), as compared with the corresponding normal diazo-ether absorption peaks at 564, 828, 995, 1219 and 1508 cm^{-1} and the parallel *p*-$NO_2C_6H_4{}^{14}N{=}^{15}NOCH_3$ fundamentals at 562, 825, 983, 1219 and 1489 cm^{-1}, respectively.[62]

6.23 Labelled Amino-acids

The 1850–3250 cm^{-1} absorption spectrum of ^{15}N-labelled glycine shows bands at 2520, 2600, 2700, 2812, 2900, 2970, 3007 and 3168 cm^{-1}, as compared with the corresponding normal glycine absorption peaks at 2520, 2610, 2700, 2824, 2900, 2970, 3007 and 3164 cm^{-1}, respectively.[65]

The main ^{15}N-induced shifts in the absorption of $^{15}ND_3^+COO^-$ appear for its 1984 (-12 cm^{-1}), 2129 (-9), 2190 (-12) and 2226 cm^{-1} (-13) bands.[65]

In the case of $[^{15}NH_3^+COOH]Cl^-$ the isotopic shifts are even smaller, the only appreciable difference in frequency between its bands in the N—H stretching region and the corresponding normal glycine hydrochloride absorption peaks being for its 2602 (-10) cm^{-1} vibration.[65]

The ^{15}N-shift for the symmetrical $^{15}NH_3^+$ bending fundamental of this hydrochloride was found to be[66] 4·5 cm^{-1}.

The 530–1700 cm^{-1} absorption spectrum of a potassium bromide pellet of ^{15}N-DL-alanine showed only four bands with an isotopic shift of more than 3 cm^{-1}, at 541 (C—C—N bending), 1145 (C—C—N out-of-phase stretching), 1234 (NH_3^+ rocking) and 1525 cm^{-1} (symmetrical NH_3^+ deformation), corresponding to the ^{14}N-alanine fundamentals at 545, 1149, 1238 and 1531 cm^{-1}, respectively.[67]

The α-$^{15}NH_3^+$ group of ^{15}N-DL-lysine and $^{15}N_2$-DL-lysine monohydrochlorides gives rise to two ^{15}N-sensitive absorption bands, at 979 (C—^{15}N stretching) and 1512–1511 cm^{-1} (symmetrical deformation) shifted by *ca.* 4 cm^{-1} each from the corresponding ^{14}N-hydrochloride band.[68] The ε-$^{15}NH_3^+$ group of the $^{15}N_2$-hydrochloride shows similar bands at 995 and 1525 cm^{-1}, respectively, as compared with the parallel ε-$^{14}NH_3^+$ group 1002 and 1529 cm^{-1} fundamentals.[68]

The symmetrical α-$^{15}NH_3^+$ bending fundamental of the ^{15}N-DL-lysine dihydrochloride appears at 1520 cm^{-1}, i.e. shifted by 7·5 cm^{-1} from the respective $^{14}NH_3^+$ frequency. In the case of its α,ε-$^{15}N_2$ isotopomer, ^{15}N-characteristic absorption bands were observed in the 650–1750 cm^{-1} region at[68] 778, 817, 929, 956, 993, 1006, 1023, 1043, 1133, 1170, 1307, 1495 (symmetrical ε-$^{15}NH_3^+$ bending) and 1520 cm^{-1} (α-$^{15}NH_3^+$ deformation), displaying isotopic shifts of 4–8 cm^{-1}.

The infrared absorption spectrum of 98 atom % ^{15}N-DL-glutamic acid shows its ^{15}N-sensitive bands at 534 (C—C—N bending), 943 (C—N stretching), 1122, 1148 (both NH_3^+ rocking) and 1510 cm^{-1} (symmetric NH_3^+ deformation), corresponding to the normal glutamic acid bands at 538, 946, 1125, 1151 and 1518 cm^{-1}, respectively.[69]

The symmetrical $^{15}NH_3^+$ bending fundamental of the ^{15}N-DL-glutamic acid hydrochloride was found at 1503 cm^{-1} while the analogous $^{15}ND_3$ band was observed at 1154 cm^{-1}, corresponding to the respective $^{14}ND_3$ frequency of 1168 cm^{-1}. The $^{15}ND_3^+$ rocking vibration of the $^{15}ND_3$-glutamic acid hydrochloride appears at 819 cm^{-1}, i.e. shifted by 5 cm^{-1} from the parallel $^{14}ND_3$ fundamental.[70]

6.24 Various Other ^{15}N-Labelled Compounds

While a normal barbituric acid solution in dimethylsulphoxide exhibits a shoulder at about 1707 cm^{-1}, a strong absorption band at 1716 cm^{-1} and a medium band at 1740 cm^{-1} (all being due to various C=O stretching modes of vibration) a similar 99 atom % ^{15}N-labelled barbituric acid solution shows its absorption peaks in this region at about[71] 1710 and 1730 cm^{-1}. Figs. 6.2 and 6.3 reproduce the absorption of the normal and ^{15}N-labelled barbituric acids, respectively, in the C=O stretching region.

The 400–4000 cm^{-1} spectrum of gaseous B—^{15}N-aminoborazine shows its ^{15}N-characteristic absorption bands at 719 (*m*; NH_2 wagging), 814 (*w*),

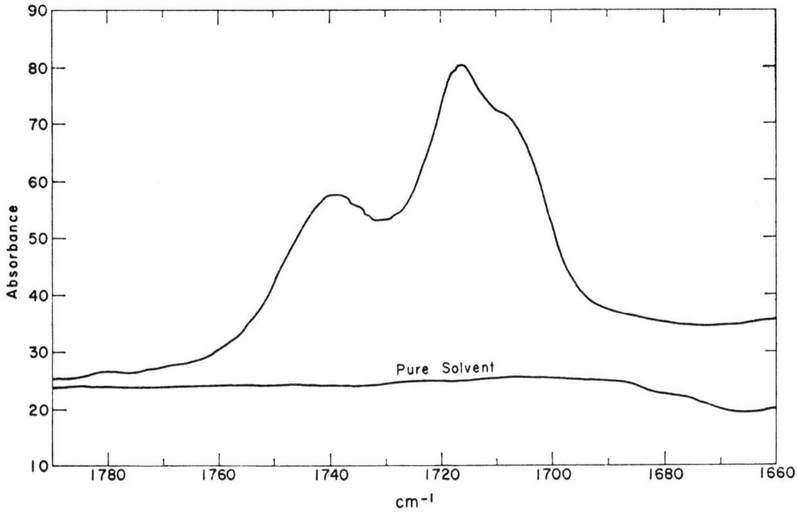

Fig. 6.2 The infrared absorption of normal barbituric acid in dimethylsulphoxide (0·00208 g/ml).

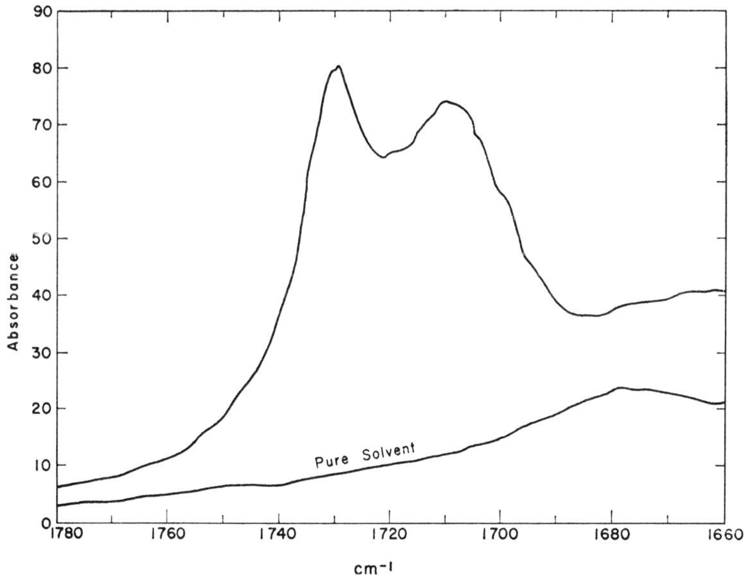

Fig. 6.3 The infrared absorption of ^{15}N-barbituric acid in dimethylsulphoxide (0·00357 g/ml; 0·105 mm cell).

1604 (*s*; NH_2 bending) and 3552 cm^{-1} (*w*; N—H stretching), corresponding to the normal compound bands at 725, 817, 1612 and 3561 cm^{-1}, respectively.[72]

The fact that ^{15}N-enriched SiD_3CN shows its $C\equiv{}^{15}N$ stretching band shifted from 2212 cm^{-1} to 2181 cm^{-1}, i.e. by 31 cm^{-1} only, was used to prove that silyl cyanides really exist in the cyanide, rather than in the isocyanide, form since this shift agrees only with the (31 cm^{-1}) value calculated for the cyanide structure but not with that (38 cm^{-1}) computed for the isocyanide isomer.[73] In the case of trimethylsilyl ^{15}N-isocyanide, in fact, the 2095 cm^{-1}—$N^+\equiv C^-$ stretching band of the normal compound was observed at 2060 cm^{-1}, in good agreement with the value of 38 cm^{-1} calculated for the ^{15}N-shift of this vibration.[74]

The $P={}^{15}N$ stretching band of a potassium bromide pellet of ^{15}N-triphenylphosphinephenylimide appears at 1324 cm^{-1} (*vs*), showing an isotopic shift of 20 cm^{-1}. Its 1033 cm^{-1} (*m*) C_6H_5—N stretching fundamental and its 559 cm^{-1} absorption peak are also ^{15}N-sensitive, the corresponding $(C_6H_5)_3P={}^{14}N-C_6H_5$ bands being shown at 1039 and 564 cm^{-1}, respectively.[75]

The 847 (mainly N—N stretching), 1053 (mainly P=N stretching) and 1549 cm^{-1} (approximately: C=N stretching) fundamentals of normal P-triphenyl-N-methylenephosphaketazine are shifted in the case of the $^{15}N_2$-isotopomer to 831, 1029 and 1530 cm^{-1}, respectively.[76] The infrared spectra of the $(C_6H_5)_3P={}^{14}N-{}^{15}N=CH_2$ and $(C_6H_5)_3P={}^{15}N-{}^{14}N=CH_2$ isotopomers display these bands at 841, 1049, 1530 and 837, 1028, 1549 cm^{-1}, respectively.[76] The absorption of various deuterated ^{15}N-labelled modifications of this ketazine was also reported.[76]

The P=N stretching band of various $RCO^{15}N=PCl_2X$ compounds (R = CF_3, CCl_3, CF_3CCl_2, C_2Cl_5, C_6H_5, *m*-Cl—C_6H_4, *m*-NO_2—C_6H_4; X = Cl, CH_3) appears in the 1272 (R = CCl_3; X = Cl) – 1386 cm^{-1} (R = CF_3, X = CH_3) region showing an ^{15}N-shift of 18–10 cm^{-1}, respectively.[77] $CF_3CCl_2{}^{15}N=PCl_3$ and $C_2Cl_5{}^{15}N=PCl_3$, also show $P={}^{15}N$ bands at 1400 and 1408 cm^{-1} (shifted by 50 and 37 cm^{-1}), respectively. Their $C-{}^{15}N$ stretching vibration varies between 1172 cm^{-1} (for R = CCl_3; X = Cl) and 1180 cm^{-1} (R = C_6H_5; X = Cl), exhibiting an isotopic shift[77] of 3–17 cm^{-1}.

The ^{15}N-sensitive fundamentals of ^{15}N-dimethyl-*p*-toluenesulphonamide were observed[78] at 945 (*m*; S—N stretching), 1183 (*w*; N—CH_3 in-phase stretching) and 1251 cm^{-1} (*w*; out-of-phase N—CH_3 stretching), corresponding to the parallel normal sulphonamide bands at 955, 1192 and 1264 cm^{-1}.

In the case of *p*-CH_3—C_6H_4—SO_2—$^{15}N=PCl_3$ the $S-{}^{15}N$ stretching absorption peak was found at 756 cm^{-1} and the P=N stretching fundamental appeared at 1183 cm^{-1}, as compared with the corresponding normal phosphineimine 763 and 1199 cm^{-1} frequencies.[75] When the chlorine atoms of this molecule are substituted by phenyl radicals, its $P{=}^{15}N{-}S$ bending, $S{-}^{15}N$ stretching and P=N stretching bands appear at 574, 780 and 1124 cm^{-1}, corresponding to the ^{14}N-compound 578 (*m*), 784 (*m*) and 1147 cm^{-1} (*vs*) vibrations.[75]

The N=N stretching, P=N stretching and asymmetrical N_3 moiety stretching fundamentals of $(C_6H_5)_3P{=}^{14}\overset{\gamma}{N}{-}^{14}\overset{\beta}{N}{=}^{14}N{-}SO_2C_6H_4CH_3$ which are shown at 980, 1230 (*vs*) and 2114 cm^{-1} are shifted to 968, 1213 and 2095 cm^{-1} when its γ-N is ^{15}N-labelled and to 977, 1212 and 2073 cm^{-1} when its β-N is similarly exchanged.[79]

The infrared absorption spectra of copper complexes of various ^{15}N-labelled aldehyde imines have been studied.[80]

The I form of ^{15}N-polyglycine was observed to exhibit its ^{15}N-characteristic infrared absorption bands at 1010 (skeletal stretching), 1210, 1503 (II amide band), 3064 (1503 overtone) and 3293 cm^{-1} (N—H stretching), as compared with the corresponding normal polyglycine bands at 1016, 1214, 1517, 3088 and 3303 cm^{-1}, respectively.[81] In the case of the II form of ^{15}N-polyglycine the ^{15}N-sensitive absorption peaks were found at 1023, 1245 (III amide band), 1541, 3061 and 3290 cm^{-1}, corresponding to the normal polyglycine bands at 1028, 1249, 1554, 3086 and 3303 cm^{-1}, respectively.[81]

The effects of ^{15}N-substitution of deoxyribonucleic acid on its absorption in the 600–1800 cm^{-1} region have also been reported.[82]

REFERENCES

1. Borodko, Yu. G., Shilova, A. K. and Shilov, A. E. (1967). *Doklady Akad. Nauk. S.S.S.R.* **176**, 1297.
2. Keck, D. B. and Hause, C. D. (1968). *J. Mol. Spectroscopy* **26**, 163.
3. Fletcher, W. H. and Begun, G. M. (1957). *J. Chem. Phys.* **27**, 579.
4. Cleare, M. J. and Griffith, W. P. (1967). *J. Chem. Soc.* (*A*). 1144.
5. Miki, E. (1968). *Bull. Chem. Soc. Japan.* **41**, 1835.
6. Miki, E., Ishimori, T., Yamatera, H. and Okuno, H. (1966). *J. Chem. Soc. Japan* (*Pure Chem. Sect.*) **87**, 703.
7. Grosso, R. P. and McCubbin, T. K. Jr. (1964). *J. Mol. Spectroscopy* **13**, 240.
8. Plíva, J. (1964). *J. Mol. Spectroscopy* **12**, 360.
9. Migeotte, M., Neven, L. and Swensson, J. (1957). *Mem. soc. roy. sci. Liège* **2**.
10. Arakawa, E. T. and Nielsen, A. H. (1958). *J. Mol. Spectroscopy* **2**, 413.
11. Olman, M. D. and Hause, C. D. (1968). *J. Mol. Spectroscopy* **26**, 241.
12. McKean, D. C. (1966). *Spectrochim. Acta* **22**, 269.
13. Shvangiradze, R. R. and Dzhamagidze, Sh. Z. (1962). *Optika i Spektroskopiya* **12**, 364.
14. Plíva, J., Špirko, V. and Papoušek, D. (1967). *J. Mol. Spectroscopy* **23**, 331.
15. Jones, L. H., Asprey, L. B. and Ryan, R. R. (1967). *J. Chem. Phys.* **47**, 3371.
16. Jones, L. H., Ryan, R. R. and Asprey, L. B. (1968). *J. Chem. Phys.* **49**, 581.
17. Morgan, H. W., Staats, P. A. and Goldstein, J. H. (1957). *J. Chem. Phys.* **27**, 1212.
18. Moore, C. B. and Rosengren, K. (1966). *J. Chem. Phys.* **44**, 4108.

19. McGraw, G. E., Bernitt, D. L. and Hisatsune, I. C. (1966). *J. Chem. Phys.* **45**, 1392.
20. Bernitt, D. L., Miller, R. H. and Hisatsune, I. C. (1967). *Spectrochim. Acta* **23A**, 237.
21. Nemes, L. (1968). *Acta Chim. Acad. Sci. Hung.* **56**, 153.
22. Bayersdorfer, L. Engelhardt, U., Hoehne, K., Fischer, J. and Jander, J. (1968). *Z. Naturforsch.* **B23**, 1602.
23. McGraw, G. E., Bernitt, D. L. and Hisatsune, I. C. (1965). *J. Chem. Phys.* **42**, 237.
24. Palm, A., Castelli, A. and Alexander, C. Jr. (1968). *Spectrochim. Acta* **24A**, 1658.
25. Yeranos, W. A. and Joncich, M. J. (1967). *Mol. Phys.* **13**, 263.
26. Monse, E. V., Kauder, L. N. and Spindel, W. (1964). *J. Chem. Phys.* **41**, 3898.
27. Begun, G. M. and Fletcher, W. H., (1960). *J. Mol. Spectroscopy* **4**, 388.
28. McDowell, R. S., Horrocks, W. D. and Yates, J. (1961). *J. Chem. Phys.* **34**, 530.
29. Maki, A. and Decius, J. C. (1959). *J. Chem. Phys.* **31**, 772.
30. Singh, K. and Banerji, N. N. (1965). *J. Chem. Soc.* 1513.
31. Theophanides, T. and Turell, G. C. (1967). *Spectrochim. Acta* **23A**, 1927.
32. Bryant, J. I. (1966). *Spectrochim. Acta* **22**, 1475.
33. Savoie, R. and Pézolet, M. (1967). *Canad. J. Chem.* **45**, 1677.
34. Sidman, J. W. (1957). *J. Amer. Chem. Soc.* **79**, 2675.
35. McGraw, G. E., Bernitt, D. L. and Hisatsune, I. C. (1967). *Spectrochim. Acta* **23A**, 25.
36. Decius, J. C. (1954). *J. Chem. Phys.* **22**, 1941.
37. Khanna, R. K., Lingscheid, J. and Decius, J. C. (1964). *Spectrochim. Acta* **20**, 1109.
38. Decius, J. C. (1955). *J. Chem. Phys.* **23**, 1290.
39. Miller, R. H., Bernitt, D. L. and Hisatsune, I. C. (1967). *Spectrochim. Acta* **23A**, 223.
40. Karelin, A. I., Kharitonov, Yu. Ya. and Rosolovskii, V. Ya. (1968). *Zhur. priklad. Spektroskopii* **8**, 256; *C.A.* (1968). **68**, 109672a.
41. Hewkin, D. J. and Griffith, W. P. (1966). *J. Chem. Soc.* (*A*) 472.
42. Buttler, M. J. and McKean, D. C. (1965). *Spectrochim. Acta* **21**, 465.
43. Randall, E. W., Ellner, J. J. and Zuckerman, J. J. (1965). *Inorg. Nucl. Chem. Letters* **1**, 109.
44. Tsuboi, M. (1960). *Spectrochim. Acta* **16**, 505.
45. Randall, E. W. and Zuckerman, J. J. (1968). *J. Amer. Chem. Soc.* **90**, 3167.
46. Mitchell, R. W., Burr, J. C. Jr. and Merritt, J. A. (1967). *Spectrochim Acta* **23A**, 195.
47. Van Senden, K. G. (1965). *Rec. Trav. chim.* **84**, 1459.
48. Maki, A. G. Plyler, E. K. and Thibault, R. (1964). *J. Opt. Soc. Amer.* **54**, 869.
49. Fritzsche, H., Schulze, W., Letsch, G. and Huebner, H. (1964). *Z. Chem. Leipzig* **4**, 105
50. Fritzsche, H., Schulze, W. and Letsch, G. (1962). *Z. Chem. Leipzig* **2**, 357.
51. Kuebler, R., Luettke, W. and Weckherlin, S. (1960). *Z. Elektrochem.* **64**, 650.
52. Weckherlin, S. and Luettke, W. (1966). *Annalen* **700**, 59.
53. Nixon, E. R. and Verderame, F. D. (1964). *J. Chem. Phys.* **41**, 1682.
54. Maki, A. G. (1965). *J. Chem. Phys.* **43**, 3193.
55. Gray, L. S. Jr., Fassel, V. A. and Kniseley, R. N. (1960). *Spectrochim. Acta* **16**, 514.
56. Moore, C. B. and Pimentel, G. C. (1964). *J. Chem. Phys.* **40**, 329.

57. Randall, E. W., Yoder, C. M. S. and Zuckerman, J. J. (1966). *Inorg. Chem.* **5**, 2240.
58. (a) Weckherlin, S. and Luettke, W. (1960). *Z. Elektrochem.* **64**, 1228.
(b) Strukov, O. G., Smirnov, S. K., Dubov, S. S. and Danilina, L. L. (1969). *Zhur. org. Khim.* **5**, 1257, for the spectra of ^{15}N-labelled benzamide and other amides.
59. Coburn, R. A. and Dudek, G. O. (1968). *J. Phys. Chem.* **72**, 1177.
60. Coburn, R. A. and Dudek, G. O. (1968). *J. Phys. Chem.* **72**, 3681.
61. Dudek, G. (1967). *J. Org. Chem.* **32**, 822.
62. Kuebler, R. and Luettke, W. (1963). *Ber. Bunsenges. phys. Chem.* **67**, 2.
63. Pinchas, S., Samuel, D. and Silver, B. L. (1964). *Spectrochim. Acta* **20**, 179.
64. Daehne, S. and Stanko, H. (1962). *Spectrochim. Acta* **18**, 561.
65. Dupuy, B. and Garrigou-Lagrange, C. (1968). *J. Chim. phys.* **65**, 450.
66. Dupuy, B. and Garrigou-Lagrange, C. (1967). *J. Chim. phys.* **64**, 1509.
67. Tsuboi, M., Takenishi, T. and Nakamura, A. (1961). *Spectrochim. Acta* **17**, 634.
68. Nakamura, A. (1965). *J. Chem. Soc. Japan* **86**, 500.
69. Takenishi, T. (1960). *J. Chem. Soc. Japan* **81**, 1380.
70. Takenishi, T. (1961). *J. Chem. Soc. Japan* **82**, 15.
71. Pinchas, S. (1966). *Spectrochim. Acta* **22**, 1889.
72. Porter, R. F. and Yeung, E. S. (1968). *Inorg. Chem.* **7**, 1306.
73. Linton, H. R. and Nixon, E. R. (1958). *Spectrochim. Acta* **10**, 299.
74. Booth, M. R. and Frankiss, S. G. *Chem. Commun.* **1968**, 1347.
75. Wiegraebe, W., Bock, H. and Luettke, W. (1966). *Chem. Ber.* **99**, 3737.
76. Bock, H., Schnoeller, M. and Dieck, H. T. (1969). *Chem. Ber.* **102**, 1363.
77. Shokol, V. A., Kisilenko, A. A. and Derkach, G. I. (1969). *Zhur. obshchei Khim.* **39**, 874.
78. Rademacher, P., Wiegraebe, W. and Luettke, W. (1967). *Chem. Ber.* **100**, 1213.
79. Bock, H. and Schnoeller, M. (1969). *Chem. Ber.* **102**, 38.
80. Dudek, E. P. and Dudek, G. (1967). *Inorg. nucl. Chem. Lett.* **3**, 241.
81. Suzuki, S., Iwashita, Y., Shimanouchi, T. and Tsuboi, M. (1966). *Biopolymers* **4**, 337.
82. Tsuboi, M., Shuto, K. and Higuchi, S. (1968). *Bull. Chem. Soc. Japan* **41**, 1821.

CHAPTER 7

Oxygen-18 Compounds

7.1 Introduction

Owing to the commercial availability of water highly enriched in oxygen-18 (obtained from the fractional distillation of ordinary water and containing up to 99 atom % oxygen-18; from the Miles Yeda Co. Ltd., Rehovoth, Israel) an extensive number of various highly ^{18}O-labelled compounds have recently been prepared and their infrared absorption spectra investigated, in certain cases down to *ca.* 200 cm^{-1}. As a result, the effect of substituting oxygen-18 for oxygen-16 on almost all the fundamentals of a series of representative molecules, belonging to diverse important chemical classes of compounds, has been evaluated. The shifts observed in their frequencies have been found to be very significant for checking the correctness of existing band assignments, as well as for the suggestion and confirmation of new ones, in such cases as those of the nitro-hydrocarbons, amino acids, metal acetylacetonates, alkali permanganates, etc.

Most of the ^{18}O-labelled molecules investigated so far contain at least one double or triple XO bond, where X = C, N, P, S, V, Mn, As or U. In such cases the potential energy of those vibration modes which involve an appreciable XO stretching is localized to a large extent (see Section 3.2) in these bonds. It is therefore possible to get an estimate of the ^{18}O-induced shift of these modes by assuming it to be not much lower than for the corresponding diatomic X—O oscillator having the same stretching frequency.

The ν^i/ν ratio of isotopic stretching frequencies of diatomic molecules being very nearly equal to $(\mu/\mu^i)^{1/2}$ (Section 3.2) one can write the following equation for this ratio in the case of the X—^{18}O and X—^{16}O oscillators:

$$\frac{\nu^i}{\nu} = \left(\frac{\mu}{\mu^i}\right)^{1/2} = \left(\frac{16 \times m_X \times (18 + m_X)}{(16 + m_X) \times 18 \times m_X}\right)^{1/2} = \frac{2}{3}\left(\frac{2(18 + m_X)}{16 + m_X}\right)^{1/2} \quad (7.1)$$

where m_X = mass of the X atom. After m_X is substituted in eqn. 7.1 by its value and the characteristic stretching frequency of the corresponding X—^{16}O group is multiplied by the resulting (calculated) ν^i/ν ratio, the stretching frequency of the diatomic X—^{18}O vibrator, and hence the respective expected ^{18}O-induced shift, can be obtained from this equation.

Table 7.1 gives some of the ^{18}O-shifts calculated by this method, for various diatomic oscillators, together with the shifts actually observed for compounds containing the respective XO groups.

TABLE 7.1 ^{18}O-Induced Isotopic Shifts of XO Stretching Bands (cm^{-1})

X	Normal group frequency[a]	Calculated v^i/v ratio	Calculated isotopic group frequency	Calculated shift	Observed shift	Reference
C	1667 [$(C_6H_5)_2CO$]	0·976	1627	40	30	1
N	1440 [$C_6H_5NO_2$][b]	0·974	1403	37	24[b]	2
P	1202 [$(C_6H_5)_3PO$]	0·963	1158	44	30	3
S	1051 [$(C_6H_5)_2SO$]	0·962	1011	40	30	4
As	880 [$(C_6H_5)_3AsO$]	0·953	839	41	40	5

[a] This is *not* the stretching frequency of the diatomic XO *molecule* but the observed absorption frequency of the $X^{16}O$ *group* (in a polyatomic molecule). Since this frequency varies a little from molecule to molecule it refers always to a certain compound, the formula of which is added in parentheses after each value.

[b] Mean value for the two N=O stretching bands.

It is obvious from Table 7.1 that all the ^{18}O-shifts calculated for the diatomic XO oscillators are in reasonable agreement with the corresponding shifts observed for the compounds which contain the respective $X^{18}O$ groups. This agreement can be seen to improve quickly as the mass of the X atom increases, its R—XO amplitude dwindles down and the vibrational energy of the XO stretching concentrates more and more in the XO group.

The fact that the observed $X{=}^{18}O$ stretching absorption bands appear generally about 20–40 cm^{-1} lower than the corresponding $X{=}^{16}O$ bands allows the former to be resolved from the latter, in the spectrum of an ^{18}O-enriched compound, even if its enrichment is not more than[6] 2% and the instrument used has only a usual resolving power.

If the labelled compound contains a chemical group which comprises more than one oxygen atom and the ^{18}O-enrichment of this group is appreciable the absorption spectrum of the compound will generally show several sets of XO stretching bands, that belonging to the normal group (if the labelling is not too high), that belonging to the possible mixed modification(s) and that due to the fully labelled compound.

Figure 7.1 reproduces the absorption of a 43 atom % ^{18}O-nitrobenzene solution in the symmetrical NO_2 stretching region. The band shown in Fig. 7.1 at 1349 cm^{-1} belongs to the (statistically calculated) 32% content of the normal nitrobenzene, the absorption manifested at 1335 cm^{-1} stems from the 49% concentration of the mixed species while that revealed at 1322 cm^{-1} is due to the 19% amount of fully labelled compound.

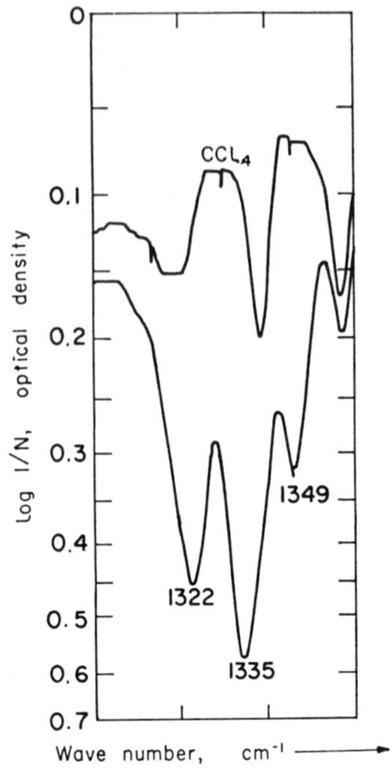

Fig. 7.1 The symmetrical NO_2 stretching band of ^{18}O-labelled nitrobenzene. (0·01 g/ml CCl_4; 0·2 mm cell.)

In the case of inorganic salts of oxyacids, however, no separate XO_n^{m-} stretching bands can usually be discerned for each of the isotopic modifications constituting the labelled sample. The reason for this is that these substances are investigated mainly in the crystalline state and under these conditions, as is well known, the infrared absorption bands of salts are usually broad, so that a merging together of corresponding bands of all the components of a labelled salt sample occurs.

Since the case of R—H groups is similar to that of multiply bonded groups, in respect to the concentration of the potential energy of their stretching vibrations in these groups (Section 3.2), one can estimate the ^{18}O-induced shift of the O—H stretching band of various alcohols etc. in precisely the same way as was done above for multiply bonded XO groups. Such estimations, however, can be expected to hold only for bands observed on samples free from appreciable hydrogen bonding, e.g. present in a highly diluted solution in a more or less inert solvent, since it has been observed[7, 8] that the strength of an ^{18}O—H····X hydrogen bond appears to be different from its value for the corresponding ^{16}O—H····X bond.

The agreement between the ^{18}O-shifts estimated by this method for phenolic or carbinolic hydrogen-bond free OH stretching bands and the respective observed values is indeed very good.

The theoretically expected relationships between the various fundamental frequencies of the ^{18}O-labelled bent XO_2, planar XO_3 and tetrahedral XO_4 anions and the corresponding frequencies of the respective normal anions (Sections 3.3, 3.4, 3.8) were confirmed in the case of a number of inorganic salts. This agreement between the expected and the observed frequencies of these labelled salts proved to be of importance for the confirmation of several controversial assignments of characteristic absorption bands belonging to the respective normal salts.

The frequencies observed for the vibrational modes of a couple of ^{18}O-labelled organic compounds, on the other hand, have been used, together with other isotopic frequency data, for carrying out a normal co-ordinate analysis of the fundamental vibrations of these compounds.

The characteristic absorption bands of various $C{=}^{18}O$ groups have also been used to follow the enzyme kinetics of the formation of labelled inosine from adenosine,[9] to clarify the mechanism of the 1-phenylallyl *p*-nitrobenzoate rearrangement to *trans*-cinnamyl *p*-nitrobenzoate[10], etc. Such tracer studies are reported in Chapter 12.

No work seems to have been done so far on the absorption of highly ^{17}O-enriched samples in the infrared region although the vibration–rotation spectrum of a number of ^{17}O-compounds has been investigated on samples with the natural abundance of oxygen-17.[11]

7.2 Oxygen-18 and Oxygen-17 Carbon Monoxide

The centre of the C—O stretching absorption band of the gaseous $^{12}C^{16}O$ molecules appears at[12] 2143·3 cm^{-1} (vacuum) hence, on basis of the proper ν^i/ν ratio given in Table 7.1, one can expect the centre of the corresponding vibration–rotation band of the gaseous $^{12}C^{18}O$ molecules to appear at about $2143 \times 0{\cdot}976 = 2092$ cm^{-1}. An investigation of the infrared absorption of a carbon monoxide gas sample of a natural isotopic abundance has indeed shown that the $^{12}C^{18}O$ molecules show the centre of this band[13] at 2092·1 cm^{-1}. Analysing the fine structure of this band yielded the values for the molecular vibrational and rotational constants[13] of $^{12}C^{18}O$, summarized in Table 7.2.

The 2-O and 3-O overtones of the $^{12}C—^{18}O$ stretching have also been investigated.[14] The isotopic shift observed for the 2-O overtone was $101{\cdot}03 \pm 0{\cdot}05$ cm^{-1}, in good agreement with the calculated value. This band was also investigated in emission in the 4000–4360 cm^{-1} region.[14]

The effect of the solid state on the $^{12}C^{18}O$ fundamental band is reviewed in Chapter 10.

TABLE 7.2 The Vibrational and Rotational Constants (cm^{-1}) of $^{12}C^{18}O$[14]

	Calculated	Observed
ω_e	2117·39	2117·32
$\chi_e\omega_e$	12·65	12·60
B_e	1·83912	1·8400 ± 0·0007
α_e	0·01630	0·01624 ± 0·00005
D	$5{\cdot}39 \times 10^{-6}$	$5{\cdot}25 \times 10^{-6}$

The stretching of $^{12}C^{18}O$ molecules was shown to be responsible for a number of lines in the infrared solar spectrum.[15]

The centre of the vibration–rotation band of gaseous $^{12}C^{17}O$ molecules was found[12] at 2116·3 cm^{-1}, almost half-way (2117·7) between the 2143·3 and the 2092·1 cm^{-1} values, observed for the $^{12}C^{16}O$ and the $^{12}C^{18}O$ molecules, respectively. This measurement was made on a carbon monoxide sample containing only the low natural abundance of 0·04%.

The $C^{18}O$ stretching frequency of ^{18}O-labelled carbon monoxide bound to haemoglobin appears at 1907·3 cm^{-1} as compared with the 1903·8 cm^{-1} value calculated for it on basis of the bound $C^{16}O$ frequency (1950·3; 1950·8) and the change in the isotopic mass. This 3·5 cm^{-1} discrepancy suggests that carbon monoxide is bound to the haemoglobin molecule by a co-ordination of its oxygen atom to the iron atom of the haemoglobin.[16]

7.3 $^{15}N,^{18}O$-LABELLED NITRIC OXIDE

The analysis of the fine structure of the fundamental and first overtone vibration–rotation absorption bands of gaseous $^{15}N^{18}O$ yielded their band origins at 1793·202/1793·405 cm^{-1} (a doublet, see Section 6.3) and 3560·761/3561·186 cm^{-1}, respectively, as compared with the 1842·82/1843·05 cm^{-1} values observed for the fundamental doublet of $^{15}N^{16}O$.[17]

The values of its other important vibration–rotation constants have also been reported.[17]

7.4 ^{18}O-LABELLED WATER MOLECULES

The triatomic water molecule belonging to the C_{2v} point group shows three fundamentals: ν_1 and ν_2 of the symmetrical A_1 species and the perpendicular ν_3 stretching vibration which belongs to the B_1 species, all of them being infrared-active.

The high resolution absorption spectrum of $H_2{}^{18}O$ vapour in the 3350–4020 cm^{-1} region was analysed to yield the values for its ν_1 and ν_3 band centres as well as its A_0, B_0, C_0 (27·535, 14·516 and 9·2437 cm^{-1}, respectively) and other rotational constants.[18, 19]

The ν_2 and ν_3 frequencies of $D_2{}^{18}O$ vapour were determined from the minimum and maximum, respectively, of the corresponding vibration–rotation bands of an 89 atom % sample.[20]

Since the symmetrical ν_1 stretching vibration of the water molecule absorbs only weakly in the infrared region its frequency was determined from the Raman spectrum of the labelled sample.[20] These frequencies agree well with the normal frequencies calculated for $D_2{}^{18}O$ either on basis of the valence forces assumption[21a] (Section 3.8) or using the general quadratic potential function,[21b] as can be seen from Table 7.3.

TABLE 7.3 The Fundamental Frequencies (cm^{-1}) of Isotopic Water Molecules[a]

Vibration	$H_2{}^{16}O$ Observed[18, 19]	$D_2{}^{16}O$ Observed[22]	$H_2{}^{18}O$ Calculated[21b]	$H_2{}^{18}O$ Observed[19]	$D_2{}^{18}O$ Calculated[21b]	$D_2{}^{18}O$ Calculated[21a]	$D_2{}^{18}O$ Observed[20b]
ν_1	3657·05	2671·69	3647	3649·68	2656	2653·4	2656·8 ± 2
ν_2	1594·59[c]	1178·33	1586	—	1166	1171·6	1169·3 ± 2
ν_3	3755·97	2788·05	3744	3741·58	2762	2767	2764 ± 5

[a] Unless otherwise indicated, in vacuum.
[b] In air.
[c] Ref. 22.

The $H_2{}^{17}O$ ν_1 and ν_3 vibrations were found at 3653·15 and 3748·36 cm^{-1}, respectively[19] (on a sample which consisted mainly of $H_2{}^{18}O$).

Vibrational bands due to $H_2{}^{17}O$ and $H_2{}^{18}O$ molecules were observed in the solar infrared spectrum[15] (due to the atmospheric content of these compounds).

7.5 Labelled Carbon Dioxide

The O═C═O molecules being linear show two infrared-active (ν_2, ν_3) and one Raman-active (ν_1) fundamental bands. The Raman band is a doublet formed by Fermi-resonance (Section 2.10) between ν_1 and one of the overtones of the ν_2 bending ($2\nu_2{}^{\circ}$).

The absorption of an ^{18}O-labelled sample of CO_2 which contained about 95% of $^{12}C^{18}O_2$ and 3% of $^{12}C^{16}O^{18}O$, in the 637–695 cm^{-1} and the 1800–5000 cm^{-1} regions, has been investigated under a very high resolution.[23–25] This study gave precise values for the band centres of the $^{12}C^{18}O_2$ ν_2 and ν_3 fundamentals and a number of its combination and overtone bands as well as for the centre of the ν_2 band of $^{12}C^{16}O^{18}O$. These values are generally in good agreement with the frequencies calculated for these centres taking into account first- and second-order corrections for the anharmonicity of the vibrations. The different rotational constants of $^{12}C^{18}O_2$ were also obtained.

Since the sample used for these measurements contained about 1% of $^{13}C^{18}O_2$ a number of vibrational and rotational constants of this species were determined as well.

The $^{12}C^{16}O^{18}O$ band centres of the ν_3 fundamental and 9 of its combination and overtone bands in the 1620–5400 cm^{-1} region were observed on a sample of 10 atom % ^{18}O-carbon dioxide.[26] These bands also include a band due to the $2\nu_1$ overtone which is infrared-*inactive* for the symmetrical CO_2 modifications since the latter, in contradistinction to $C^{16}O^{18}O$, do possess a symmetry centre.

The centre of the $^{13}C^{16}O^{18}O$ ν_3 band as well as various rotational constants of these mixed species were also determined on this sample, the results agreeing with the values calculated theoretically for all of these constants.

The frequencies obtained for the centres of the fundamental bands of the various ^{18}O-labelled CO_2 modifications are summarized in Table 7.4.

TABLE 7.4 The Fundamental Frequencies (cm^{-1}) of Isotopic CO_2 Molecules

Vibration	$^{12}C^{16}O_2$ (ref. 27)	$^{12}C^{18}O_2$ (ref. 24)	$^{13}C^{18}O_2$ (ref. 24)	$^{12}C^{16}O^{18}O$ (ref. 26)	$^{13}C^{16}O^{18}O$ (ref. 27)
ν_1	—	—	—	^{a,b}	^{a}
ν_2	667·3 (vacuum)	657·3	638·6	662·6	^{a}
ν_3	2349·3	2313·97	2247·24	2332·16	2265·98

^{a} Not measured.
^{b} $2\nu_1$ = 2500·7.

It is interesting to note (see Table 7.4) that the frequency of the $^{12}C^{16}O^{18}O$ fundamentals is in each case almost exactly half-way between the corresponding values of $^{12}C^{16}O_2$ and $^{12}C^{18}O_2$. In the case of ν_2 the isotopic shifts are 4·7 versus 10·0 cm^{-1} for $^{12}C^{18}O_2$ while for ν_3 these are 17 and 35 cm^{-1}, respectively.

The detailed investigation of the infrared absorption spectra of the various isotopic modifications of CO_2 helped to spot a mistake in the attribution of the individual bands of the Fermi-resonance ν_1–$2\nu_2°$ diad at about 1285 and 1388 cm^{-1}. Formerly, the lower band was assigned a higher degree of $2\nu_2$ character and a lower degree of ν_1 character while the higher band was assigned inversely. This resulted in obtaining isotopic values for the Fermi-resonance perturbation constant W_e (which enters in the calculation of the band centres of many combination bands) that were incompatible with theory. Thus, e.g. instead of the calculated *decrease* of about 4% in the value of W_e on going over from $C^{16}O_2$ to $C^{18}O_2$ an *increase* of about 4% in its value was found.[24, 25]

When, however, the assignment of the ν_1–$2\nu_2^0$ bands is inverted[28] and $2\nu_2^0$ is taken as the higher frequency of the resonating pair then the isotopic relations regarding W_e are satisfied.[29]

Certain P transitions of the $^{12}C^{18}O_2$ molecules from the first excited level of the ν_3 vibration (00°1) to the first excited level of the symmetrical ν_1 stretching (10°0) or the second excited level of the ν_2 bending (02°0) were found[30] to give rise to laser action. For the P (18) branch of the first band the emitted frequency is 1070·6 cm^{-1} (vacuum) as compared with 945·9 cm^{-1} for the corresponding normal frequency, the very great isotopic shift of this difference band being due to the Fermi resonance existing in the ν_1 level.

7.6 ^{18}O-Nitrous Oxide

The linear NNO molecule shows three infrared-active bands of which only the strongest one, that at about 2223·5 cm^{-1} (vacuum) in the case of the normal modification, was observed for the natural abundance content of $^{14}N_2{}^{18}O$. The centre of this ν_3, asymmetrical stretching, band was observed[31] at 2219·68 cm^{-1} as compared with the calculated value of 2219·6 cm^{-1}. That the ^{18}O-isotopic shift for this band is so low (about 4 cm^{-1} only) is the result of the fact that most of the energy of this vibration is centred in the multiple N—N bond and not in the N—O bond which has a partial single bond character. This makes the ν_3 stretching of the nitrous oxide molecule practically an internal azide group stretching as is also evident from its frequency which is very near to the 2150 cm^{-1} frequency of the azide radical.[32] The vibrational–rotational ν_3 band of $^{15}N_2{}^{18}O$ was recorded and analysed among the other 15, combination and overtone bands observed on a sample of $^{15}N_2{}^{18}O$ which was investigated in the 200–5000 cm^{-1} region.[33,34] The frequency of its centre was taken as 2147·59 cm^{-1} as compared with the value of 2146·3 cm^{-1} calculated from the vibrational data obtained on the $^{14}N_2{}^{16}O$ and $^{15}N^{14}N^{16}O$ species.

The ν_1 vibration of $^{15}N_2{}^{18}O$ is found at[33] 1225·98 cm^{-1} while its rotational constant B_{000} has the value of 0·38186 cm^{-1} as compared with the value of 0·3818 cm^{-1} calculated for it.

The various (harmonic) force constants of the nitrous oxide molecule were evaluated on basis of all the isotopic spectra observed for it.[33, 34]

7.7 ^{18}O-Labelled Nitrosyl Halides

The study of the infrared absorption of gaseous ^{18}O-labelled (and ^{15}N, ^{18}O-labelled) nitrosyl fluoride,[35] nitrosyl chloride[36] and nitrosyl bromide[37] yielded the location of their main absorption bands, their fundamental harmonic vibrational frequencies and the quadratic valence force constants of the nitrosyl halide molecules. Their observed fundamental bands are given in Table 7.5.

TABLE 7.5 The Fundamental Bands (cm^{-1}) Observed for Various Isotopic Nitrosyl Halides

Band	Description	^{16}ONF	^{18}ONF	$^{16}ON^{35}Cl$	$^{18}ON^{35}Cl$	$^{16}ONBr$	$^{18}ONBr$
ν_1	N=O stretching	1843·5	1795·6	1799·7	1749·0 1752·7[a]	1799·0	1751·5 (*vs*)
ν_2	Bending	765·8	757·8	595·8	588·4	542·0	—
ν_3	N—X stretching	519·9	511·3	331·9	324·9	266·4	—

[a] In Fermi resonance.

7.8 LABELLED SULPHUR DIOXIDE

The SO_2 molecule belonging to the $\boldsymbol{C_{2v}}$ point group shows three fundamental absorption bands in the infrared spectrum.

A gaseous oxygen-18 enriched sulphur dioxide sample which contained 60 mole % $S^{16}O_2$, 35 mole % $S^{16}O^{18}O$ and 5 % $S^{18}O_2$ showed these bands[38] as summarized in Table 7.6.

TABLE 7.6 The Observed Absorption Frequencies (cm^{-1}) of Isotopic SO_2 Molecules

Vibration	$S^{16}O_2$	$S^{16}O^{18}O$	$S^{16}O^{18}O$ (calcul.)[39]	$S^{18}O_2$
ν_1	1152 ± 1	1122	1133	Not observed
ν_2	518·0 ± 0·5	506·8	511	Not observed
ν_3	1360·5 ± 0·5	1341·1	1328	1316
$\nu_1 + \nu_3$	2497 ± 0·5	2451	—	Not observed

The appreciable discrepancy between the observed $S^{16}O^{18}O$ frequencies and those calculated for them (see Table 7.6) is the result of neglecting some of the potential constants in their calculation (since there are not enough spectral data for their determination).[39]

The ν_3 frequencies observed for carbon tetrachloride solutions of $S^{16}O_2$, $S^{16}O^{18}O$ and $S^{18}O_2$ appear at[40] 1345, 1327 and 1302 cm^{-1}, respectively, while cyclohexane solutions show them at about 1344, 1330 and 1305 cm^{-1}.

7.9 ^{18}O-LABELLED FOUR-ATOMIC MOLECULES

Gaseous ^{18}O-labelled nitrous acids[41] exhibit their *cis* and *trans* form bands (see Section 6·9) as reported in Table 7.7.

The 350–3000 cm^{-1} absorption of solid fluorine $^{18}O_2$-peroxide (at 77°K) shows its fundamentals at 362 (symmetric bending), 444 (antisymmetric O—O—F bending), 586 (antisymmetric O—F stretching), 595 (symmetric

TABLE 7.7 The Fundamentals (cm^{-1}) of the ^{18}O-Labelled Nitrous Acids[41]

Form	Vibration	$H^{16}ON^{18}O$	$H^{18}ON^{16}O$	$H^{18}ON^{18}O$
cis	ν_2	1606	1643	1606
	ν_4	846	840	840
	ν_5	582	593	576
	ν_6	638	631	631
trans	ν_1	3588	3571	3571
	ν_2	1666	1702	1666
	ν_3	1258	1251	1251
	ν_4	785	779	779
	ν_5	570	570	564
	ν_6	540	537	537

O—F stretching) and 1239 cm^{-1} (O—O) stretching), corresponding to the $F^{16}O_2F$ bands at 369, 457, 615, 621 and 1306 cm^{-1}, respectively.[42]

7.10 LABELLED PHOSPHORUS OXYCHLORIDE

The $POCl_3$ molecule belonging to the C_{3v} point group (see Fig. 7.2) possesses six, infrared-active, fundamentals of which the symmetrical P—Cl

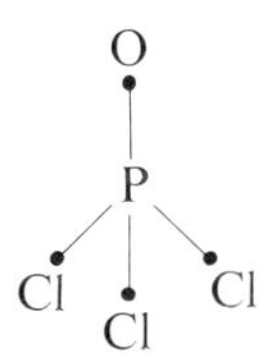

Fig. 7.2 The structure of $POCl_3$.

stretching ν_1, the P=O stretching ν_2 and the symmetrical P—Cl bending ν_3 frequencies form the A_1 species. The remaining ν_4, ν_5 and ν_6 vibrations belong to the doubly degenerate E species.

The absorption of a $POCl_3$ sample which contained 97 atom % oxygen-18 and 0·5 atom % oxygen-17 was investigated in solution, in carbon tetrachloride or cyclohexane, in the 200–4000 cm^{-1} region[43] which contains all but one (ν_6) of these fundamentals. The results of this measurement are given in Table 7.8.

It can be seen from Table 7.8 that the isotopic shift of the ν_2, P=O stretching vibration of $POCl_3$ amounts to 45 cm^{-1}, this value being in close agreement with the isotopic shift of 48 cm^{-1} expected for a 1310 cm^{-1} stretching frequency of a diatomic P=O molecule (cf. Table 7.1).

Applying the Teller-Redlich product rule (Section 3.4) to the A_1 species of the $POCl_3$ vibrations, one can write:

$$\omega_1^i\omega_2^i\omega_3^i/\omega_1\omega_2\omega_3 = [(16/18) \times 155{\cdot}4/153{\cdot}4]^{1/2},$$

TABLE 7.8 The Observed Infrared Absorption Bands (cm^{-1}) of $P^{16}OCl_3$ and $P^{18}OCl_3$

Vibration	$P^{16}OCl_3$ (CCl_4)	$P^{18}OCl_3$ (CCl_4)	$P^{16}OCl_3$ (C_6H_{12})	$P^{18}OCl_3$ (C_6H_{12})
ν_1			482	476
ν_2	1306	1261	1310	1265
ν_3			264	263
ν_4	586	589	587	585
ν_5			339	333
$2\nu_2$	2636^b			
$2\nu_4$	1169	~1180	1176	
$\nu_1 + \nu_2^a$			1790	
$\nu_1 + \nu_4$	~1060	~1060		

[a] By mistake this combination vibration is designated in ref. 43 as $\nu_2 + \nu_4$.
[b] Observed on the pure vapour of $POCl_3$.

which after substitution by the values computed[43] for the various ω's from the corresponding ν values of Table 7.8, and calculation gives 0·9496 = 0·9489. This agreement can be taken as proof for the correctness of the above given assignment of the various $POCl_3$ fundamentals.

7.11 ^{18}O-ENRICHED NICKEL CARBONYL

The tetrahedral $Ni(CO)_4$ molecule which belongs to the highly symmetrical T_d point group can be expected to show four infrared-active fundamental bands, due to the ν_5 CO stretching vibration, the ν_7 Ni—CO stretching frequency and the two asymmetrical bending fundamentals ν_3 and ν_6. In fact, however, only three fundamental bands are observed in the infrared absorption spectrum since ν_3 and ν_6 coincide.

The infrared spectrum of a gaseous nickel carbonyl sample which was made up of about 80% of $Ni(C^{18}O)_4$ and 20% of $Ni(C^{16}O)(C^{18}O)_3$ was also found[44, 45] to show only three fundamental bands. Using the Raman spectrum results for this sample the Raman-active ν_1 frequency of gaseous $Ni(C^{18}O)_4$, its ω_1 frequency (from the ν_1 and the $\nu_1 + \nu_5$ and $2\nu_5$ Raman frequencies) and from it its ω_2 frequency (by applying the Teller-Redlich rule to the isotopic $\omega_1\omega_2$ products) were all calculated. These results were then used in computing some of the $Ni(CO)_4$ force constants. An infrared absorption investigation made on a nickel carbonyl sample of lower ^{18}O-enrichment gave results for the various CO stretching frequencies of the $Ni(C^{16}O)(C^{18}O)_3$ and $Ni(C^{16}O)_2(C^{18}O)_2$ species. All the data relating to the isotopic species measured so far are summarized in Table 7.9.

It seems interesting that in the case of the mixed isotopic modifications studied a number of the absorption bands corresponding to the ν_5 band of

TABLE 7.9 Fundamental Vibration Frequencies (cm^{-1}) of Gaseous Isotopic Nickel Carbonyls

Vibration	$Ni(C^{16}O)_4$	$Ni(C^{16}O)_2(C^{18}O)_2$	$Ni(C^{16}O)(C^{18}O)_3$	$Ni(C^{18}O)_4$
ν_1	2131	2115	2104·7	2084
ν_2	367·5			354·4
ν_3	459			454
ν_4	63·6			60·6
ν_5	2057	(2058, 2028, 2011)[a]	(2039, 2011)[a]	2010·6 ± 0·1
ν_6	459			454
ν_7	422			417·4 ± 0·2
ν_8	74			69·5

[a] The ν_5 vibration is split in this case owing to the lower symmetry of this modification; the values cited are for the vibrations corresponding to ν_5 in this instance.

the homogeneous species have almost exactly the same frequency as that of the ν_5 band of $Ni(C^{16}O)_4$ or of $Ni(C^{18}O)_4$. It is further noteworthy that the isotopic shift of about 47 cm^{-1} observed both for the ν_1 and for the ν_5 bands of $Ni(C^{18}O)_4$ agrees well with the value of 51 cm^{-1} estimated for it on basis of the diatomic CO oscillator approximation (cf. Section 7.1). Hence it is evident that the ν_1 and ν_5 fundamentals of the nickel carbonyl molecules are almost "pure" CO vibrations.

The 400–550 cm^{-1} spectra of the ^{18}O-labelled nickel carbonyl complexes of the general type $Ni(CO)_{4-n}L_n$ [L = PF_3, $P(OCH_3)_3$, $P(CH_3)_3$, $As(CH_3)_3$, $Sb(C_2H_5)_3$] made it possible to recognize and assign the symmetrical Ni—CO stretching and degenerate Ni—C—O bending fundamentals of tri-substituted nickel carbonyl complexes.[46]

7.12 LABELLED SALTS

A number of ^{18}O-labelled salts of the more important oxyacids have been prepared and their infrared absorption investigated. Their infrared spectra are interesting in connection with the assignment of the absorption bands observed for the corresponding normal salts as well as from the point of view of their own qualitative and quantitative analysis.

Since salts in the solid state are made up of ions held in place by strong electrostatic and other lattice forces, their infrared absorption is the result of the characteristic vibrations of the ions which constitute them as modified by the forces and symmetry governing their lattice. Although the infrared-active characteristic frequencies of the isolated ions (as observed, e.g. in the Raman spectrum of their aqueous solutions) are usually, more or less, reproduced by the infrared spectra of the corresponding salts a number of phenomena give evidence to the effect of the lattice on the characteristic vibrations of salts. First of all, their infrared absorption bands are usually very broad since being, as a rule, measured on solid particles, their exact frequency, in the case of each vibration, depends upon the number of ions vibrating as a unit (as a result of the strong oscillator coupling existing in the lattice); hence a whole region of frequencies is absorbed together in each case.

The originally internal ion frequencies also tend to combine with various lattice modes of vibration and this fact brings about both a further broadening of the fundamental vibrational bands and the appearance of new combination bands.

The crystalline unit cell symmetry of a salt is very often much lower than that of the ions which constitute it hence it happens frequently that vibrations which are infrared-inactive for the symmetrical isolated ions become active for the less symmetrical crystalline unit cells. This difference in symmetry between the unit cell and the ions also splits very often degenerate bands into a number of branches since their degeneracy is lost in the crystal. A further splitting of ionic fundamental frequencies occurs sometimes in the crystalline salt as a result of a vibration coupling between two or more similar ions belonging to the same cell, the vibration of which can be either in-phase or out-of-phase.

It must finally be mentioned here that there often exist also specific interactions between neighbouring ions in the lattice, such as arise from the formation of partial covalent bonds, hydrogen-bonding, etc. These interactions modify considerably the frequency and absorption intensity of the original ionic vibrations.

As a result of the effect of the lattice forces on the ionic frequencies the relationships to be expected in the case of isolated ionic oscillators (cf. Section 3.4), between the fundamental frequencies of the labelled ions and the corresponding frequencies of the respective normal ions often do not hold for the labelled salts.

7.13 Potassium ^{18}O-cyanate

The infrared absorption spectrum of dilute solutions of potassium cyanate in single potassium halide crystals shows the natural abundance content $N{\equiv}C{-}^{18}O^-$ fundamental bands, at 150°K, at 1176 (ν_1 in Fermi resonance with $2\nu_2$) and 2162 (ν_3) cm^{-1} as compared with the corresponding $NC^{16}O^-$ absorption peaks at 1206, 1293 (ν_1; $2\nu_2$) and 2170 cm^{-1}, respectively.[47]

The ν_3 frequency of the $N^{13}C^{18}O^-$ ion in such a solution was observed at[47] 2105 cm^{-1}.

7.14 Labelled Silver Nitrite

The bent nitrite ion 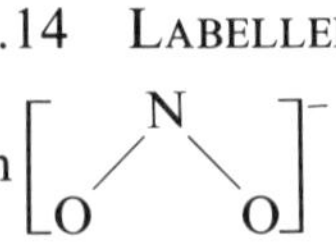 (point group: C_{2v}) can be expected to show three fundamental infrared absorption bands and the absorption spectrum of a Nujol mull of normal sodium nitrite exhibits indeed three such bands, at *ca.* 828 (ν_2), 1261 (ν_3) and 1328 cm^{-1} (ν_1). The Nujol mull of normal silver nitrite shows, however, *four* bands, at about 829 (weak), 846, 1251 (strong) and 1374 cm^{-1}, the second one also being due to the ν_2 bending vibration which is split in the case of the silver nitrite, as well as in the case of the potassium salt, as a result of lattice forces.[48]

A Nujol mull of *well* over 60 atom % silver ^{18}O-nitrite displayed these bands at 807, 827, 1227 (broad and strong) and 1365 (*sh*) cm^{-1} besides a band at about 847 cm^{-1} which belongs to the normal silver nitrite content of the sample.[48] Since $AgN^{18}O_2$ is probably the major component of the measured labelled sample the normal frequencies of the $N^{18}O_2^-$ ion in this case are taken to be about 1365 (ν_1), 817 (mean of 807 and 827; ν_2) and 1227 (ν_3) cm^{-1} as compared with the corresponding values of 1374, 838 and 1251 cm^{-1} for the normal ion in this lattice.

Taking the ONO ion angle as 126° and applying the isotopic relations (eqn. 3.8 and 3.9) holding for the C_{2v} triatomic oscillator, one gets the following expected ratios for the $N^{18}O_2$ zero-order fundamental frequencies:

$$\omega_3^i/\omega_3 = 0{\cdot}9801; \quad \omega_1^i\omega_2^i/\omega_1\omega_2 = 0{\cdot}9267.$$

Substituting the observed absorption band frequencies (cf. Section 3.3) given above for the corresponding zero-order frequencies in these equations gives, however:

$$\omega_3^i/\omega_3 = 1227/1251 = 0{\cdot}9808;$$

$$\omega_1^i\omega_2^i/\omega_1\omega_2 = 1365 \times 817/(1374 \times 838) = 0{\cdot}9685.$$

Thus it is seen that while the application of eqn. 3.8 gives a result which agrees excellently with the observed ratio, eqn. 3.9 is totally unsatisfactory for the prediction of the $\omega_1^i\omega_2^i/\omega_1\omega_2$ ratio of silver nitrite. This fact is no doubt the result of the strong effect of the lattice forces on the symmetrical ν_1 and ν_2 vibrations which is also evident in the splitting of the ionic ν_2 fundamental in the silver nitrite crystal. The known tendency of the silver cation to form partly covalent bonds with anionic oxygen atoms[49] probably also contributes to the discrepancy between the expected and the observed $\omega_1^i\omega_2^i/\omega_1\omega_2$ ratio for this silver salt.

7.15 Labelled Uranyl Ions of Solid $K_3UO_2F_5$

The normal uranyl ion UO_2^{2+} which is probably linear and symmetrical[50] and hence possesses only one infrared-active stretching vibration (ν_3), the other stretching vibration (ν_1) being only Raman-active, was shown, in the case of the $K_3^+[UO_2F_5]^{3-}$ salt, to absorb at about[50] 863 cm^{-1}.

A sample of 38 atom % $^{18}O—K_3UO_2F_5$ was observed to display also bands at about[51] 773 and 822 cm^{-1}. The 822 cm^{-1} band belongs to the $U^{18}O_2^{2+}$ ion while the ν_3 band of the $U^{16}O^{18}O$ ion, which is expected at about 845 cm^{-1}, seems to be masked by the normal uranyl band.[50] The 773 cm^{-1} absorption must be assigned to the "symmetrical" ν_1 stretching of the $U^{16}O^{18}O^{2+}$ ion, which is now infrared-active as the result of the asymmetry introduced by the unhomogeneous labelling of the ion.[50]

The valence vibrations of the $U^{16}O^{18}O^{2+}$ ion in other uranyl complexes have also been reported.[52]

7.16 ^{18}O-Enriched Potassium Nitrate

The planar NO_3^- ion belongs to the $\boldsymbol{D_{3h}}$ point group and therefore possesses four fundamental vibrations, the most symmetrical of which, ν_1 (in-phase stretching of the three N—O bonds), only, being infrared-inactive. Their frequencies in aqueous solutions are located at 1050 (ν_1), 831 (ν_2), 1390 (ν_3) and 720 (ν_4) cm^{-1}.

The infrared absorption spectrum of crystalline potassium nitrate, as that of a number of other crystalline nitrates, however, shows also the ν_1 band, at about 1049 cm^{-1}. It exhibits the ν_2 symmetrical deformation fundamental at about 828 cm^{-1} and the ν_4 degenerate bending absorption peak at about[48] 716 cm^{-1}. The degenerate ν_3 stretching vibration of nitrate salts is often split considerably. Its mean value in the case of potassium nitrate was taken as[48] 1371 cm^{-1}.

An 84 atom % potassium ^{18}O-nitrate sample displayed its ν_1, ν_2 and ν_4 bands at about 1028, 817 and 705 cm^{-1}, respectively. The centre of its ν_3 absorption was taken[48] to be at 1362 cm^{-1}.

When the relationships expected to hold for isotopic $\boldsymbol{D_{3h}}$ XY_3 oscillators (eqns. 3.33, 3.34 and 3.36) are applied to the NO_3^- isotopic fundamentals of potassium nitrate the following isotopic ratios are obtained for the fully labelled $N^{18}O_3^-$ species:

$$\nu_1^i/\nu_1 = 0{\cdot}943; \qquad \nu_2^i/\nu_2 = 0{\cdot}987; \qquad \nu_3^i\nu_4^i/\nu_3\nu_4 = 0{\cdot}931.$$

These values can be compared with the following ratios observed for the labelled sample which is mainly a mixture of the fully labelled modification with the $N^{16}O^{18}O_2^-$ species:

$$\nu_1^i/\nu_1 = 1028/1049 = 0{\cdot}980; \qquad \nu_2^i/\nu_2 = 817/828 = 0{\cdot}987;$$
$$\nu_3^i\nu_4^i/\nu_3\nu_4 = 1362 \times 705/(1371 \times 716) = 0{\cdot}977.$$

It can be seen from these results that while the symmetrical out-of-plane deformation (ν_2) of the NO_3^- ion is not much affected by the lattice forces, its ν_1 and at least one of the degenerate vibrations ν_3 and ν_4 are strongly influenced by them.

7.17 Labelled Sodium Sulphite

The pyramidal SO_3^{2-} ion belongs to the C_{3v} point group and can be expected to show four, infrared-active, fundamentals. The location of its in-phase S—O stretching vibration, ν_1, at 967 cm^{-1}, its symmetrical bending vibration, ν_2, at 620 cm^{-1} and its degenerate bending frequency, ν_4, at 469 cm^{-1} follows from its Raman spectrum.[53] The frequency of its degenerate S—O stretching mode, ν_3, being at about[54] 1164 cm^{-1}, rather than at 933 cm^{-1} as assumed earlier,[53] is suggested by the infrared absorption spectra of the sodium, potassium and barium sulphites and by the frequency of the corresponding ν_3 stretching vibration of the SO_3 molecule. The band occurring occasionally at *ca.* 933 cm^{-1} can be due to the $2\nu_4$ overtone.

The infrared absorption spectrum of normal sodium sulphite showed these fundamentals at about[54] 965, (ν_1), 630 (ν_2), 1164 (ν_3) and 494 (ν_4) cm^{-1}.

A Nujol mull of 57 atom % sodium ^{18}O-sulphite (which contained about 32% of fully labelled sulphite, 42% of $Na_2S^{16}O^{18}O_2$ and 18% of $Na_2S^{16}O_2{}^{18}O$) exhibits its in-phase S—O stretching band at 951 cm^{-1} and its symmetrical bending absorption peak at 620 cm^{-1}. The ν_3 and ν_4 vibrations of the normal sulphite split in this case into several branches[54] as a result of the lower symmetry of the $Na_2S^{16}O^{18}O_2$ species which can be taken as the "weighted mean component" of the labelled sample (the broad absorption peaks of which being probably very near to those of this "mean component").

If the mean frequency of the ^{18}O-enriched sulphite absorption branches corresponding to the normal ν_3 and ν_4 vibrations, respectively, are considered, one obtains the following values for comparison with the normal ν_1, ν_2, ν_3 and ν_4 fundamentals: 951, 620, 1152 and 482 cm^{-1}. These results give the following observed isotopic ratios for the $S^{16}O_3^{2-}$ and $S^{16}O^{18}O_2^{2-}$ oscillators:

$$``\nu_1^i\nu_2^i"/\nu_1\nu_2 = 951 \times 620/(965 \times 630) = 0{\cdot}970;$$
$$``\nu_3^i\nu_4^i"/\nu_3\nu_4 = 1152 \times 482/(1164 \times 494) = 0{\cdot}966.$$

Equations 3.39 and 3.40 for the isotopic frequency ratios of an XY_3 pyramidal oscillator give the following values, respectively, for the corresponding ratios of the $S^{18}O_3^{2-}$ modification: 0·922, 0·917. For the $S^{16}O^{18}O_2^{2-}$ species, however, one can expect values which are nearer to unity by about a third of the respective differences, i.e. in the vicinity of 0·948 and 0·945 (see e.g. Table 7.1 for the ν_3 frequency of $C^{16}O^{18}O$ which is very nearly half-way between those of $C^{16}O_2$ and $C^{18}O_2$; see also Section 3.3.). This fair agreement between the observed and the expected isotopic ratios seems to support the assignment of the 1164 cm^{-1} band to the ν_3 vibration of normal sodium sulphite.

7.18 BaS^{16}O^{18}O$_3$

The isolated tetrahedral SO_4^{2-} ion shows its four, Raman-active, fundamentals at about 981 (ν_1), 451 (ν_2), 1104 (ν_3) and 613 (ν_4) cm^{-1}. Of these vibrations only the triply degenerate S—O stretching, ν_3, and the S—O bending ν_4, are infrared-active.

The infrared absorption spectrum of crystalline barium sulphate shows, however,[55] an in-phase S—O stretching (ν_1) band at about 981 cm^{-1} besides ν_3 absorptions at 1076, 1112 and 1187 cm^{-1} and ν_4 bands at 611 and 637 cm^{-1}.

An 87% $BaS^{16}O^{18}O_3$ sample shows these bands at[55] 961, 1059, 1095, 1189, 604 and 634 cm^{-1}, respectively, with an additional shoulder at 1072 cm^{-1}.

The calculated frequencies for the isolated $^{32}S^{18}O_4^{2-}$ ion are[56] 927 (ν_1), 424 (ν_2), 1074 (ν_3) and 582 (ν_4) cm^{-1}.

7.19 Labelled Phosphates

The isolated PO_4^{3-} anion, being very similar to the sulphate ion, also shows four fundamentals, at 980 (ν_1), 363 (ν_2), 1082 (ν_3) and 515 (ν_4) cm^{-1}, only two of which being infrared-active.

A Nujol paste of normal barium phosphate shows, however, infrared absorption peaks due to all the PO_4^{3-} fundamentals,[57] at 980 (ν_1), 367 and 392 (ν_2), 1082 (ν_3), 435, 489, 540 and 590 cm^{-1} (ν_4).

A mull of 75% barium ^{18}O-phosphate, (which can be calculated to be composed of

$$5\%\ Ba_3(P^{16}O_3{}^{18}O)_2, \quad 21\% Ba_3(P^{16}O_2{}^{18}O_2)_2,$$
$$42\%\ Ba_3(P^{16}O^{18}O_3)_2 \quad \text{and} \quad 31\%\ Ba_3(P^{18}O_4)_2)$$

exhibited these bands at 937 (ν_1 of the weighted mean species of this sample, $P^{16}O^{18}O_3^{3-}$), 1037 (ν_2), 526 and 558 (ν_3), 976 (ν_4), 361 and 392 (ν_5) and 426 and 515 (ν_6) cm^{-1}. The higher number of the absorption bands in this case is the result of the lower symmetry of the unhomogeneous main labelled species, which splits, e.g. the normal degenerate 1082 cm^{-1} band into the 976 and 1037 cm^{-1} branches.

The infrared absorption of an aqueous solution of 89 atom % ^{18}O-labelled K_3PO_4 displayed a weak band at 948 cm^{-1} and a strong band at 1050 cm^{-1}, corresponding to the normal phosphate aqueous solution bands at 938 (*vw*) and 1012 (*s*) cm^{-1}, respectively.[58] These bands must be assigned to the symmetric ν_1 and asymmetric ν_3 P—O stretching fundamentals of $P^{16}O_4^{3-}$ and

$P^{18}O_4^{3-}$, accordingly.[58] The fact that ν_1 although theoretically infrared-inactive gives rise to weak infrared bands in these solutions is probably the result of random fluctuations in the number and identity of the molecules hydrogen bonded to the four oxygen atoms of the PO_4^{3-} ions, which bring about a weak statistical asymmetry in the structure of the solvated PO_4^{3-} ions.

The astonishing observation that ^{18}O-labelling the hydrated PO_4^{3-} ions *increases* their ν_1 stretching frequency by about 10 cm^{-1} and their ν_3 stretching fundamental by *ca.* 38 cm^{-1} has been attributed to a weaker hydrogen bonding of the water molecules to the ^{18}O-labelled ions which more than compensates for the ^{18}O-induced isotopic shifts of these vibrations to lower frequencies.[58]

Similar phenomena have also been observed for crystalline *hydrated* ^{18}O-labelled sodium orthovanadate and its solutions in D_2O. Thus, while powdered $Na_3V^{16}O_4 \cdot 14H_2O$ exhibits its ν_3 absorption peaks at 822 and 866 cm^{-1} its 80 atom % ^{18}O-labelled modification displays them at about[59] 830 and 867 cm^{-1}. In the case of the orthovanadate solutions in D_2O, which contain mainly the DVO_4^{2-} ion species, the effect of ^{18}O-labelling is even more pronounced[59] *increasing* its asymmetric V—O stretching absorption band from 854 to 870 cm^{-1}.

7.20 Labelled Potassium Permanganate

The tetrahedral MnO_4^- ion also belonging to the $\boldsymbol{T_d}$ point group possesses again four normal modes of vibration, only ν_3 and ν_4 of which being infrared-active.

Crystalline potassium permanganate, however, can be expected to show infrared absorption bands due to all of the MnO_4^- fundamentals since the site symmetry of this ion in such crystals is low.[60]

This salt indeed shows infrared bands at[61] 844 (ν_1), 385 (ν_2), 910 (ν_3) and 407 (ν_4) cm^{-1}. The assignment of its ν_1, ν_2 and ν_3 vibration modes, however, has until recently been doubtful.[61, 62] This assignment has finally been proved by the infrared absorption of a potassium bromide pellet of 86 atom % potassium ^{18}O-permanganate (55% $KMn^{18}O_4$) which displayed its fundamentals at 806, 368, 873 and 386 cm^{-1}, respectively. Applying eqn. 3.19 and eqn. 3.20 to this case gives the following calculated isotopic ratios for the $Mn^{18}O_4^-$ ions:

$$\nu_1^i/\nu_1 = 0{\cdot}943; \qquad \nu_2^i/\nu_2 = 0{\cdot}943; \qquad \nu_3^i\nu_4^i/\nu_3\nu_4 = 0{\cdot}918.$$

Substituting the various ν's in these ratios by the observed frequencies, on basis of the assignment given above, results in the following observed isotopic ratios:

$$\nu_1^i/\nu_1 = 0{\cdot}953; \qquad \nu_2^i/\nu_2 = 0{\cdot}956; \qquad \nu_3^i\nu_4^i/\nu_3\nu_4 = 0{\cdot}910.$$

Taking into account the fact that the investigated sample also contained about 36% of the $KMn^{16}O^{18}O_3$ modification and the effect of the lattice forces on the MnO_4^- vibrations this agreement between the observed and the calculated ratios can be accepted as very good and therefore as supporting strongly the assignment used in these calculations.

A D_2O solution of 83·3 atom % ^{18}O-labelled sodium permanganate exhibited its ν_3 Mn—^{18}O stretching band at 878 cm^{-1} as compared with the 917 cm^{-1} frequency of the corresponding Mn—^{16}O fundamental.[63] The X—^{18}O stretching band of this aqueous solution thus behaves normally and shows an ^{18}O-induced shift to *lower* frequencies comparable to that observed for KBr pellets of crystalline isotopic sodium[63] (from 904 to 870 cm^{-1}) or potassium permanganates[61] (from 910 to 873 cm^{-1}).

D_2O solutions of 84·5 atom % ^{18}O-labelled sodium orthoarsenate also display a normal ^{18}O-isotopic shift for their ν_1 and ν_3 $As^{18}O_4^{3-}$ stretching bands which appear at 808 (*sh*) and 822 (*s*) cm^{-1} (with an additional $As^{16}O^{18}O_3^{3-}$ band (*m*) at 854 cm^{-1}), respectively,[63] as compared with the corresponding As—^{16}O fundamentals at 834 (*sh*) and 865 (*s*) cm^{-1}. These shifts parallel the ^{18}O-induced shift observed for the crystalline sodium arsenate absorption band which moves[63] from 851 to 804 cm^{-1}.

The reason for the normal behaviour of the hydrated $Mn^{18}O_4^-$ and $As^{18}O_4^{3-}$ ions, while the hydrated $P^{18}O_4^{3-}$ and ^{18}O-vanadate ions exhibit an *increase* in their stretching frequencies, relative to the respective normal ion fundamentals, seem to lie in the low extent to which the MnO_4^- and AsO_4^{3-} stretching vibrations are affected by the hydration of these ions.[63]

Organic Compounds

7.21 ^{18}O-ENRICHED ALCOHOLS AND PHENOLS

The infrared absorption of only a few ^{18}O-labelled alcohols or phenols seems to have been investigated so far, these being ^{18}O-phenol[64] (74 atom %), ^{18}O-2,4,6-triphenylphenol,[65,66] ^{18}O-triphenylmethyl carbinol[67] (78 atom %) and propylene-1,2-diol[68] (29 atom %).

Since such molecules usually tend to associate, forming hydrogen bonds with each other, and such associated molecules show diffuse O—H bands, these substances have as a rule been investigated in dilute carbon tetrachloride, chloroform or cyclohexane solutions where they exist as monomers. Table 7.10 collects the absorption frequencies of $C_6H_5{}^{18}OH$, $(C_6H_5)_3C^{18}OH$ and $^{18}O—C_3H_8O_2$ for which significant isotopic shifts ($\Delta\nu \geqslant 3$ cm^{-1}) have been observed. Their present assignment, based in a number of cases on the magnitude of these, ^{18}O-induced, isotopic shifts, is included in the Table together with the previous (literature) assignment, as far as this differs from the present assignment.

TABLE 7.10 Absorption Frequencies (cm^{-1}) Differing Significantly for the R—^{18}OH Compounds from the Corresponding R—^{16}OH Values (R = C_6H_5,[64] $(C_6H_5)_3C$,[67] CH_3—CHOH—CH_2[68])

Normal frequency	Labelled compound frequency	Present assignment	Previous assignment
		C_6H_5OH	
272	277	Out-of-plane C—OH bending	
320	315	Out-of-plane O—H bending	
400	389	In-plane C—OH bending	
510	506	Out-of-plane, X-sensitive, ring deformation	
536	530	In-plane, symmetrical X-sensitive, ring deformation	
608	604	In-plane, unsymmetrical ring deformation	
812	802	In-plane, symmetrical X-sensitive ring deformation	
827	841	Out-of-plane, X-sensitive, C—H bending	
1152	1155	In-plane C—H bending	
1256	1250	C_{ar}—OH stretching	
1313		510 + 812	In-plane C—H deformation
	1302	506 + 802	
1333		320 + 1000	530 + 814
	1330	315 + 1000	
1472		Ring stretching	
	1455, 1480	Resonance of ~ 1470 ring stretching with 224 + 1250	
3620	3610	O—H stretching	
		$(C_6H_5)_3COH$	
582	572	C—OH bending (in the C—O—H plane)	
920	917	C—H bending	
1018	1006	C—OH stretching	
1038	1035	C—H bending	
1085	1073	Another C—OH stretching	C—H bending
1298	1294	In-plane O—H bending	
1689	1685	920 + ~ 765	
3621	3609	O—H stretching	
		$CH_3CHOHCH_2OH$	
3638	3642	—CH_2^{16}O—H stretching	
~3490	~3480	Internally hydrogen bonded O—H stretching	

It can be seen from Table 7.10 that substituting ^{18}O for ^{16}O in the hydroxylic group lowers its stretching frequency by 10–12 cm^{-1}. This value is in perfect agreement with expectation since the vibrational energy of this stretching is localized, practically, only in the O—H group (cf. Section 4.1), hence one can treat this vibration as belonging to a diatomic O—H molecule and apply to it eqn. 7.1 which gives, in this case,

$$\frac{\nu^i}{\nu} = \frac{2}{3}\left(\frac{2(18 + 1)}{16 + 1}\right)^{1/2} = 0{\cdot}9966$$

and $\Delta\nu = \nu - \nu^i = (1 - 0{\cdot}9966)\nu = 0{\cdot}0034{\cdot}3620\ cm^{-1} = 12\ cm^{-1}$. The data of Table 7.10 also show that the C—OH stretching bands of the R—OH compounds undergo an ^{18}O-induced isotopic shift of 6–12 cm^{-1}. This fact makes possible the identification of such bands by their observed ^{18}O-shifts, as has recently been done for the 1085 cm^{-1} band of triphenylmethyl carbinol[67] which was previously assigned to a C—H bending vibration.

The absence of any ^{18}O-shift for the 1330 cm^{-1} band of this carbinol has shown its previous assignment to an in-plane O—H bending vibration to be wrong since such a vibration must display some shift as a result of ^{18}O-labelling. This band must therefore belong to a C—C stretching vibration.

The CH_3—O—C stretching bands of ^{18}O-2,4,6-triphenylanisole appear at[65,66] 984 and 1216 cm^{-1} as compared with the respective CH_3—^{16}O—C values of 1008 and 1228 cm^{-1}.

7.22 Labelled Aldehydes

^{18}O-Enriched aldehydes, which contain the C=^{18}O group, can, according to Table 7.1, be expected to show C=O stretching bands at frequencies lower by about 30 cm^{-1} from the corresponding band frequencies of their normal analogues. It was indeed found[6] that a number of *ca.* 50 atom % ^{18}O-aldehydes exhibited their C=^{18}O stretching bands (in tetrahydrofuran solution) at the positions indicated in Table 7.11, which are, as a rule, lower by about 30 cm^{-1} from the respective normal C=O frequencies.

7.23 ^{18}O-Enriched Ketones

The infrared absorption spectra of ^{18}O-enriched ketones have been studied more thoroughly than those of the labelled aldehydes which have been measured only in the C=O stretching region. 2,4-Dimethyl-3-pentanone containing about 35 atom % ^{18}O was investigated (in carbon disulphide solution) in the 850–4000 cm^{-1} region.[69] Its only bands for which an ^{18}O-isotopic shift was observed were those due to the C=O stretching vibration, which moved from 1712 to 1681 cm^{-1}, and the C=O bending, C—C stretch-

TABLE 7.11 The C═O Stretching Frequencies (cm^{-1}) of ^{18}O-Labelled Aldehydes[6]

Compound	Structure	$C{=}^{16}O$ Frequency	$C{=}^{18}O$ Frequency	Difference
Acetaldehyde	CH_3CHO	1720	1692	28
Benzaldehyde	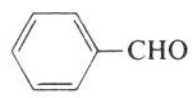	1703	1675	28
1-Naphthaldehyde	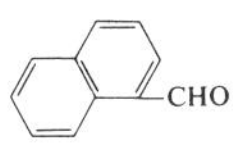	1692	1664	28
2-Naphthaldehyde	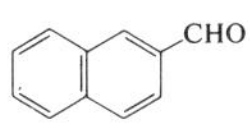	1697	1668	29
9-Anthraldehyde	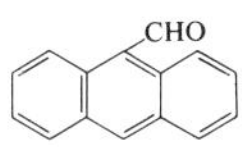	1676	1650	26
9-Phenanthraldehyde	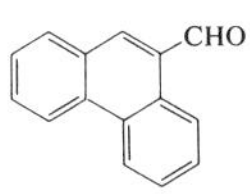	1691	1661	30
3-Indole aldehyde	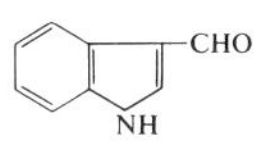	1667	1641	26
Chlorophyl b	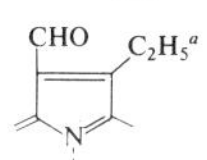	1665	1637	28

[a] Partial structure.

ing fundamental (see Fig. 7.3), which shifted from 1203 to 1192 cm^{-1}, respectively.

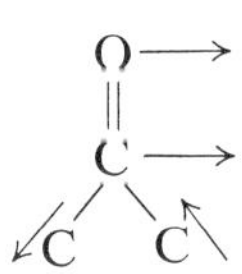

Fig. 7.3 The ketone vibrational mode which gives rise to the absorption band at about 1200 cm^{-1}.

The observed $C{=}^{18}O$ stretching shift, amounting here also to *ca.* 30 cm^{-1} (1712–1681 cm^{-1}) while the calculated value for the corresponding diatomic $C^{18}O$ oscillator is about 41 cm^{-1} (see Table 7.1), is in harmony with the normal co-ordinate analysis of the acetone molecule vibrational modes.[70] According to this analysis the 1720 cm^{-1} acetone vibration has a *ca.* 80 % C═O stretching and about 20% C—C stretching character. Hence, the ^{18}O-shift to be expected of this band is $0{\cdot}80 \times 41 = 33$ cm^{-1}.

A comparison of the infrared absorption of a carbon tetrachloride solution of 89 atom % ^{18}O-benzophenone with the respective normal ketone absorption has also shown[1] that the only significant differences in the absorption of the isotopic benzophenones in the 370–3500 cm^{-1} region are in the C=O stretching vibration, which moves from 1667 to 1637 cm^{-1}, and the 640 cm^{-1} C=O bending frequency, which goes over to 637 cm^{-1} in the case of the labelled ketone.

Table 7.12 reports the frequency of further ketonic C=^{18}O stretching bands,[7] measured in piperidine solutions.

TABLE 7.12 The C=^{18}O Stretching Frequency (cm^{-1}) of Various Ketones

Compound	Frequency	Isotopic shift	Compound	Frequency	Shift
Acetone	1678	30	Cyclohexanone	1678	29
Cyclobutanone	1747	33	Fluorenone	1685	31
Cyclopentanone	1702	33	Indanone	1683	30
Cyclopent-2-enone	1708	31	Ethylaceto-acetate	1684[a]	27

[a] The value for the ketonic carbonyl group.

^{18}O-Labelled cyclopent-2-enone solutions in carbon tetrachloride exhibit their C=O stretching band at 1689 cm^{-1} as compared with the corresponding normal ketone fundamental[71] at 1718 cm^{-1}. Such ^{18}O-2,3-dimethylcyclohex-2-enone solutions display it at 1636 cm^{-1}, corresponding to the respective C=^{16}O stretching 1670 cm^{-1} frequency.[71] The values for these absorption peaks in the case of other solvents have also been reported.[71]

^{18}O-2-Cyclopentylidenecyclopentanone solutions in various solvents exhibit its main C=^{18}O stretching band between 1679 (in $CHBr_3$) and 1694 cm^{-1} (in *n*-hexane) showing an ^{18}O-shift of 19–23 cm^{-1}. Their C=C stretching fundamental which has a *ca.* 30% C=O stretching character appears in the 1621–1636 cm^{-1} span displaying an isotopic shift of[71] 9–11 cm^{-1}.

The ^{18}O-sensitive absorption bands of ^{18}O-tropone appear[72] at 698 (ring deformation), 740 (*sh*), 1138, 1570 (in CCl_4; C=O stretching), 1755, 2910 and 3130 cm^{-1} (2 × 1570), corresponding to the normal tropone bands at 708, 745, 1141, 1594, 1760, 2915 and 3175 cm^{-1}, respectively.

The 1643 and 1861 cm^{-1} fundamentals of cyclohexane solutions of diphenylcyclopropenone, which are probably due to an interaction between its C=O and C=C stretching vibrations, move in the case of its ^{18}O-labelling[73] to 1633 and 1837/1850 cm^{-1} (a doublet due to Fermi resonance). For its carbon tetrachloride solutions they appear at 1626 and 1838 cm^{-1}, respectively.[73]

The 600–1800 cm^{-1} absorption spectrum of 85 atom % ^{18}O-tropolone solutions (in $CHCl_3$ or CS_2) shows its ^{18}O-sensitive bands at 1206, 1243, 1256 (C—O stretching character), 1294, 1433, 1467 and 1536 cm^{-1} (C=O valence vibration character) corresponding to the normal compound bands at 1209, 1247, 1263, 1300, 1437, 1470 and 1547 cm^{-1}, respectively.[74]

The 400–1700 cm^{-1} absorption of 87 atom % ^{18}O-labelled bistropolonato copper (II) has also been reported.[74]

The ^{18}O-induced shift of about 30 cm^{-1} observed for the *ca.* 1570 cm^{-1} band of "1-phenylazo-2-naphthol" and "2-phenylazo-1-naphthol", which actually exist in the hydrazo-form (see Fig. 7.4), has been used as proof that

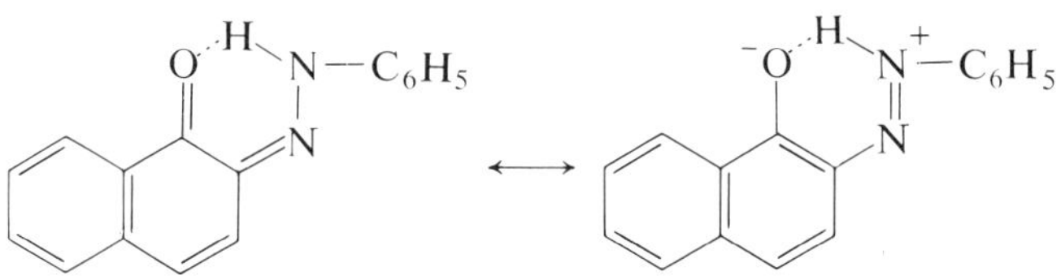

I II

Fig. 7.4 The structure of "2-phenylazo-l-naphthol".

this band is really due to the C=O stretching vibration of these compounds,[75] the frequency of which is lowered from its usual value, as a result of a resonance such as that existing between the I and II forms of "2-phenylazo-1-naphthol" (Fig. 7.4).

7.24 Labelled Compounds Containing More Than One Ketonic Group

^{18}O-Enriched compounds containing two or more ketonic groups adjacent one to the other or conjugated one with the other were also investigated.

2,3-Butanedione containing about 50 atom % ^{18}O, the major component of which is $CH_3C^{16}OC^{18}OCH_3$, dissolved in carbon tetrachloride showed two sharp absorption bands of similar intensity,[68] at 1688 and 1721 cm^{-1}. Since the last frequency is also the C=O absorption frequency of the normal diketone and the 1721–1688 = 33 cm^{-1} shift is the one to be expected of the fully labelled diketone (by analogy to the isotopic shifts recorded in Table 7.12) it must be concluded that the C=O absorption peaks of the mixed isotopic diketone also fall at, or very near to, these locations, i.e. *this modification behaves in this respect, practically, as if it were a mixture of the two isotopically homogeneous species*. It seems that these bands of the mixed diketone correspond, respectively, to the in-phase and out-of-phase C=O stretching modes while their frequencies, in the case of an homogeneous diketone, are very near one to the other; in fact one can discern a shoulder at about 1714 cm^{-1} on top of the normal diacetyl band at about[68] 1721 cm^{-1}.

Ninhydrin,

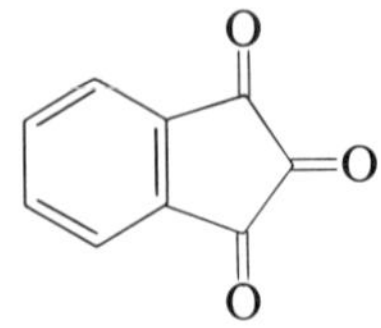

which normally absorbs at about 1714 and 1739 cm^{-1}, was observed[7] to show a band at about 1692 cm^{-1} after an ^{18}O-enrichment of about 50%. This band is probably the result of a merging together of a number of absorption peaks due to various ^{18}O-labelled modifications of ninhydrin. The difference between the mean $C{=}^{16}O$ frequency of ninhydrin (1714 + 1739)/2 = 1726 cm^{-1}) and its $C{=}^{18}O$ frequency (1692) being 34 cm^{-1} is, however, very near to the usual ketonic ^{18}O-shift of 27–33 cm^{-1}.

The infrared absorption of $^{18}O_2$-*p*-benzoquinone and D_4-$^{18}O_2$-*p*-benzoquinone (belonging to the $\boldsymbol{D_{2h}}$ point group), of *ca.* 80 atom % enrichment,[76, 77] in the 300–4000 cm^{-1} region was investigated as part of a study of the spectra of seven isotopic modifications of *p*-benzoquinone, aimed at the arrival at the correct assignment of all of its vibrations.

Table 7.13 reports the frequencies of those quinone infrared absorption bands which showed a significant ^{18}O-shift, together with their assignment.[78]

TABLE 7.13 The ^{18}O-Shifted Infrared-Active Fundamental Frequencies[a] (cm^{-1}) of Isotopic *p*-Benzoquinone[78]

Normal Quinone	$^{18}O_2$-Compound	D_4-$^{18}O_2$-Quinone	Assignment
409	394	387	B_{3u} C=O bending
505	502	412	B_{1u} C=O out-of-plane bending
728	718	630	B_{2u} ring bending
944	941	864	B_{2u} C—C stretching
1354	1351	1158	B_{2u} C—H bending
(1666[b])	1634	1625	B_{2u} C=O stretching

[a] For CCl_4 solutions.
[b] Calculated from the actual Fermi-resonating frequencies at 1656 and 1668 cm^{-1}.

It can be seen from Table 7.13 that the out-of-phase C=O stretching vibration of *p*-benzoquinone also shows an ^{18}O-shift of *ca* 30 cm^{-1}, similarly to the usualy $C{=}^{18}O$ shift of ordinary ketones (see Table 7.12).

7.25 ENRICHED BENZOYL CHLORIDE

The only ^{18}O-labelled acid halogenide to be investigated so far seems to be benzoyl chloride.[79] When the absorption of a 90 atom % ^{18}O-benzoyl chloride carbon tetrachloride solution is compared with that of a normal benzoyl chloride solution in the 830–3500 cm^{-1} region, it is seen that the only significant difference between them is that while the normal chloride shows a very strong band at 1777 cm^{-1} and strong bands at 1736 and 1203 cm^{-1}, the labelled compound shows a strong absorption peak at 1747 cm^{-1} and very strong bands at 1711 and 1199 cm^{-1}.

The 1777/1736 cm^{-1} doublet of the normal compound is the result of Fermi resonance between the $C{=}^{16}O$ stretching frequency, originally at about 1765 cm^{-1}, and the overtone of the 875 cm^{-1} vibration, to be expected at about 1748 cm^{-1}. In the case of the labelled chloride the original $C{=}^{18}O$ frequency can be anticipated to appear at about 1730 cm^{-1} while the overtone of the 872 cm^{-1} vibration is expected at about 1740 cm^{-1}. Fermi resonance between these vibrations is therefore again possible and is responsible for the resulting 1747/1711 cm^{-1} doublet. Since, however, the *lower*, 1711 cm^{-1}, band possesses now the larger C=O character (being nearer to the original 1730 cm^{-1} $C{=}^{18}O$ stretching frequency) this band is now *stronger* than the upper one (at 1747 cm^{-1}).

The 4 cm^{-1} ^{18}O-shift of the 1203 cm^{-1} band is analogous to the similar (albeit larger) ^{18}O-shift of the 1203 cm^{-1} band of 2,4-dimethyl-3-pentanone (Section 7.23). This fact suggests that the former is due to a similar vibrational mode (see Fig. 7.3).

7.26 Labelled Acids

In the case of carboxylic acids two groups can be labelled by oxygen-18, the hydroxylic and the carbonylic groups. Since the energy of the stretching of each of these groups is highly localized in itself one can expect $^{18}O_2$-labelled acids to show isotopic shifts very similar to those of ^{18}O-labelled alcohols on one hand and to those of ^{18}O-ketones on the other. The necessity, however, to use highly diluted chloroform solutions, in order to break down the strongly hydrogen bonded acid dimers, makes it very difficult to investigate any but the strongest band of the monomeric carboxylic acids, i.e. that due to their CO stretching vibration.

Such solutions of *ca.* 90 atom % ^{18}O-benzoic acid show their CO band at[79] 1688 cm^{-1}, while normal benzoic acid under these conditions absorbs at 1715 cm^{-1}. In carbon tetrachloride, however, even a concentration of 0·03 g per litre showed a strong labelled dimeric acid band, at 1670 cm^{-1}, near a weaker monomeric acid band at 1707 cm^{-1}, as compared with the corresponding normal values of 1690 and 1739 cm^{-1}, respectively. These values give an ^{18}O-isotopic shift of 32 cm^{-1} for the monomeric benzoic acid carbon tetrachloride solution as compared with the value of 27 cm^{-1} for the respective chloroform solution.

Carbon tetrachloride solutions of labelled dimeric benzoic acid exhibit ^{18}O-sensitive absorption peaks in the 33–500 cm^{-1} region only at 270 (*vs*;

O—C=O bending) and 380 (w) cm^{-1}, corresponding to the 279 and 388 cm^{-1} bands of the normal dimeric benzoic acid.[80] Cyclohexane solutions of $(CH_3C^{18}O_2H)_2$ show characteristic far infrared bands at 175 (s) and 461 cm^{-1} (vs), as compared with the 179 and 478 cm^{-1} values, respectively, observed for normal dimeric acetic acid solutions.[80]

60 Atom % ^{18}O-labelled sodium deuteroformate (in a KBr pellet) absorbs at 732 ($C^{18}O_2^-$ bending), 750 ($C^{16}O^{18}O^-$ bending), 911 (out-of-plane C—D deformation), 1005 (D—$C^{18}O_2^-$ bending), 1009 (D—$C^{16}O^{18}O^-$ bending), 1289 (symmetric $C^{18}O_2^-$ stretching), 1310 (symmetric $C^{16}O^{18}O^-$ stretching), 1577 (asymmetric CO_2 stretching) and 2132 cm^{-1} (C—D stretching). The fundamentals of the $DC^{16}O_2^-$ ions appear at 766 (CO_2^- bending), 914 (C—D deformation), 1013 (C—D bending), 1332 (symmetric CO_2^- stretching), 1590 (asymmetric CO_2^- stretching) and[81] 2135 cm^{-1} (C—D stretching).

7.27 Crystalline 75 Atom % ^{18}O-α-Glycine

The infrared absorption of ^{18}O-labelled glycine, which as all the amino acids exists as a dipolar ion, in this case as $NH_3^+—CH_2—CO_2^-$, is very significant in respect to the assignment of those fundamental vibrations of glycine which are localized mainly in the oxygen atoms.

A powdered sample of one of the three crystalline modifications of glycine, which contained about 75 atom % ^{18}O was investigated in the 400–4000 cm^{-1} region.[82]

Figure 7.5 shows the absorption of this sample in the 650–1650 cm^{-1} region.

Table 7.14 gives the frequency of the normal α-glycine bands which are

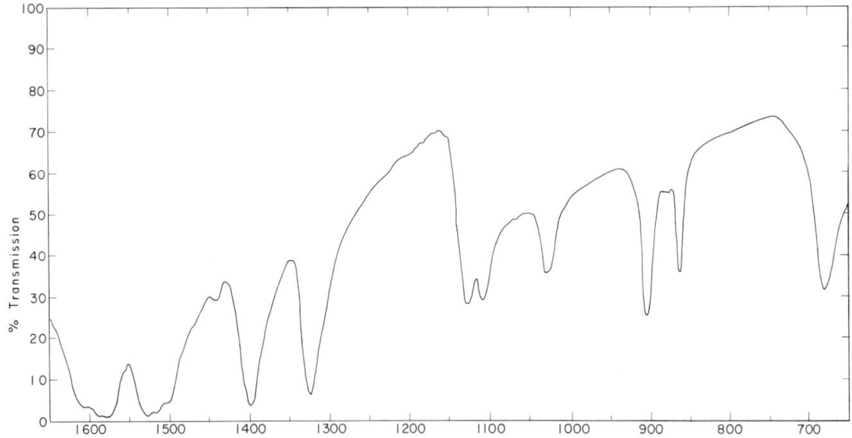

Fig. 7.5 Infrared absorption of 75 atom % ^{18}O-labelled glycine between 650 and 1650 cm^{-1}.

shifted significantly by ^{18}O, together with the corresponding ^{18}O-α-glycine frequencies and their assignments. The previous assignments of these bands, made on basis of a normal co-ordinate analysis of the glycine modes of vibration,[83] are also added as far as they differ from the new assignments suggested in most cases by the isotopic shifts observed for ^{18}O-α-glycine.

TABLE 7.14 The Main ^{18}O-Shifted Bands (cm^{-1}) of Normal and ^{18}O-α-Glycine[82]

Normal frequency	Labelled frequency	^{18}O-Shift	Present assignment	Previous assignment[83]
608	601	7	CO_2^- wagging	
700	685	15	CO_2^- rocking	CO_2^- scissoring
894	870	24	CO_2^- scissoring	C—C stretching
1334	1324	10	CH_2 wagging with a C—CO_2^- stretching character	CH_2 wagging
1413	1401	12	CO_2^- symmetrical stretching	
1604	1586, 1574	18; 30	CO_2^- antisymmetrical stretching	

The 1586 and 1574 cm^{-1} bands belong, respectively, to the $C^{16}O^{18}O^-$ and $C^{18}O_2^-$ species present in the labelled sample.

Assuming that the $-CO_2^-$ oscillator is harmonic and practically isolated from the rest of the glycine molecule it is possible to estimate the antisymmetrical $-C^{18}O_2$ stretching frequency by applying here eqn. 3.8 which, taking the OCO angle as 125·5°, gives in this case:

$$\frac{\nu^{(i)2}}{1604^2} = \frac{12 \times 16(12 + 2 \times 18 \times 0{\cdot}8141^2)}{12 \times 18(12 + 2 \times 16 \times 0{\cdot}8141^2)} = 0{\cdot}9598.$$

Hence $\nu^i = 1604(0{\cdot}9598)^{1/2} = 1571\ cm^{-1}$, in good agreement with the observed value of 1574 cm^{-1}.

The fact that there appears only one band due to the symmetrical CO_2^- stretching of the labelled glycine seems to be the result of a merging together of the two $C^{16}O^{18}O^-$ and $C^{18}O_2^-$ absorption peaks expected in the 1400 cm^{-1} region.

The ^{18}O-shift of 10 cm^{-1} observed for the 1334 cm^{-1} band strongly suggests that the vibration responsible for this band possesses a very appreciable C—CO_2^- stretching character which makes the frequency of this vibration quite sensitive to ^{18}O-labelling.

Assuming such a C—CO_2^- stretching vibration to be mainly localized in this group, one can get a rough estimate of the ^{18}O-shift of such a vibration by applying here also eqn. 3.6, taking the CO_2 group as one unit with isotopic masses of 12 + 2 × 16 = 44 and 12 + 2 × 18 = 48, respectively. One thus obtains:

$$\frac{\nu^i}{\nu} = \left(\frac{\mu}{\mu^i}\right)^{1/2} = \left(\frac{\frac{1}{12} + \frac{1}{48}}{\frac{1}{12} + \frac{1}{44}}\right)^{1/2} = (0{\cdot}98)^{1/2} = 0{\cdot}99.$$

Hence, the shift $\nu - \nu^i = \nu(1 - 0{\cdot}99) = 1334 \times 0{\cdot}01 = 13\ \text{cm}^{-1}$, in reasonable agreement with the observed shift of 10 cm^{-1} for this band.

This result proves that the 894 cm^{-1} band which shows an ^{18}O-shift of 24 cm^{-1} cannot be due to the C—C stretching frequency as assigned before,[83] but must belong to the CO_2^- bending vibration which is more localized in the oxygen atoms and as such can be expected to be much more sensitive to the ^{18}O-labelling.

7.28 ^{18}O-ENRICHED ESTERS

The first study of the infrared absorption of ^{18}O-*enriched* compounds seems to have been carried out on an ester.[10] This ester which had its carbonyl group oxygen atom enriched by about 9 atom % ^{18}O was cinnamyl *p*-nitrobenzoate which showed its $C{=}^{18}O$ stretching band, in a carbon tetrachloride solution, at about 1700 cm^{-1} as compared with the 1730 cm^{-1} frequency of the corresponding normal ester. The ^{18}O-induced shift of the esteric CO stretching band is thus equal to that of the ketonic CO (see Table 7.12) frequency.

A more detailed investigation has been carried out on a methyl (^{18}O-carbonyl) benzoate sample of 84 atom % enrichment.[79] Table 7.15 gives the ^{18}O-shifts observed for the main ^{18}O-sensitive bands of methyl benzoate in the 660–4000 cm^{-1} region.

TABLE 7.15 The Frequency (cm^{-1}) of the Main ^{18}O-Sensitive Bands of Methyl Benzoate[79]

Normal compound	^{18}O-Labelled compound	Shift	Assignment
675	671	4	Out-of-plane CO_2CH_3 bending
707	702	5	CO_2CH_3 rocking[a]
822	808	14	In-plane CO_2CH_3 bending[b]
1111	1108	3	=C—O—C symmetrical stretching
1277	1273	4	=C—O—C asymmetrical stretching
1727	1696	31	C=O stretching

[a] This assignment seems now more probable than that given in ref. 79, since the rocking of the CO_2^- group has now also been shown (Table 7.14) to absorb at *ca.* 700 cm^{-1}.

[b] This interpretation of the 822 cm^{-1} band appears now more plausible than that of ref. 79, in the light of the recent assignment of the 894 cm^{-1} band of glycine to the CO_2^- scissoring.[82]

The $C{=}^{18}O$ stretching frequency of a transition complex formed during the ester cleavage of phenyl benzoate was also observed at[84] 1696 cm^{-1}.

7.29 LABELLED AMIDES

The carbonyl group of amides forms part of the $-\overset{\overset{\displaystyle O}{\|}}{C}-NR_1R_2$ system which is usually strongly in resonance with the polar form $^{-}O-\underset{|}{C}=N^{+}R_1R_2$ where the C—O linkage is no more a double bond. Hence the ^{18}O-induced shift of the stretching vibration of this group can be expected to be appreciably lower than that of the corresponding band in the case of ketones and esters. Table 7.16 shows that this is really the case for monomeric amide molecules (in dilute solutions).

TABLE 7.16 The CO Stretching Frequency (cm^{-1}) of Isotopic Amides

Compound	Solvent	$C={}^{16}O$ frequency	$C={}^{18}O$ frequency	Shift	Reference
Dimethylformamide	CCl_4	1691	1667	24	—
Propionamide	$CHCl_3$	1680	1660	20	85
Benzamide	$CHCl_3$	1675	1655	20	86
Benzamide	CCl_4	1690	1666	24	86
N-Methylbenzamide	CCl_4	1673	1649	24	87
NN-Dimethylbenzamide	CCl_4	1640	1616	24	88

Besides the C=O band, the "II amide" band of benzamide also exhibits a remarkable ^{18}O-shift, its frequency moving from 1584 cm^{-1} for the normal amide to 1574 cm^{-1} for the labelled compound.[86] This unexpected shift suggests that the vibrational mode responsible for the II band of primary amides affects considerably also the position of the oxygen atom[86] and not only the N—H bonds as previously assumed.

The "III amide" band of benzamide, on the other hand, does not exhibit any shift from its 1358 cm^{-1} frequency as a result of ^{18}O-labelling although Miyazawa *et al.*[89] have assumed that the "III amide" band of formamide (at 1310 cm^{-1}) is in fact due to a symmetrical stretching of the N—C=O group.

The fact that the ^{18}O-isotopic shift of the C=O stretching band is the same for a primary, for a secondary and for a tertiary amide (as can be seen from Table 7.16) shows that the localization of this vibration in the CO group in the case of the primary and secondary amides is not appreciably lower than that in the case of the tertiary amides. This conclusion is in contradiction[88] to the assumption that the primary amide C=O stretching mode is strongly coupled with the N—H bending vibration or that the secondary amide CO stretching mode is coupled with that of the C—N stretching.[89,90]

The shifting of the 1267 cm^{-1} band of normal NN-dimethylbenzamide to

1261 cm^{-1}, in the case of the labelled modification, is in harmony with the analogous shifts of the similar, C=O bending, C—C stretching, ketonic band at *ca.* 1200 cm^{-1} (see Section 7.23 and Fig. 7.3) and benzoyl chloride band at 1203 cm^{-1} (Section 7.25).

Chloroform solutions of over 67 atom % ^{18}O-labelled 2-pyridone display their C=^{18}O stretching band at 1651 cm^{-1} and their II amide band at 1538 cm^{-1}. For 1-deutero-2-pyridone solutions in carbon tetrachloride, the ^{18}O-isotopic shift of the main CO stretching absorption peak amounts to 14 cm^{-1} while the C=^{18}O fundamental of labelled 1-methyl-2-pyridone (in $CHCl_3$) appears at 1572 cm^{-1} as compared with the corresponding 1583 cm^{-1} C=^{16}O frequency.[91,92] The ^{18}O-1-methyl-2-pyridone also displays ^{18}O-sensitive bands at 868 ($\Delta\nu = 8$) and 1655 cm^{-1} ($\Delta\nu = 4$).

^{18}O-4-Pyridone shows characteristic bands at 1381 ($\Delta\nu = 9$) and 1512 cm^{-1} ($\Delta\nu = 4$) which were observed in the case of ^{18}O-1-methyl-4-pyridone at 1392 and 1555 cm^{-1}, respectively.[93]

7.30 ^{18}O-Enriched Ureas

The infrared absorption of labelled urea as well as N,N′-dicyclohexyl and N,N′-di-*p*-tolylurea of 80–90 atom % ^{18}O-enrichment has been investigated quite extensively.[94–96]

Since these compounds are not soluble in the non-polar solvents usually used in infrared spectroscopy their spectra have been measured either in the solid phase (present as Nujol mulls, alkali halide pellets, fine powders or single crystals) or on solutions in dimethylsulphoxide or dimethylsulphoxide-carbon tetrachloride mixtures.

Figure 7.6 shows the absorption of powdered 87% ^{18}O-urea in the C=O stretching region. Figure 7.7 shows the respective absorption of a solution of this compound in a carbon tetrachloride-dimethylsulphoxide mixture (Fig. 7.7 gives, however, the *absorbance* of the solution, not its % transmission as Fig. 7.6).

It is immediately seen from these figures that going over from the solid phase to a solution brings about an inversion in the relative intensities of the absorption bands above 1600 cm^{-1}; while the lower band (at 1624 cm^{-1}) of solid ^{18}O-urea is much stronger than its upper one (at 1672 cm^{-1}) in the solution spectrum, the first band is much weaker (at 1618 cm^{-1}) than the upper band (at 1671 cm^{-1}). The same behaviour has also been observed for the normal urea which shows these bands at about 1602, 1626 and 1682 cm^{-1} for the powdered sample and at 1617 and 1686 cm^{-1} for its solution.[94]

This phenomenon is the result of a resonance which occurs in the solid phase between the original C=O stretching frequency (at about 1635 cm^{-1} for the normal urea and at about 1620 cm^{-1} for the ^{18}O-labelled urea) and the symmetrical NH_2 bending vibration (to be expected at about 1650 cm^{-1}

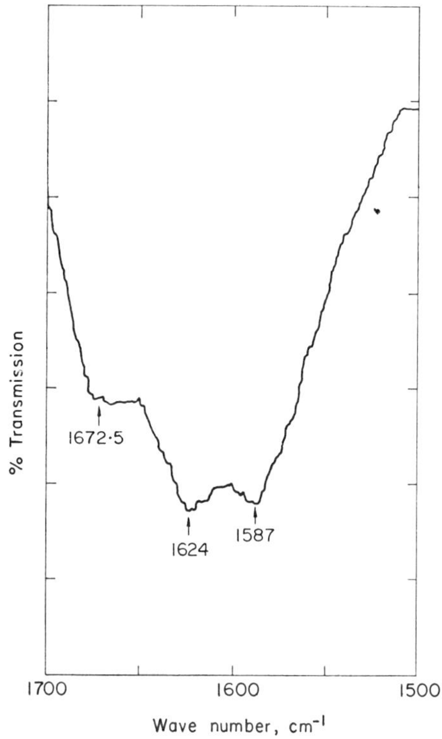

Fig. 7.6 The absorption of powdered ^{18}O-urea in the carbonyl stretching region.

for the normal compound and at about 1640 cm^{-1} for the labelled modification). This resonance brings about the observed band pairs at about 1602 and 1682 cm^{-1} in the case of the normal substance and at about 1587 and 1672 cm^{-1} in the case of the labelled species. In both cases the upper band has a larger symmetrical NH_2 bending character than the lower band (which is more a C=O stretching absorption) and hence is relatively weak.

In the solution phase, however, the original C=O stretching frequency of both of the isotopic ureas can be expected to be much higher than in the solid phase since their C=O group is strongly hydrogen bonded to their NH_2 groups in the latter case while it is practically free from hydrogen bonds in solution. The expected solution C=O stretching frequency of normal urea is therefore at about 1675 cm^{-1} while it is at about 1655 cm^{-1} for the ^{18}O-urea. The original symmetrical NH_2 bending vibration of both the normal and the labelled urea, in solution, can on the other hand be assumed to appear already at *ca.* 1630 cm^{-1}, since their NH_2 groups are less hydrogen bonded in solution and hydrogen bonding is known to *increase* the bending frequency of such groups. The original C=O stretching vibrations of urea solutions are

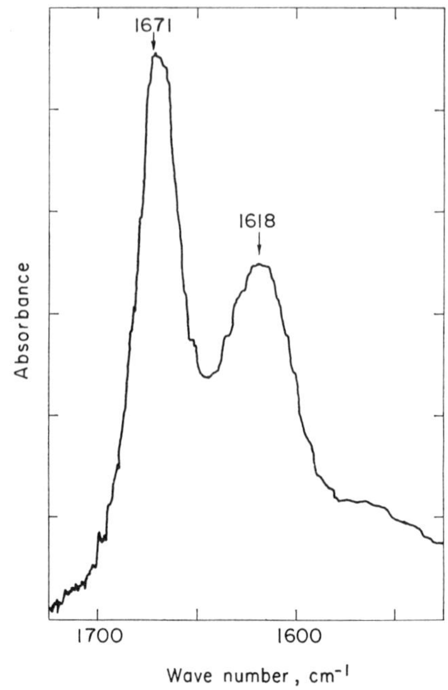

Fig. 7.7 The absorption of a solution of ^{18}O-urea in the mixed solvent in the C=O stretching region.

thus higher in frequency than their symmetrical NH_2 bending frequency and when these modes of vibration interact again it is now the *higher* band that has in each case a stronger C=O stretching character than the other resulting band. This means that the *higher* band of the urea solutions absorption spectrum in the double bond region must be stronger than its lower band since a C=O stretching band is usually the strongest band in the spectrum of such compounds.[94]

The solid phase urea band at about 1625 cm^{-1} belongs to the asymmetrical NH_2 bending mode and as such is almost totally insensitive to ^{18}O-labelling. In the case of the solution spectra it no doubt merges together with the lower band of the Fermi-resonating C=O stretching–symmetrical NH_2 bending vibrations.[94]

The normal urea powder band at 1150 cm^{-1} shows an ^{18}O-induced shift of 10 cm^{-1}; this fact is in harmony with its assignment[97] to the ν_3 mode of vibration calculated to have a 35% CO stretching character and therefore an expected ^{18}O-shift of 9·7 cm^{-1}.

The 576 cm^{-1} solid phase urea band merges together in the case of the labelled urea, with its 558 cm^{-1} absorption peak since it is due to an in-plane deformation vibration of the N—CO—N moiety[97] and as such must shift appreciably to lower frequencies on ^{18}O-labelling.

The absorption in the 830–3700 cm^{-1} region of a 90 atom % ^{18}O—N,N′-dicyclohexylurea solution in a 3:1 mixture of carbon tetrachloride and dimethylsulphoxide showed a significant ^{18}O-shift only for the C=O stretching band which moves from 1668 cm^{-1} for the normal compound to 1647 cm^{-1}. In the case of a pure dimethylsulphoxide solution the corresponding frequencies were observed at 1660 and 1627 cm^{-1}, respectively.[96]

In the solid phase, the C=O stretching frequencies for the normal and ^{18}O-labelled dicyclohexylurea appear at 1628 and 1611 cm^{-1}, respectively.[95]

Since the solid *p*-toluenesulphonate of the ^{18}O-labelled dicyclohexylurea shows the 1669 cm^{-1} band of the corresponding normal toluenesulphonate at the same position this band cannot belong to a carbonyl group. Hence it was concluded that very probably the dicyclohexylurea is not N-protonated by acids but O-protonated. The 1437 to 1427 cm^{-1} and 1135 to 1129 cm^{-1} ^{18}O-shifts, observed for this sulphonate, are therefore likely to be due to in-plane O—H bending and C—OH stretching vibrations, respectively.[95] A similar conclusion about the protonation of amides has been drawn from the fact that the *ca.* 1707 cm^{-1} band frequency of the solid hydrohalides of propionamide remains practically unaltered even for an ^{18}O-enrichment of 50 atom % or over.[85]

The absorption spectrum of a carbon tetrachloride-dimethylsulphoxide solution of 80 atom % ^{18}O—N,N′-di-*p*-tolylurea in the 700–3700 cm^{-1} region showed two significant isotopic shifts. The first of these concerns the C=O stretching band which moves from 1708 cm^{-1} for the normal ditolylurea to 1688 cm^{-1}. The second shift is in the position of the 1197 cm^{-1} band of the normal compound which moves to 1192 cm^{-1} in the case of the ^{18}O-ditolylurea. This band seems to be due to the symmetrical N—CO—N stretching vibration which also affects the position of the carbonyl group.[96]

7.31 6-^{18}O-Purine Nucleosides

6—^{18}O-Purine nucleosides can in fact be considered as a special case of labelled lactams or, more generally, of ^{18}O-enriched amides, as is evident, e.g. from Fig. 7.8 which represents the structure of inosine. Since amides have been shown above (Section 7.29) to have C=^{18}O stretching frequencies which are lower by about 20 cm^{-1} from the corresponding C=^{16}O frequencies it can thus be expected that such nucleosides will show ^{18}O-induced shifts of 15–20 cm^{-1} for their C=O stretching bands. In accordance with this expectation it has indeed been observed[9] that while normal inosine in D_2O shows its C=O stretching band at about 1675 cm^{-1}, ^{18}O-labelled inosine exhibits its C=^{18}O absorption at *ca.* 1660 cm^{-1}. In the case of the 5′-monophosphate of this nucleoside the C=O bands appear at 1672 and 1659

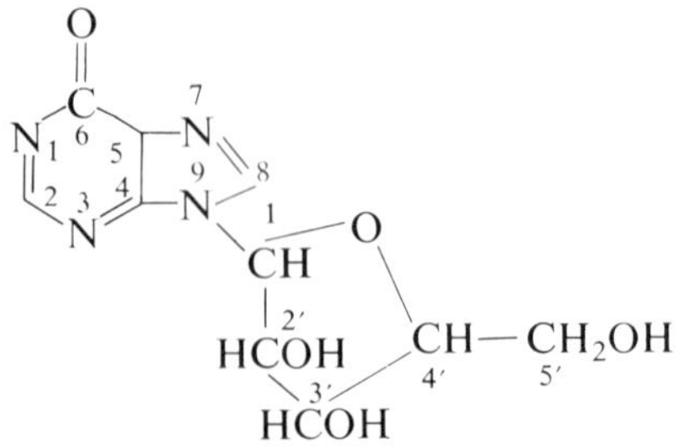

Fig. 7.8 The structure of inosine.

cm^{-1}, respectively, while acetonitrile solutions of isotopic 2′,3′,5′-tri-O-acetyl-inosines show the corresponding absorption peaks at about[98] 1707 and 1690 cm^{-1}.

Guanosine in D_2O, on the other hand, shows these bands at 1665 and 1652 cm^{-1}, respectively, the relatively low difference between these isotopic frequencies suggesting that the guanosine CO stretching mode is strongly coupled to other vibrational modes as a result of an interaction of this group with other groups of this compound. Since normal and ^{18}O-guanosine solutions in basic D_2O exhibit bands, in this region, only at 1576, 1591 and 1571, 1591 cm^{-1}, correspondingly, it is evident that they do not contain C=O groups, hence the negative charge of the guanosine anion is localized on the oxygen atom.[98]

7.32 Labelled Nitro Compounds

Similarly to the ionized carboxyl group, the nitro group also possesses two equal X⋯O bonds the angle between which is very nearly 126° (in glycine 125·5°; in nitromethane[99] 127°). Since the nitrogen atom is similar to the carbon atom in its mass, X—O bond-strength, etc. one can therefore anticipate to find a great extent of analogy between the various characteristic vibration frequencies of the different isotopic species of the nitro group and the corresponding frequencies of the ionized carboxyl group. Table 7.17 shows that when the isotopic CO_2^- group fundamentals of glycine[82] are compared with the respective NO_2 frequencies of nitromethane[99] and nitrobenzene[2] this anticipation is generally fulfilled.

The only case where the analogy between the isotopic CO_2^- frequencies and the corresponding NO_2 frequencies does not hold can be seen from Table 7.17 to be that of the rocking vibrations of these groups. These discrepancies can be traced back to the effect of hydrogen bonding on the rocking frequencies of the glycine CO_2^- groups on one hand and to the large mass of the benzene ring on the other.

The difference between the isotopic shift of the $N^{18}O_2$ asymmetrical stretching vibration of nitromethane (38 cm^{-1}) and that of nitrobenzene

TABLE 7.17 The Frequency (cm^{-1}) of the Characteristic Vibrations of Various Isotopic CO_2^- and NO_2 Groups

Vibration type	$C^{16}O_2^{-\,a}$	$N^{16}O_2$ $PhNO_2^{\,b}$	$N^{16}O_2$ $MeNO_2^{\,b}$	$C^{16}O^{18}O^-$	$N^{16}O^{18}O$ $PhNO_2^{\,c}$	$N^{16}O^{18}O$ $MeNO_2^{\,d}$	$C^{18}O_2^-$	$N^{18}O_2$ $PhNO_2^{\,c}$	$N^{18}O_2$ $MeNO_2^{\,d}$
Wagging	608	532	?	601	520	?	?	508	?
Rocking	700	417	?	?	414	?	685	408	?
Scissoring	894	850	914[e]	?	838	900[e]	870	824	884[e]
(N)CO_2—X stretching	1334	1109	1095[e]	?	1105	1089[e]	1324	?	?
Symmetric stretching	1413	1349	1375	1401	1335	1357	?	1322	?
Asymmetric stretching	1604	1531	1564	1586	1518	1545	1574	1510	1526

[a] The values (for the powdered glycines) are taken from Table 7.14.
[b] For a CCl_4 solution.
[c] For a CCl_4 solution of a 43 atom % ^{18}O-enriched sample.
[d] For a 58 atom % sample.
[e] This assignment is now more probable than that given in ref. 99 in light of the new assignments made in ref. 2.

(21 cm^{-1}) is probably the result of the conjugation existing between the nitro group and the phenyl ring in nitrobenzene. This conjugation can be expected to diminish the effect of increasing the mass of the oxygen atoms, on the stretching frequencies of the conjugated nitro group, which is now a less-isolated oscillator. The fact that the ^{18}O-isotopic shift for the symmetrical NO_2 stretching of $CH_3N^{16}O^{18}O$ is equal to about 18 cm^{-1} while the corresponding shift for $C_6H_5N^{16}O^{18}O$ amounts only to *ca.* 13 cm^{-1} is also in accordance with this expectation.[2]

The ^{18}O-induced shift of 38 cm^{-1} of the asymmetrical NO_2 stretching frequency of nitromethane can be seen to be very near to the 37 cm^{-1} shift calculated for a diatomic NO oscillator of a 1440 cm^{-1} stretching frequency (Table 7.1).

Applying the isotopic relation (eqn. 3.8) holding for the ω_3's of isotopic $\boldsymbol{C_{2v}}$ triatomic oscillators to the values observed for the $CH_3N^{16}O_2$ and $CH_3N^{18}O_2$ asymmetrical NO_2 stretching bands gives: $\omega_3^i/\omega_3 = 1526/1564 = 0{\cdot}976 = 0{\cdot}981$ (calculated), this agreement supporting the assignment of these bands.

A rough estimate of the symmetrical NO_2 stretching frequency of $CH_3N^{16}O^{18}O$ can be obtained by assuming it to be very near to the stretching frequency of a diatomic N—X oscillator, where X = ^{16}O + ^{18}O, and applying Hooke's law (Section 3.2). This gives[99] a v^i/v ratio of 0·986 as compared with the observed value of 1357/1375 = 0·987.

The large observed ^{18}O-induced shift of the 850/914 cm^{-1} band, of the nitro-compounds investigated so far, to 824/884 cm^{-1}, proved[2] that these bands cannot be assigned, as had been done before,[100] to C—NO_2 stretching modes but must instead be attributed to the respective NO_2 scissoring vibrations.

7.33 Organic P=^{18}O Compounds

It has been seen above (Section 7.10) that labelling phosphorus oxychloride with oxygen-18 shifts its P=O stretching band at about 1310 cm^{-1} by about 45 cm^{-1}, as compared with the value of 48 cm^{-1} calculated for this shift on basis of the diatomic model. Similarly, investigation of the infrared absorption spectrum of carbon tetrachloride solutions of normal and 86 atom % ^{18}O-triphenylphosphine oxide has shown[3] that ^{18}O-labelling of this compound brings about a decrease of 30 cm^{-1} in its P=O stretching band, which appears normally at 1202 cm^{-1}, while no other really significant change is observed in the rest of its absorption in the 660–3700 cm^{-1} region.

The fact that 87 atom % ^{18}O-triphenylphosphine oxide which was recrystallized from normal water did not show a significant decrease in the intensity of its P=^{18}O stretching band was used to prove that the hydrate of this compound does not have the dihydroxy-structure.[3]

The P=^{18}O stretching band of ^{18}O-dipropylphosphite (actually: $(C_3H_7O)_2HP{=}O$) was observed[101] at 1258 cm^{-1} while the corresponding normal isotopomer absorbs at 1216 cm^{-1}.

7.34 Compounds with S = ^{18}O Groups

The infrared spectrum of 81 atom % ^{18}O-diphenylsulphoxide in the 670–3700 cm^{-1} region differs significantly from that of the corresponding normal sulphoxide only in the vicinity of the S=O stretching band. The frequency of this band in the case of a chloroform solution of the labelled compound is 1009 cm^{-1} as compared with the respective normal value of 1041 cm^{-1}. In carbon tetrachloride solutions the original S=O stretching frequencies are somewhat higher (since hydrogen bonding is no longer possible between the S=O groups and the solvent molecules) and can therefore enter into resonance with the aromatic ring vibrations at *ca.* 1090 (in the normal case) and[4] 1025 cm^{-1} (in the labelled compound).

The effect of ^{18}O-labelling on the SO_2 asymmetrical stretching vibration of sulphones is complicated by the fact that this vibration usually gives rise to a band triplet.[4] In the case of normal diphenylsulphone in a carbon tetrachloride solution this vibration is responsible for absorption peaks at 1295, 1309 and 1327 cm^{-1} while a similar solution of an 80 atom % ^{18}O-enriched sulphone showed these bands at 1282 and 1297 cm^{-1}, the lowest band of the triplet, which for the normal sulphone is its weakest branch, being probably masked in the case of the labelled sulphone by the tail absorption of the neighbouring bands. These results give for this vibration a mean ^{18}O-shift of *ca.* 28 cm^{-1} as compared with the value of 32 cm^{-1} given above for the S=O stretching frequency of diphenylsulphoxide in a chloroform solution.

The symmetrical SO_2 stretching frequency of the normal and fully labelled sulphone chloroform solutions appears at 1162 cm^{-1} and 1138 cm^{-1}, respectively.[5]

The absorption spectrum of a highly enriched sample of

$$C_6H_5 \cdot CH_2 \cdot S^{16}O^{18}O \cdot C_6H_4CH_3$$

in solution in chloroform shows the following bands in the SO_2 stretching region[102] at 1316 ($S^{16}O_2$ asymmetrical stretching), 1304 (asymmetrical stretching), 1299 ($S^{16}O^{18}O$ asymmetrical stretching), 1284 (asymmetrical stretching), 1160 ($S^{16}O_2$ symmetrical stretching), 1147 ($S^{16}O^{18}O$ symmetrical stretching), 1135 ($S^{18}O_2$ symmetrical stretching) and 1119 (aromatic) cm^{-1}.

7.35 Labelled Triphenylarsine Oxide

Triphenylarsine oxide, $(C_6H_5)_3As{=}O$ is the arsenic analogue of triphenylphosphine oxide (see Section 7.33). In this case, however, the much larger mass of the arsenic atom (75), as compared with that of phosphorus (31), moves only very little during the X=O stretching vibration. As a result, the As=O stretching vibration can be assumed to be localized to a very high degree in the As=O group. The fact that 90 atom % ^{18}O-triphenylarsine oxide in a chloroform solution shows the As=O stretching band[5] at about 840 cm^{-1} (while its normal position is at 880 cm^{-1}) is in harmony with this assumption since the As=^{18}O stretching frequency expected on the basis of a diatomic As=O model was shown above to be 839 cm^{-1} (Table 7.1).

7.36 Metal ^{18}O-Acetylacetonates

The correct assignment of the infrared absorption bands of the metal acetylacetonates (see Fig. 7.9) has recently[103, 104] been established by comparing the absorption spectrum of the powdered Cr(III) and Mn(III) derivatives of over 90 atom % ^{18}O-acetylacetone with that of the corresponding normal complexes. Figures 7.10 and 7.11 show the Cr(III) complexes spectra in the

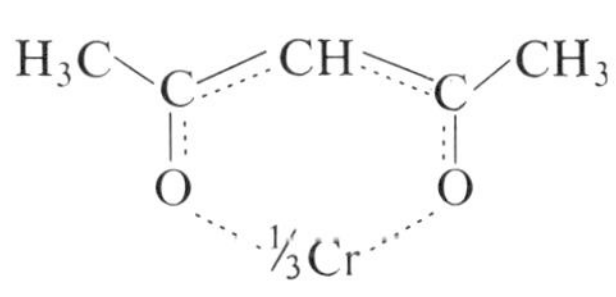

Fig. 7.9 The structure of Cr(III) acetylacetonate.

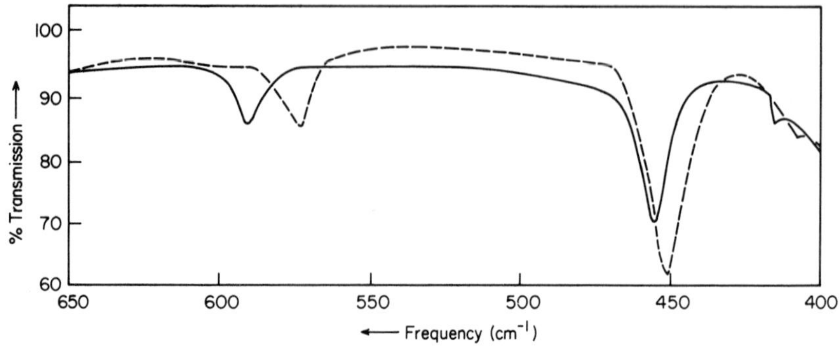

Fig. 7.10 Infrared spectra of normal and ^{18}O-Chromium III acetylacetonate in the 400–650 cm^{-1} region.
Normal ———. Labelled – – – –.

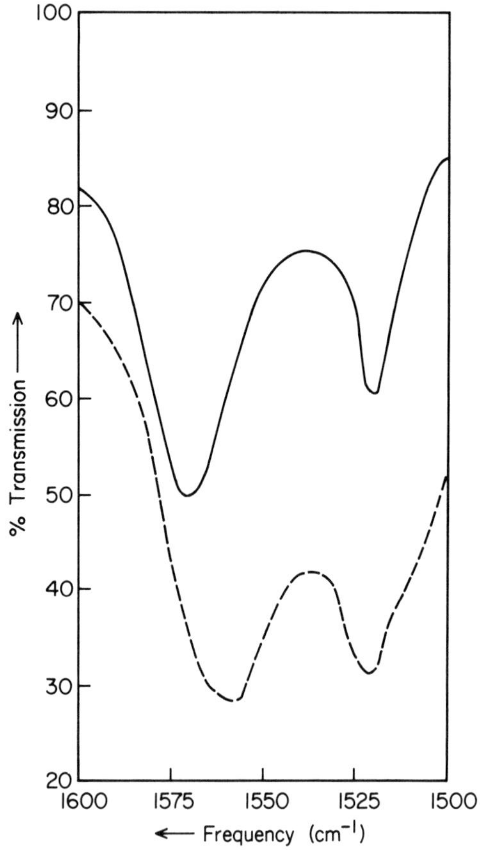

Fig. 7.11 Infrared spectra of normal and ^{18}O-Chromium III acetylacetonate in the 1500–1600 cm^{-1} region.
Normal ———. Labelled – – – –.

400–650 and 1500–1600 cm^{-1} regions, respectively, while Table 7.18 summarizes the main results obtained thus together with the previous assignment[105] of the ^{18}O-shifted bands (based on Urey–Bradley Force Field normal co-ordinate analysis) and that indicated by the observed ^{18}O-shifts.

TABLE 7.18 ^{18}O-Induced Isotopic Shifts of the Main Cr(III) Acetylacetonate Absorption Bands in the 400–1650 cm^{-1} Region

Normal band (cm^{-1})	Shift	Previous assignment[105]	Indicated assignment[104]
416	8	Out-of-plane deformation	O—Cr—O bending
456	5	Cr—O stretching	C—CH_3 bending + Cr—O stretching
592	19	Out-of-plane deformation	Cr—O stretching
661	3	C—CH_3 bending + Cr—O stretching	Out-of-plane deformation
678	11	Ring deformation + Cr—O stretching	(No change)
1380	?	CH_3 asymmetrical bending	CH_3 symmetrical bending
1425	1	CO stretching + CH bending	CH_3 asymmetrical bending
1519	0	C=O stretching	C=C stretching
1569	13	C=C stretching	Partial C=O stretching

The study of the Mn(III) complexes gives similar results although the spectra of the latter are more complicated as a result of a distortion of the octahedral configuration of the oxygen ligands about the central atom, in this case.

The ^{18}O-shift of *ca.* 19 cm^{-1} observed for the Cr—O stretching frequency is quite near to the value of 25 cm^{-1} which can be calculated for the ^{18}O-induced shift of a 592 cm^{-1} stretching frequency of a diatomic Cr—O oscillator. This fact strongly supports the assignment of the 592 cm^{-1} band to the Cr—O stretching fundamental, given in Table 7.18 as its "indicated assignment", over the previous assignment.

The ^{18}O-acetylacetonate infrared spectra measured lately,[106, 107] are similar to the corresponding spectra of the labelled Cr(III) complex, thereby proving again the present assignment of the normal acetylacetonate bands.

REFERENCES

1. Laulicht, I. and Pinchas, S. (1963). *Israel J. Chem.* **1**, 404.
2. Pinchas, S., Samuel, D. and Silver, B. L. (1964). *Spectrochim. Acta* **20**, 179.
3. Halmann, M. and Pinchas, S. (1958). *J. Chem. Soc.*, 3264.
4. Pinchas, S. Samuel, D. and Weiss-Broday, M. (1962). *J. Chem. Soc.*, 3968.

5. Bernstein, J., Halmann, M., Pinchas, S. and Samuel, D. (1964). *J. Chem. Soc.*, 821.
6. Byrn, M. and Calvin, M. (1966). *J. Amer. Chem. Soc.* **88**, 1916.
7. Pinchas, S. and Meshulam, E. (1970). *Chem. Commun.* 1147.
8. Pikman, H. and Pinchas, S. (1970). *J. Inorg. Nuclear Chem.* **32**, 2441.
9. Howard, F. B. and Miles, H. T. (1964). *Biochem. Biophys. Res. Commun.* **15**, 18.
10. Braude, E. A. and Turner, M. (1958). *J. Chem. Soc.*, 2404.
11. Mills, I. M. and Thompson, H. W. (1953). *Trans. Faraday Soc.* **49**, 224.
12. Ewing, G. E. and Pimentel, G. C. (1961). *J. Chem. Phys.* **35**, 925.
13. Mills, I. M. and Thompson, H. W. (1953). *Trans. Faraday Soc.* **49**, 224.
14. Plyler, E. K. Benedict, W. S. and Silverman, S. (1952). *J. Chem. Phys.* **20**, 175.
15. Migeotte, M., Neven, L. and Swensson, J. (1957). *Mem. soc. roy. sci. Liège No. 2.*
16. Alben, J. O. and Caughey, W. S. (1968). *Biochemistry, N.Y.* **7**, 175.
17. Griggs, J. L. Jr., Rao, K. N., Jones, L. H. and Potter, R. M. (1967). *J. Mol. Spectroscopy* **22**, 383.
18. Fraley, P. E., Thesis (The Ohio State University, 1966; University Microfilms Ann Arbor, No. 66–10, 011).
19. Fraley, P. E., Rao, K. N. and Jones, L. H. (1969). *J. Mol. Spectroscopy* **29**, 312.
20. Pinchas, S., Halmann, M. and Stoicheff, B. P. (1959). *J. Chem. Phys.* **31**, 1692.
21. (a) Urey, H. C. (1947). *J. Chem. Soc.*, 567.
(b) Gompertz, G. and Orville-Thomas, W. J. (1959). *J. Phys. Chem.* **63**, 1331.
22. Benedict, W. S., Gailar, N. and Plyler, E. K. (1956). *J. Chem. Phys.* **24**, 1139.
23. Chakerian, C. Jr. (1964). *Diss. Abs.* **25** (5), 2775.
24. Chakerian, C. Jr. and Eggers, D. F. Jr. (1965). *J. Chem. Phys.* **43**, 757.
25. Chakerian, C. Jr. and Eggers, D. F. Jr. (1968). *J. Mol. Spectroscopy* **27**, 59.
26. Berney, C. V. and Eggers, D. F. Jr. (1964). *J. Chem. Phys.* **40**, 990.
27. Herzberg, G. (1945). "Infrared and Raman Spectra of Polyatomic Molecules". p. 274. D. Van Nostrand Company, Princeton, N.J.
28. Amat, G. and Pimbert, M. (1965). *J. Mol. Spectroscopy* **16**, 278.
29. Gordon, H. R. (1966). *Diss. Abs.* **26**, 6781.
30. Wieder, I. and McCurdy, G. B. (1966). *Phys. Rev. Letters* **16**, 565.
31. Plyler, E. K., Tidwell, E. D. and Maki, A. G. (1964). *J. Res. Natl. Bur. Std.* (*U.S.*). **68A**, 79.
32. Milligan, D. E., Brown, H. W. and Pimentel, G. C. (1965). *J. Chem. Phys.* **25**, 1080.
33. Griggs, J. L. Jr., Rao, K. N., Jones, L. H. and Potter, R. M. (1965). *J. Mol. Spectroscopy* **18**, 212; (1968) **25**, 34.
34. Mantz, A. W., Rao, K. N., Jones, L. H. and Potter, R. M. (1969). *J. Mol. Spectroscopy* **30**, 513.
35. Jones, L. H., Asprey, L. B. and Ryan, R. R. (1967). *J. Chem. Phys.* **47**, 3371.
36. Jones, L. H., Ryan, R. R. and Asprey, L. B. (1968). *J. Chem. Phys.* **49**, 581.
37. Laane, J. L., Jones, L. H., Ryan, R. R. and Asprey, L. B. (1969). *J. Mol. Spectroscopy* **30**, 485.
38. Polo, S. R. and Wilson, M. K. (1954). *J. Chem. Phys.* **22**, 900.
39. Venkateswarlu, K. and Mariam, S. (1965). *Indian J. pure appl. Phys.* **3**, 117.
40. Lichtin, N. N., Laulicht, I. and Pinchas, S. (1964). *Inorg. Chem.* **3**, 537.
41. McGraw, G. E., Bernitt, D. L. and Hisatsune, I. C. (1966). *J. Chem. Phys.* **45**, 1392.
42. Loos, K. R., Goetschel, C. T. and Campanile, V. A. (1968). *Chem. Commun.*, 1633.

43. Laulicht, I., Pinchas, S., Sadeh, D. and Samuel, D. (1964). *J. Chem. Phys.* **41**, 789.
44. Bouquet, G. and Bigorgne, M. (1965). *Compt. rend.* **261**, 2865.
45. Jones, L. H., McDowell, R. S. and Goldblatt, M. (1968). *J. Chem. Phys.* **48**, 2663.
46. Bouquet, G., Loutellier, A. and Bigorgne, M. (1968). *J. Mol. Structure* **1**, 211.
47. Maki, A. and Decius, J. C. (1959). *J. Chem. Phys.* **31**, 772.
48. Anbar, M., Halmann, M. and Pinchas, S. (1960). *J. Chem. Soc.*, 1242.
49. Pauling, L. (1940). "The Nature of the Chemical Bond", p. 243. Cornell Univ. Press, Ithaca, U.S.A.
50. Jones, L. H. (1964). *J. Opt. Soc. Amer.* **54**, 1283.
51. Kharitonov, Yu. Ya. and Buslaev, Y. A. (1963). *Optika i Spektroskopiya* **14**, 586.
52. Vdovenko, V. M., Mashirov, L. G. and Suglobov, D. N. (1969). *Doklady Akad. Nauk. S.S.S.R.* **185**, 824.
53. Simon, A. and Waldmann, K. (1955). *Z. physik. Chem.* (Leipzig) **204**, 235.
54. Petreanu, E., Pinchas, S. and Samuel, D. (1963). *J. Chem. Soc.*, 2822.
55. Spencer, B. (1959). *Biochem. J.* **73**, 442.
56. Vojta, G. and Koch, D. (1964). *Z. phys. Chem.* (*Leipzig*) **225**, 295.
57. Petreanu, E., Pinchas, S. and Samuel, D. (1965). *J. Inorg. Nucl. Chem.* **27**, 2519.
58. Pinchas, S. and Sadeh, D. (1968). *J. Inorg. Nucl. Chem.* **30**, 1785.
59. Pinchas, S. (1969). *J. Chem. Phys.* **51**, 2284.
60. Mueller, A. and Krebs, B. (1965). *Z. Naturforsch.* **20b**, 1127.
61. Pinchas, S. Samuel, D. and Petreanu, E. (1967). *J. Inorg. Nucl. Chem.* **29**, 335.
62. Mueller, A. and Krebs, B. (1966). *Z. Naturforsch.* **21b**, 3.
63. Pinchas, S. (1969). *Israel J. Chem.* **7**, 5.
64. Pinchas, S., Sadeh, D. and Samuel, D. (1965). *J. Phys. Chem.* **69**, 2259.
65. Dimroth, K., Berndt, A. and Volland, R. (1966). *Chem. Ber.* **99**, 3040.
66. Dimroth, K. and Berndt, A. (1968). *Chem. Ber.* **101**, 2515.
67. Pinchas, S., Samuel, D. and Wasserman, I. (1967). *Spectrochim. Acta* **23A**, 1307.
68. Pinchas, S. and Cohen, J. Unpublished results.
69. Karabatsos, G. J. (1960). *J. Org. Chem.* **25**, 315.
70. Francis, S. A. (1951). *J. Chem. Phys.* **19**, 942.
71. Junge, H. (1968). *Spectrochim. Acta* **24A**, 1965.
72. Junge, H. (1968). *Spectrochim. Acta* **24A**, 1951.
73. Krebs, A. and Schrader, B. (1967). *Annalen* **709**, 46.
74. Junge, H. (1968). *Spectrochim. Acta* **24A**, 1957.
75. Lippert, E., Samuel, D. and Fischer, E. (1965). *Ber. Bunsenges. phys. Chem.* **69**, 155.
76. Becker, E. D., Ziffer, H. and Charney, E. (1963). *Spectrochim. Acta* **19**, 1871.
77. Ziffer, H., Charney, E. and Becker, E. D. (1965). *J. Chem. Phys.* **42**, 914.
78. Becker, E. D., Charney, E. and Anno, T. (1965). *J. Chem. Phys.* **42**, 942.
79. Pinchas, S., Samuel, D. and Weiss-Broday, M. (1961). *J. Chem. Soc.*, 2382.
80. Statz, G. and Lippert, E. (1967). *Ber. Bunsenges. phys. Chem.* **71**, 673.
81. Junge, H. and Musso, H. (1968). *J. Chem. Soc.* (*B*)., 389.
82. Laulicht, I., Pinchas, S., Samuel, D. and Wasserman, I. (1966). *J. Phys. Chem.* **70**, 2719.
83. Suzuki, S., Shimanouchi, T. and Tsuboi, M. (1963). *Spectrochim. Acta* **19**, 1195.
84. Mitton, C. G. and Schowen, R. L. (1968). *Tetrahedron Lett.* **55**, 5803.
85. Stewart, R., Muenster, L. J. and Edward, J. T. (1961). *Chem. and Ind.*, 1906.
86. Pinchas, S., Samuel, D. and Weiss-Broday, M. (1961). *J. Chem. Soc.*, 1668.
87. Pinchas, S., Samuel, D. and Weiss-Broday, M. (1961). *J. Chem. Soc.*, 2666.
88. Pinchas, S., Samuel, D. and Weiss-Broday, M. (1961). *J. Chem. Soc.*, 3063.

89. Miyazawa, T., Shimanouchi, T. and Mizushima, S. I. (1956). *J. Chem. Phys.* **24**, 408.
90. Price, W. C. and Fraser, R. D. B. (1953). *Proc. Roy. Soc.* **141B**, 66.
91. Coburn, R. A. and Dudek, G. O. (1968). *J. Phys. Chem.* **72**, 1177.
92. Keller, G. H., Bauer, L. and Bell, C. L. (1968). *Canad. J. Chem.* **46**, 2475.
93. Coburn, R. A. and Dudek, G. O. (1968). *J. Phys. Chem.* **72**, 3681.
94. Laulicht, I., Pinchas, S., Petreanu, E. and Samuel, D. (1965). *Spectrochim. Acta* **21**, 1487.
95. Stewart, R. and Muenster, L. J. (1961). *Canad. J. Chem.* **39**, 401.
96. Lapidoth, A., Pinchas, S. and Samuel, D. (1963). *J. Chem. Soc.*, 1128.
97. Yamaguchi, A., Miyazawa, T., Shimanouchi, T. and Mizushima, S. I. (1957). *Spectrochim. Acta* **10**, 170.
98. Howard, F. B. and Miles, H. T. (1965). *J. Biol. Chem.* **240**, 801.
99. Halmann, M. and Pinchas, S. (1960). *J. Chem. Soc.*, 1246.
100. Green, J. H. S., Kynaston, W. and Lindsey, A. S. (1961). *Spectrochim. Acta* **17**, 486.
101. Aksnes, G. and Aksnes, D. (1964). *Acta Chem. Scand.* **18**, 1623.
102. Stirling, C. J. M. (1963). *J. Chem. Soc.*, 5741.
103. Pinchas, S. and Laulicht, I. (1966). *Israel J. Chem.* **4**, 48pp.
104. Pinchas, S., Silver, B. L. and Laulicht, I. (1967). *J. Chem. Phys.* **46**, 1506.
105. Nakamoto, K. (1963). "Infrared Spectra of Inorganic and Coordination Compounds". p. 216–220. Wiley, New York.
106. Musso, H. and Junge, H. (1966). *Tetrahedron Lett.*, 4003.
107. Junge, H. and Musso, H. (1968). *Spectrochim. Acta* **24A**, 1219.

CHAPTER 8

Various Labelled Compounds

8.1 Isotopic Lithium Compounds

While sodium and potassium carbonates do not show any absorption peak in their 400–650 cm^{-1} spectrum natural abundance (92·6 atom % ^{7}Li) lithium carbonate was observed to exhibit fairly strong bands at about 416 and 497 cm^{-1} with shoulders at 397 and 435 cm^{-1} (see Fig. 8.1). These bands are due to the stretching vibrations of the LiO_4 tetrahedra present in the lithium carbonate crystals.[1,2]

When the absorption of $^{6}Li_2CO_3$ is investigated in this region, these peaks are found at 447 and 531 cm^{-1} (with shoulders at 414 and 469 cm^{-1}), respectively, thus displaying *positive* isotopic shifts of 31 and 34 cm^{-1}, correspondingly, as a result of the *decrease* in the mass of the lithium atoms.[1,2] These shifts agree very well with the value of 33 cm^{-1} calculated for the ν_1 symmetrical stretching fundamental of an isolated tetrahedral $^{6}LiO_4$ oscillator on basis of a 416 cm^{-1} frequency for the respective $^{7}LiO_4$ case (see eqn. 3.19).

Similar ^{6}Li—O stretching bands were observed for 95 atom % $^{6}Li_3PO_4$ at[3] 403, 452, 479 and 517 cm^{-1}, corresponding to the $^{7}Li_3PO_4$ bands at 385, 431, 460 and *ca.* 500 cm^{-1}. Significant ^{6}Li-shifts were also found for the $^{6}Li_3PO_4$ bands at 330 ($\Delta\nu = +23$), 554 ($\Delta\nu = +38$) and 601 cm^{-1} ($\Delta\nu = +9$).

In the case of Li_3VO_4 the ^{6}Li-characteristic absorption peaks appeared at 366, 397, 430, 463, 490 and 537 cm^{-1}, as compared with the parallel 355, 387, 415, 443, 463 and 512 cm^{-1} $^{7}Li_3VO_4$ frequencies.[3]

$^{6}Li_3AsO_4$ exhibits bands at 322, 400, 424, 474, 531 and *ca.* 567 cm^{-1}, corresponding to the $^{7}Li_3AsO_4$ bands at 306, 368, 402, 447, 507 and *ca.* 553 cm^{-1}, respectively.[3]

The ^{6}Li-characteristic absorption peaks of $^{6}Li_2MoO_4$ appear at 300, 309, 345, 415, *ca.* 450 and 490 cm^{-1}, as compared with the parallel $^{7}Li_2MoO_4$ bands at 294, 305, 340, 391, 432 and 464 cm^{-1}, respectively,[2] while the main ^{6}Li-shifts for $^{6}LiWO_4$ were observed for its 452 ($\Delta\nu = +19$) and 497 cm^{-1} ($\Delta\nu = +26$) bands.[2]

The 290–1600 cm^{-1} spectrum of $^{6}LiFeCr_4O_8$ was also reported.[2]

The 400–3500 cm^{-1} absorption of gaseous $C_2H_5{}^{6}Li$ shows its CH_2—^{6}Li

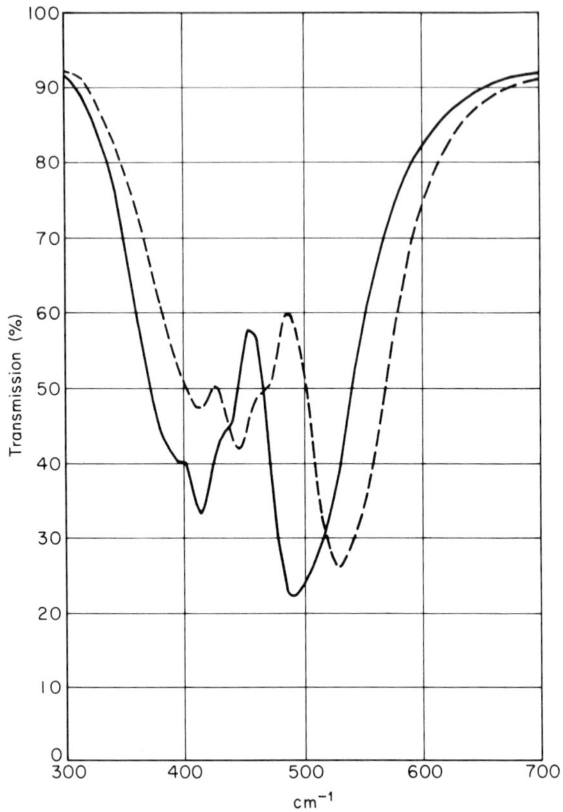

Fig. 8.1 Isotopic shift in the infrared spectrum of Li_2CO_3 in the low frequency region. Curve ——— $^7LiCO_3$. Curve ----- $^6Li_2CO_3$.

stretching band at about 508 cm^{-1}, as compared with the 528 cm^{-1} frequency observed for the respective CH_2—^7Li stretching vibration.[4] In the case of its crystalline state the isotopic effect of ^6Li on this band is lower as a result of an accompanying change in its structure occurring upon its crystallization.[4]

8.2 Labelled Boron Halides

The homogeneous boron trihalides which are planar and belong to the D_{3h} point group possess four normal modes of vibration: ν_1 (symmetrical stretching), ν_2 (symmetrical deformation), ν_3 (degenerate stretching) and ν_4 (degenerate bending) of which only ν_1 is infrared-inactive.

Gaseous natural abundance boron trifluoride, containing about 19 atom % ^{10}B and 81 atom % ^{11}B, displays its infrared-active fundamentals at[5,6] 720

(ν_2 of $^{10}BF_3$), 691 (ν_2 of $^{11}BF_3$), 1504·7 (ν_3 of $^{10}BF_3$), 1453·5 (ν_3; $^{11}BF_3$), 482 (ν_4; $^{10}BF_3$) and 480·4 cm^{-1} (ν_4; $^{11}BF_3$). Liquid BF_3, however, shows also its theoretically-forbidden ν_1 band, at 880 cm^{-1} (unaffected by the mass of the boron atom since this atom is not participating in this vibration), due to the self-association of the liquid BF_3 molecules. This fact also results in a shifting of their ν_2, ν_3 and ν_4 fundamentals to 688, 1488 and 477 cm^{-1} in the case of the $^{10}BF_3$ species and to 656, 1433 and 477 cm^{-1} for the $^{11}BF_3$ isotopomer.[6] Solid $^{10}BF_3$ exhibits its normal vibrations at 878 (ν_1), 665 (ν_2), 1476 (ν_3) and 477 (ν_4) cm^{-1}, for the same reason.[6]

The analysis of the highly resolved ν_2 vibration–rotation band of gaseous $^{10}BF_3$ yielded the value of 0·3447 cm^{-1} for its B_0 constant, as compared with the 0·3450 cm^{-1} value found for the respective $^{11}BF_3$ constant.[7] A similar analysis of its ν_3 band resulted in a value of 0·3527 cm^{-1} for the $^{10}BF_3$ B_0 constant and in the evaluation of A_0 and other of its rotational constants.[8] The centre of its ν_3 band was thus determined to be at 1502·5 cm^{-1}, corresponding to the respective $^{11}BF_3$ 1449 cm^{-1} frequency.[9]

The analysis of the $2\nu_3$ vibration–rotation absorption of $^{10}BF_3$ and $^{11}BF_3$ was also carried out.[10]

The infrared spectra of the complexes formed by $^{10}BF_3$ and $^{11}BF_3$ with methyl fluoride,[11] dimethyl ether,[12] diethyl ether,[12] tetrahydrofuran,[12] esters,[13] ammonia,[14] dimethyl sulphide[15,16] and dimethyl selenide[16] were investigated, mainly, in order to use the spectral data thus obtained to calculate the theoretical equilibrium constant of the isotopic exchange between these complexes and boron trifluoride.[12] In the case of the isotopic BF_3 complexes with diethyl ether the ^{10}B-sensitive bands appear at 504 (BF_3 deformation), 678 (B—O stretching), 758 (symmetric B—F stretching), 887 (C—O stretching), 1210 and 1252 cm^{-1} (both: antisymmetric B—F stretching), as compared with the corresponding $^{11}BF_3$ 502, 664, 756, 879, 1163 and 1210 cm^{-1} absorption peaks, respectively.[12]

Difluoroborane possesses six normal vibrations, three of which were observed to be ^{10}B-sensitive.[17] Table 8.1 reports their frequencies for HBF_2 and DBF_2.

TABLE 8.1 The ^{10}B-Sensitive Fundamental Vibrations (cm^{-1}) of HBF_2 and DBF_2[17]

Vibration	Approximate description	$H^{10}BF_2$	$H^{11}BF_2$	$D^{10}BF_2$	$D^{11}BF_2$
ν_1	Symmetrical B—F stretching	1193	1162	1147	1125
ν_3	Symmetrical B—F bending	555	550	551[a]	544
ν_6	Out-of-plane deformation	945	924	806·5	787

[a] Calculated.

The ν_2, ν_3 and ν_4 fundamentals of $^{10}BCl_3$ appear at about 480, 993·7 and 243 cm^{-1}, corresponding to the $^{11}BCl_3$ 460, 954·2 and 243 cm^{-1} frequencies, respectively.[5, 18] These values have been used in the calculation of the force constants of the boron trichloride molecules.[18]

The 525–3600 cm^{-1} absorption of the complexes formed by $^{10}BCl_3$ and $^{11}BCl_3$ with various esters has also been investigated.[13]

The observed fundamentals of $H^{10}BCl_2$ appeared at 762 (ν_1; symmetric B—Cl stretching), 2617 (ν_2; B—H stretching), 1100 (ν_4; asymmetric B—Cl stretching) and 795 cm^{-1} (ν_6; out-of-plane deformation) as compared with the corresponding $H^{11}BCl_2$ bands at 740, 2617, 1089 and 784 cm^{-1}, respectively.[19] The $D^{10}BCl_2$ and $D^{11}BCl_2$ ν_6 peak is displayed at 661 and 645 cm^{-1}, respectively.[19] An analysis of the fine structure of this band in case of 98·6 atom % $H^{11}BCl_2$ was found to support its assignment as due to the out-of-plane deformation vibration of dichloroboranes.[20]

The fundamental frequencies observed for the isotopic dichloroboranes are in excellent agreement with those calculated for them.[19]

Liquid $^{10}BBr_3$ exhibits its main infrared absorption bands at 390 (ν_2; *m–s*) and 845 cm^{-1} (ν_3; *vvs*), corresponding to the $^{11}BBr_3$ 372 and 802 cm^{-1}, fundamentals respectively,[21] while gaseous $^{10}BBr_3$ shows its ν_3 vibration at[5] 856 cm^{-1}.

The $H^{10}BBr_2$ and $D^{10}BBr_2$ ν_4 band was observed at 1048 and 901 cm^{-1}, as compared with the parallel $H^{11}BBr_2$ and $D^{11}BBr_2$ vibrations at 1036 and 870 cm^{-1}, respectively.[22]

The ν_2 and ν_3 normal frequencies of $^{10}BI_3$ (in solution) give rise to infrared absorption bands at 320 (*w*) and 724 cm^{-1} (*vvs*), corresponding to the $^{11}BI_3$ ν_2 and ν_4 frequencies of 305 and 692 cm^{-1}, respectively.[21]

8.3 Other Inorganic Isotopic Boron Compounds

96 Atom % ^{10}B-metaboric acid dispersed in a solid argon matrix absorbs at *ca.* 2061 (*w*) and 2128 (*s*) cm^{-1} while 97 atom % $H^{11}BO_2$ shows its *strong* band at 2060 cm^{-1} and its *weak* absorption peak at 2128 cm^{-1}. This means that the B=O stretching band of $H^{10}BO_2$ is located at about 2128 cm^{-1} while its parallel ^{11}B=O fundamental has a frequency of[23] 2060 cm^{-1}.

$H^{10}BO_2$ also shows absorption peaks[23] at 493 and 536 cm^{-1}, as compared with the respective $H^{11}BO_2$ values of 477 and 518 cm^{-1}.

The infrared absorption of vapours of lithium, sodium and cesium 96 atom % ^{10}B-metaborates exhibits bands at 610 and 2000 cm^{-1}, corresponding to the respective 600 and 1935 cm^{-1} $^{11}BO_2^-$ fundamentals.[24]

The frequency of the ν_2 (bending) and ν_3 (antisymmetric stretching) normal vibrations of various isotopic modifications of the linear and symmetric BO_2^- ions, when isolated in alkali halide crystals,[25] is reported in Table 8.2.

TABLE 8.2 The Infrared Absorption Bands (cm^{-1}) of Isotopic BO_2^- Ions in Alkali Halide Crystals[25]

Band	Alkali halide	$^{10}B^{16}O_2^-$	$^{11}B^{16}O_2^-$
ν_2	KBr	607	587
	RbCl	610	590
ν_3	NaCl	2068	1996
	KBr	2029	1959
	RbCl	2035	1965
$\nu_3 + \nu_2 - \nu_2$	KBr	2019	1950

$^nB_2O_3$ probably exists in the bipyramidal $\boldsymbol{D_{3h}}$ structure since it displays no B═O stretching bands.[23] It must therefore exhibit only three infrared-active fundamentals: an in-phase B—O stretching, an out-of-phase B—O stretching and an O—B—O bending mode of vibration. Concentrated solid argon matrices of natural abundance and 96 atom % $^{10}B_2O_3$ show bands at 1325 (*w*), 1345 (*w*), 1420 (*m*), 1442 (*m*) and 1368, 1442 (*vw*) and 1465 (*w*) cm^{-1}, respectively[23] which stem from the stretching fundamentals of $^{10}B_2O_3$ (1368, 1465), $^{10}B^{11}BO_3$ (1345, 1442) and $^{11}B_2O_3$ (1325, 1420).

The 400–5000 cm^{-1} absorption of 96 atom % $H_3{}^{10}BO_3$ in a potassium bromide pellet exhibits its main bands at 545 (*s*; out-of-plane BO_3 bending), 629 (*sh*), 668 (*s*; in-plane BO_3 bending), 690–870 (a very broad and strong band due to an O—H torsion), 882 (*w*; sharp), 1117 (*sh*), 1195 (*s*; sharp; B—O—H bending), 1227 (*sh*), 1418 (*sh*), 1490 (*vs*; B—O stretching), 2249 (*m*), 2340 (*w*), 2480 (*w*) and 3150 cm^{-1} (*s*; *b*; O—H stretching), corresponding to the normal boric acid bands at 544, 639, 671, 690–870, 882, 1117, 1183, 1220, 1428, 2240, 2330, 2475 and 3150 cm^{-1}, respectively.[26] The 33–500 cm^{-1} spectra of a $H_3{}^{10}BO_3$ Nujol mull (and mulled D_3BO_3) at room temperature and at $-170°C$ were also investigated.[27]

The investigation of the 250–4000 cm^{-1} spectra of gaseous $H_2{}^{10}B_2O_3$ and $H_2{}^{11}B^{10}BO_3$ (as well as the spectra of $H_2B_2{}^{18}O_3$, $H_2B_2{}^{16}O^{18}O_2$ and $D_2B_2O_3$ and the 700–950 cm^{-1} absorption of the deuterated $^{11}B^{10}B$-species) indicated that the isotopically pure modifications of this compound (see Fig. 8.2) belong to the $\boldsymbol{C_{2v}}$ point group.[28] The fundamentals of $H_2{}^{10}B_2O_3$

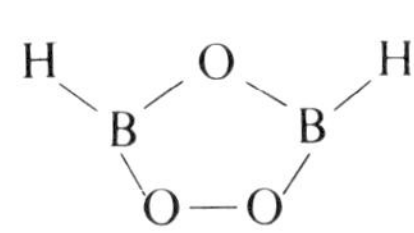

Fig. 8.2 The structure pf $H_2B_2O_3$.

were observed at 1238 (*w*; ν_2), 1146 (*vw*; ν_3), 968 (*w*; ν_4), 842 (*vw*; ν_5), 781 (*w*; ν_6), 899 (*vs*; ν_9), 431 (ν_{10}), 2680 (*s*; ν_{11}), 1419 (*vs*; ν_{12}), 1192 (*vs*; ν_{13}), 982 (*m*; ν_{14}) and 748 (*m*; ν_{15}), as compared with the $H_2{}^{11}B_2O_3$ normal frequencies of

1201, 1114, 961, 833, 776, 883, 426, 2653, 1375, 1169, 979 and 741 cm^{-1}, respectively.[28]

Isotopically pure diboranes (see Fig. 4.5) probably belong to the $\boldsymbol{D_{2h}}$ point group. Their main infrared-active fundamentals in the case of $^{10}B_2H_6$, $^{11}B_2H_6$ and $^{10}B_2D_6$ are given in Table 8.3.[29, 30]

TABLE 8.3 The Main Infrared-Active Fundamentals (cm^{-1}) of $^{10}B_2H_6$, $^{11}B_2H_6$ and $^{10}B_2D_6$[29, 30]

Vibration	Approximate description	$^{11}B_2H_6$	$^{10}B_2H_6$	$^{10}B_2D_6$
ν_8	Asymmetrical B—H(D) stretching	2612	2625	1999 (*s*)
ν_9		1859	1869	
ν_{10}		368	368	262
ν_{13}	Symmetrical B—H′(D′)[b] stretching	1905[a]	1920[a]	1491 (*m*)
ν_{14}	$BH_2(D_2)$ wagging	972	977·5	728 (*s*)
ν_{16}	Symmetrical $BH_2(D_2)$ stretching	2517	2528	1857 (*vs*)
ν_{17}	Asymmetrical B—H′(D′) stretching	1601	1606	1205 (*vvs*)
ν_{18}	Deformation	1174	1181	881 (*s*)

[a] Corrected for Fermi resonance.
[b] H′(D′)=bridging H(D).

A diborane sample containing 43·6 atom % ^{11}B, (having a $^{11}B^{10}BH_6$ content of *ca.* 49%) was observed to exhibit its ν^{16} band at 2525 cm^{-1} and its ν_{18} fundamental split into three branches, due to its $^{11}B_2$, $^{11}B^{10}B$ and $^{10}B_2$ components.[30] In the case of $^{11}B_2D_6$ this fundamental appears at 875·9 cm^{-1}.

The Urey–Bradley force field and normal frequencies of various ^{10}B and ^{11}B diboranes havė also been calculated.[31]

The 250–3000 cm^{-1} absorption of gaseous $H_3{}^{10}B_3O_3$, $H_3{}^{11}B_3{}^{18}O_3$ and various deuterated boroxines supports the assumption that the boroxine molecule is not planar.[32] The fundamentals of $H_3{}^{10}B_3O_3$ were observed at 537, 1025, 1228, 1428 and 2624 cm^{-1}, as compared with the corresponding $H_3{}^{11}B_3O_3$ bands at 530, 990, 1213, 1389 and 2620 cm^{-1}, respectively.[32]

The main ^{10}B-sensitive 250–4000 cm^{-1} absorption bands of a thin solid film of 96 atom % of ^{10}B-trifluoroboroxine at −150°C were found at 731 (*s*; *b*; ring puckering), 838 (*w*; ring deformation), 896 (*w*; *b*), 1403 (*m*; *sh*; ring stretching), 1422 (*m–s*; *b*; combination), 1468 (*s*; *vb*; ring stretching), 1545 (*m–w*) and 1781 cm^{-1} (combination), corresponding to the absorption peaks of a 89 atom % ^{11}B-labelled sample at 711, 826, 903, 1373, 1400, 1436, 1535 and 1776 cm^{-1}, respectively.[10, 33] B-enriched trichloroboroxine exhibits these bands at 675 (*w*; *sh*; ring puckering of the $^{10}B—O—^{11}B—O—^{10}B$ ring),
O

686 (*s*; ring puckering of the $^{10}B_3O_3$ ring), 760 (*m*), 771 (*m*; both: B—Cl stretchings), 786 (*m*; overtone), 815 (*m–w*; ring deformation), 1125 (*w*), 1192 (*m–s*; combination), 1220 (*s*; *b*), 1385 (*s*; *b*) and 1414 cm^{-1} (*m*; *sh*; all three: ring stretching), as compared with the parallel ^{11}B-enriched sample bands at 683, 664, 755, 766, 782, —, —, 1161, 1183, 1345 and 1390 cm^{-1}, correspondingly.[33]

The $\nu_{B\text{-}10}-\nu_{B\text{-}11}$ isotopic shifts in the case of the mixed halide boroxine $^{10}B_3O_3FCl_2$ absorption bands at 440, 770, 787, 1239 and 1394 cm^{-1} were observed to be 6, 8, 4, 42 and 42 cm^{-1}, respectively. Analagous shifts for the $^{10}B_3O_3F_2Cl$ fundamentals at 462 (ring deformation), 855 (ring deformation), 954 (B—F stretching), 1427 (B—F stretching) and 1465 cm^{-1} (ring stretching) have been determined as 5, 8, 4, 53 and 52 cm^{-1}, correspondingly.[34]

The zero-order (harmonic) normal frequencies of the $^{10}BH_4^-$, $^{10}BD_4^-$, $^{11}BH_4^-$ and $^{11}BD_4^-$ borohydride ions have been calculated after their force constants were first determined.[35]

^{10}B-Enriched potassium tetrafluoroborate exhibits its $^{10}BF_4^-$-sensitive absorption bands in the 375–4000 cm^{-1} range at 542 (ν_4), 1115, 1135, 1158 (ν_3 branches), 1305 and 1312 cm^{-1} (both: $\nu_1 + \nu_4$), corresponding to the respective natural abundance KBF_4 absorption peaks at[36] 536, 1107, 1128, —, 1312 and 1330 cm^{-1}.

A strontium tetrahydroxoborate, $Sr(B(OH)_4)_2$, ("strontium borate hydrate") sample which contained 80 atom % ^{10}B displayed (as a mull or in a KBr pellet) its out-of-phase B—O stretching bands at 936 (*vvs*) and 1001 (*sh*) cm^{-1} and its B—O—H bending absorption peaks at 1106 (*m*) and 1249 (*s*) cm^{-1}, as compared with the ^{11}B-compound bands at 916, 976, 1096 and 1240 cm^{-1}, respectively.[37]

In the case of pinnoite ("magnesium borate hydrate"), which is made up of magnesium and $(HO)_3B—O—B(OH)_3^{2-}$ ions, the ^{10}B-sensitive absorption bands of a Nujol mull of an 85 atom % ^{10}B-sample appeared at 806 (*m*; out-of-plane B—O—H deformation), *ca.* 886 (*sh*), *ca.* 926 (*sh*), 976 (*vs*), 1055 (*s*; all four: antisymmetric B—O stretchings), 1225 (*s*) and 1307 cm^{-1} (*s*; both B—O—H bendings), in parallel to the 97 atom % ^{11}B-compound absorption peaks at 801, 876, 896, 956, 1037, 1217 and 1293 cm^{-1}, correspondingly.[38]

The 400–1800 cm^{-1} spectra of 88 atom % ^{10}B-labelled "potassium (1:1:2·66) and (1:1:0·66) borate hydrates", which are actually polyborates with free OH (or, in the deuterated salts, OD) groups, have also been reported.[39]

The sodium and potassium metathioborates are composed of planar, cyclic $(SBS^-)_3$ anions which belong to the D_{3h} point group. The 45–4000 cm^{-1} absorption of Nujol mulls of about 90 atom % $Na_3{}^{10}B_3S_6$ and $K_3{}^{10}B_3S_6$ was investigated in connection with the assignment of the meta thioborate fundamentals.[40] $K_3{}^{10}B_3S_6$ was found to show its main absorption bands at 47 (*m*), 63 (*m*), 95 (*s*), 142 (*m*; all four: lattice bands), 195 (*m*;

B—S rocking), 244 (*m*; ring angle bending), 915 (*s*; ring stretching) and 936 cm^{-1} (*vs*; B—S stretching), as compared with the parallel $K_3{}^{11}B_3S_6$ bands at 47, 62, 93, 142, 195, 241, 874 and 896 cm^{-1}, respectively. The spectra of the isotopic sodium metathioborates are usually fairly similar to those of the corresponding potassium salts.[40]

8.4 Organic Boron Compounds

The asymmetrical ^{10}B—C stretching vibration of gaseous triethylborane and D_{15}-triethylborane appears at about 1135 cm^{-1}, as compared with the 1120 and 1110 cm^{-1} values for the corresponding ^{11}B—C fundamental of these compounds, respectively.[41]

Liquid phenylboron dichloride exhibits its (natural abundance ^{10}B-modification content) C—$^{10}BCl_2$ out-of-plane deformation band at 567 cm^{-1} and its C—^{10}B stretching fundamental at 1250 cm^{-1} while the corresponding ^{11}B-frequencies are shown at 552 and 1230 cm^{-1}, respectively.[42]

Liquid, natural abundance, phenylboracyclopentane shows its characteristic ^{10}B-isotopomer absorption bands at 578 (*m*; hetero-ring out-of-plane deformation), 605 (*m*; in-plane B-ring deformation), 655 (*sh*; B-ring stretching), 932 (*s*; B-ring breathing), 1050 (*sh*; B-ring stretching), 1082 (*m*; B—$C_{aromatic}$ stretching) and 1111 cm^{-1} (*m*; B-ring antisymmetrical stretching), corresponding to the ^{11}B-compound bands at 568, 601, 643, 925, 1046, 1075 and 1104 cm^{-1}, respectively.[44]

The antisymmetric ^{10}B—O stretching fundamental of $DB(OC_2H_5)_2$ was found at 1445 cm^{-1}, as compared with the 1425 cm^{-1} frequency of the parallel ^{11}B—O vibration.[43]

The ^{10}B-sensitive absorption bands of gaseous $H^{10}B(OC_2H_5)_2$ were observed at 917 (*m–s*; out-of-plane B—H deformation), 1242 (symmetrical B—O stretching), 1348 (*vvs*; CH_2 wagging) and 1451 cm^{-1} (*vvs*; asymmetrical B—O stretching), as compared with the corresponding $H^{11}B(OC_2H_5)_2$ bands at 897, 1232, 1342 and 1429 cm^{-1}, respectively.[43] In the case of $H^{10}B(O\text{-i-}C_3H_7)_2$ the ^{10}B-characteristic absorption peaks[43] appear at 919 (*m*; out-of-plane B—H deformation), 1105 (*s*; in-plane B—H bending), 1445 (*vs*; B—O stretching) and 2490 cm^{-1} (*s*; B—H stretching).

The antisymmetric $^{10}BO_2$ stretching fundamental of D_4-1,3-dioxa-2-chloroboracyclopentane was observed at 1460 cm^{-1} (*s*) while the respective $^{11}BO_2$ vibration was shown at[46] 1423 cm^{-1}.

The 400–4000 cm^{-1} absorption spectrum of gaseous methyl ^{10}B-borate, which belongs to the C_{3h} point group, exhibits its ^{10}B-sensitive bands at 709 (*vs*; out-of-plane B—O deformation) and 1385 cm^{-1} (*vs*; B—O stretching), as compared with the corresponding normal borate absorption peaks at 667 and 1363 cm^{-1}, respectively.[47,48] Perdeuteration of methyl ^{10}B-borate shifts its ^{10}B—O stretching fundamental[49] to 1435 cm^{-1} (*s*; an additional shoulder at 1442 cm^{-1}) while $(CD_3O)_3{}^{11}B$ displays its B—O stretching band at 1391 cm^{-1}.

Ethyl ^{10}B-borate shows its characteristic absorption peaks at 700 (*m*; out-of-plane B—O deformation), 1242 (*vw*; symmetric B—O stretching), 1350 (*vvs*; CH_2 wagging) and 1445 cm^{-1} (*vvs*; B—O asymmetric B—O stretching), corresponding to the ^{11}B-borate bands at 673, 1232, 1342 and 1425 cm^{-1}, respectively.[43] In the case of $(i\text{-}C_3H_7O)_3{}^{10}B$ the ^{10}B-sensitive vibrations were observed at 699 (*m–w*; BO_3 deformation), 959 (*m–s*; CH_3 rocking), 1344 (*vs*; C—H bending) and 1433 cm^{-1} (*vvs*; B—O stretching), as compared with the parallel $(i\text{-}C_3H_7O)_3{}^{11}B$ bands at 664, 954, 1335 and 1400 cm^{-1}, respectively.[45]

The asymmetric stretching fundamentals of the BO_4 groups of potassium bis(*o*-phenylenedioxy)borate and sodium dihydroxy-(*o*-phenylenedioxy)-borate were assigned on the basis of the infrared absorption spectra of their *ca.* 96 atom % ^{10}B-labelled modifications.[50] The 400–4000 cm^{-1} absorption of mulls of the labelled sodium salt displayed its ^{10}B-characteristic bands at 834 (B—O stretching), 894 (C—O stretching), 1011, 1038, 1209, 1243 and 1272 cm^{-1}, as compared with the ^{11}B-salt frequencies of 816, 888, 1000, 1025, 1196, 1224 and 1260 cm^{-1}, respectively.[50]

The 200–3300 cm^{-1} absorption of 93 atom % ^{10}B-enriched, normal and deuterated borine carbonyls was investigated in the gas as well as in the solid phase.[51] Table 8.4 reports the fundamental frequencies of the gaseous isotopic species present in these compounds.[51, 52]

TABLE 8.4 The Fundamentals (cm^{-1}) of $^{10}BH_3CO$, $^{11}BH_3CO$, $^{10}BD_3CO$ and $^{11}BD_3CO$[51, 52]

Vibration	Approximate description	$^{10}BH_3CO$	$^{11}BH_3CO$	$^{10}BD_3CO$	$^{11}BD_3CO$
ν_1 (A_1)	B—H(D) stretching	2385	2379	1695	1679
ν_2	C=O stretching	2166	2165	2169	2168
ν_3	$BH_3(D_3)$ bending	1083	1073	~883	~862
ν_4	B—C stretching	707	691	630	625
ν_5 (E)	B—H(D) stretching	2456	2441	1852	1837
ν_6	$BH_3(D_3)$ bending	~1100	1100	802	801
ν_7	Unsymmetrical bending	819	809	718	708
ν_8	Symmetrical bending	314	313	266	266

D_9-Trimethoxyboroxine solutions in CCl_4 exhibited its ^{10}B—O stretching bands[49] at 1425 (*s*), 1443 (*sh*) and 1460 (*sh*) cm^{-1}, corresponding to the respective ^{11}B—O absorption peak at 1403 cm^{-1}.

The ^{10}B—N dative bond stretching fundamental of various Lewis complexes of the general structure $R_3N—BX_3$ (R = H,D,CH_3; X = H,D,F,Cl, Br) was found in the 700–800 cm^{-1} range.[53]

The ^{10}B-sensitive bands of gaseous $(CH_3)_2N^{10}BCl_2$ appeared at 539 (*s*; Cl_2B—N out-of-plane deformation), 975 and 982 (both: *vvs*; asymmetric BCl_2 stretching), 1153 (*s*; asymmetric C_2N stretching), 1423 (*s*; CH_3 bending) and 1535 cm^{-1} (*vvs*; B—N stretching coupled with a CH_3 deformation), corresponding to the parallel ^{11}B-compound bands at 522, 942, 950, 1143, 1417 and 1525 cm^{-1}, respectively.[54]

The infrared absorption of ^{10}B-enriched $(CH_3)_2NBBr_2$ has also been investigated.[55]

Liquid $(C_2H_5)_2N^{10}BCl_2$ exhibits its ^{10}B-characteristic absorption peaks[54] at 467 (*m–w*), 540 (*m–s*), 583, 845 (*m*; symmetric NC_2 stretching), 907, 964 (both: *s*; asymmetric BCl_2 stretching), 1150 (*s*), 1282 (*s*), 1470 and 1514 cm^{-1} (*s*; B—N stretching).

In the case of $[(CH_3)_2N]_2{}^{10}BCl$ the ^{10}B-sensitive bands are shown at[54] 608 (*s*; out-of-plane $ClBN_2$ deformation), 918 (*s*; symmetric NC_2 stretching), 1079 and 1155 (both: *s*; asymmetric NC_2 stretching), 1359 (*s*; symmetric BN_2 stretching), 1407 (*s*; CH_3 bending) and 1552 cm^{-1} (*s*; asymmetric BN_2 stretching). The parallel ^{11}B-aminoborane displays[54] its symmetric BN_2 stretching vibration at 1354 cm^{-1} and its asymmetric one at 1535 cm^{-1}.

The infrared absorption of $[(C_2H_5)_2N]_2{}^{10}BCl$ shows ^{10}B-characteristic bands at 592, 605, 848, 1092, 1144, 1282 and 1522 cm^{-1}; for $^{10}B[N(CH_3)_2]_3$ such bands appear at 691 (*m–w*; out-of-plane BN_3 deformation), 917 (*m*; symmetric NC_2 stretching), 1130 (*s*; asymmetric NC_2 stretching), 1393 (*vs*; asymmetric B—N stretching) and 1516 cm^{-1} (*vs*; CH_3 bending), as compared with the corresponding ^{11}B-isotopomer vibrations of 664, 909, 1117, 1384 and 1508 cm^{-1}, respectively.[54]

The absorption spectrum of highly enriched $^{10}B[N(C_2H_5)_2]_3$ has also been reported.[54]

The infrared absorption of 80 atom % ^{10}B-borazine was used in the reassignment of the ^{11}B-borazine fundamentals at 718 and 918 cm^{-1} as being due to its a_2'' species. These bands are shown by the (gaseous) ^{10}B-isotopomer as triplets at 708, 723, 736 and 914, 928, 940 cm^{-1}, respectively.[56]

The 400–4000 cm^{-1} spectra of 94 atom % ^{10}B-labelled N-trimethylborazine, its $(CD_3)_3$ modification, their B-trichloro-derivatives, $B_3(CD_3)_3N_3(CH_3)_3$, $B_3(C_6H_5)_3N_3(CD_3)_3$ and its perdeuterated isotopomer were all measured.[57] These compounds exhibit their B—N stretching bands in the ranges of 1287–1335 (weak) and 1385–1503 cm^{-1} (strong) with other ^{10}B-characteristic (occasional) bands near 660, 675 and 970 cm^{-1} (the last one due to a B—Cl stretching).

The 400–1500 cm^{-1} absorption of 85 atom % ^{10}B-1,3,5-trimethyl-2,4,6-trichloroborazine is characterized by strong isotopic shifts for all its main bands, brought about by an appreciable coupling between its various vibrational modes. These bands[58] were observed at 675 (*m*; shifted by 14 cm^{-1}), 682 (*vs*; 13), 688 (*sh*; 12), 987 (*m*; 13), 991 (*vs*; 12; both: N—C stretchings),

1091 (*s*; 6), 1109 (*sh*; 6; CH_3 rocking), 1289 (*sh*), 1298 (*m*; 16; B—N stretching), 1406 (14; B—N stretching) and 1466 cm^{-1} (14; CH_3 bending).

The ^{10}B—N stretching fundamental of 1,3-dioxa-2-dimethylaminoboracyclopentane[46] appears at 1610 (*m*) cm^{-1} while its $(CD_2)_2$-isotopomer has been found to show it at 1599 (*m*) cm^{-1} and its perdeuterated modification exhibits it at 1556 (*s*) cm^{-1} (with a shoulder at 1563 cm^{-1}).

The alkyl and aryl esters of thioboric acid were found to display their asymmetric $^{10}BS_3$ stretching band at 942 ± 10 cm^{-1}, as compared with the respective 914 cm^{-1} $^{11}BS_3$ frequency.[59]

8.5 LABELLED SILICON COMPOUNDS

The 300–2100 cm^{-1} absorption spectra of a 45 atom % ^{29}Si- and a 30 atom % ^{30}Si-enriched gaseous SiF_4 sample exhibited, mainly, the bands given in Table 8.5.

TABLE 8.5 The Main Absorption Bands[a] (cm^{-1}) of the Isotopic SiF_4 Molecules[60]

Band	Approximate description	$^{28}SiF_4$	$^{29}SiF_4$	$^{30}SiF_4$
ν_3	Out-of-phase stretching	1031·8 (*s*)	1022·9	1014·4
ν_4	Deformation	389·35 (*s*)	387·8	386·35
$\nu_1 + \nu_3$	Combination	1825·7 (*m*)	1816·9	1808·1
$\nu_1 + \nu_4$	Combination	1190·4 (*m*)	1188·4	1186·8

[a] Measured in vacuum.

The force field and Coriolis coupling constants of these isotopic molecules have also been determined.[61,62]

Oxygen containing silicon, investigated under high resolution at liquid helium temperature, was observed to exhibit absorption peaks at 1120·09, 1130·89, 1132·72, 1134·51 and 1136·43 cm^{-1}, due partly to asymmetrical stretching vibrations of the ^{28}Si—O—^{28}Si, ^{28}Si—O—^{29}Si, ^{29}Si—O—^{29}Si, ^{28}Si—O—^{30}Si and similar isotopic groupings present in this substance as a result of the (4·7 atom % ^{29}Si and 3·1 atom % ^{30}Si occurring alongside the 92 atom % ^{28}Si of the) natural isotope content of the silicon.[63]

8.6 LABELLED SULPHUR COMPOUNDS

The vibrational–rotational $3\nu_3$ overtone band of natural abundance CS_2 was investigated under high resolution to yield also the 4556·553 cm^{-1} frequency for the centre of the $3\nu_3$ band of $C^{32}S^{34}S$ as well as a number of its vibrational and rotational constants.[64] Liquid carbon disulphide exhibits its $C^{32}S^{34}S$ $\nu_3 + 2\nu_1$ band at[65] 2793 cm^{-1}.

Gaseous 47 atom % $^{34}SO_2$ has been observed to show its ν_1 and ν_3 fundamentals at 1146·0 and 1344·2 cm^{-1}, respectively, as compared with the parallel $^{32}SO_2$ 1151·2 and 1360·8 cm^{-1} bands.[66] The ν_2 frequency of $^{34}SO_2$ has been calculated to have the value of 512·3 cm^{-1}. The rotational constants of this isotopomer have also been reported.[66] In the case of $^{35}SO_2$ these vibrations appear at[66] 1142·0 (ν_1), 510·6 (ν_2; calculated) and 1336·5 cm^{-1} (ν_3).

The rotational spectra of $^{34}S^{16}O^{18}O$ and $^{34}S^{18}O_2$ in the first ν_2 excited state have been recorded and the rotational constants of the $^{33}S^{16}O_2$ and $^{34}S^{16}O_2$ modifications, both in the ground and in the first ν_2 excited state, have been evaluated.[67]

The two infrared-active fundamentals of natural isotopic abundance SF_6 (containing about 7 atom % ^{36}S) have been observed at 611·2 (ν_4) and 930·5 (ν_3) cm^{-1} for the $^{34}SF_6$ species and at 614·5 (ν_4) and 947·9 (ν_3) cm^{-1} for the $^{32}SF_6$ isotopomer.[68] The Coriolis constants for these fundamentals of $^{32}SF_6$ and $^{34}SF_6$ have also been determined[68] and the fact that the values calculated for these constants from the observed isotopic frequencies using the theory of Moret–Bailly[69] agree well with the values obtained by conventional methods demonstrates that for a heavy spherical top molecule the isotopic fundamentals provide an accurate means of determining the Coriolis constants.[68]

The infrared absorption of crystalline 37 atom % ^{34}S-labelled potassium thiocyanate exhibited its $^{14}N^{12}C^{34}S^-$ ν_3 fundamental at 739·2 cm^{-1}, corresponding to the respective $^{14}N^{12}C^{32}S^-$ band at 748·7 cm^{-1}, while a ^{13}C-labelled sample of this salt showed its $^{14}N^{13}C^{34}S^-$ ν_3 vibration at 733·1 cm^{-1}. The other fundamentals of these ions were overlapped by the very close absorption peaks due to the corresponding $NC^{32}S^-$ vibrations.[70]

The fundamental vibrations of $^{34}SO_4^{2-}$ and $^{36}SO_4^{2-}$ were calculated to have the values of 983, 450, 1089, 608 and 983, 450, 1074, 605 cm^{-1}, as compared with the corresponding $^{32}SO_4^{2-}$ vibrations of 983 (ν_1), 450 (ν_2), 1105 (ν_3) and 611 (ν_4) cm^{-1}, respectively.[71]

8.7 Isotopic Chlorine Compounds

Since the natural abundance of ^{37}Cl amounts to 24·6% the vibration–rotation spectrum of gaseous HCl shows all its absorption lines doubled, each line appearing separately both for $H^{35}Cl$ and for $H^{37}Cl$. The fundamental H—^{37}Cl lines were observed to be shifted by 2·01 cm^{-1} from the respective $H^{35}Cl$ lines, centred around[72] 2885.9775 cm^{-1}. In the case of the first overtone lines, centred for $H^{35}Cl$ around 5667.9841 cm^{-1}, the $H^{37}Cl$ frequencies were found to be decreased by 4·00 cm^{-1}, while in the 3—O band they were measured to be lower by[73] 5·834 cm^{-1} from the corresponding $H^{35}Cl$ lines of the 8346.7816 cm^{-1} band.[72]

A mixture of solid HCl and 70% DCl (at -195°K) was found to show a strong band at 2885·2 cm^{-1} and a weak band at 2883·0 cm^{-1} due to the H—Cl

abundance phenylboron dichloride[42] appears at 535 (w) cm^{-1} while the respective frequency of its $^{35}Cl_2$-species is displayed at 552 cm^{-1}.

8.8 Other Isotopic Compounds

Natural abundance germanium difluoride, suspended in a solid neon matrix at 5°K, showed its isotopic ν_3 stretching bands at 653·0, 655·0, 656·8 and 658·9 cm^{-1} for its ^{76}Ge (7·8%), ^{74}Ge (36·5%), ^{72}Ge (27·4%) and ^{70}Ge (20·5%) species, respectively, while the parallel isotopic ν_1 stretching fundamentals appeared at 683·3, 685·0, 686·5 and 688·5 cm^{-1}, correspondingly. Natural abundance germanium tetrafluoride under these conditions exhibits similar absorption peaks at 796·7, 799·3, 802·0 and 805·0 cm^{-1}, respectively.[82]

The isotopic band centre separations for a number of near infrared (overtone and combination) absorption bands of natural abundance hydrogen selenide have been determined. These $\Delta\nu_0$ values ($\Delta\nu_0 = \nu_{H_2{}^iSe} - \nu_{H_2{}^{80}Se}$, $i = 76, 77, 78, 82$) have also been estimated theoretically.[83] The rotational constants of $H_2{}^{78}Se$ have been found to be $A = 8{\cdot}1757$, $B = 7{\cdot}7271$ and $C = 3{\cdot}9025$ cm^{-1}, as compared with the respective 8·1703, 7·7272 and 3·9013 cm^{-1} values for $H_2{}^{80}Se$.[83]

The vibration–rotation spectra of natural isotopic abundance SeF_6 in the regions of its fundamentals have also been reported.[84]

The fundamental frequencies of $^{76}SeO_4^{2-}$ and $^{82}SeO_4^{2-}$ have been calculated.[85]

The C—Br stretching band of natural isotopic abundance D_3-methyl bromide was found to show the centres of its $CD_3{}^{79}Br$ and $CD_3{}^{81}Br$ absorptions at 576·72 and 575·51 cm^{-1}, respectively.[86]

The fundamental harmonic frequencies calculated for the $^{16}O^{14}N^{79}Br$ and $^{16}O^{14}N^{81}Br$ isotopomers of nitrosyl bromide are 1832·4 (N=O stretching; ω_1), 548·1 (bending; ω_2), 269·7 (N—Br stretching; ω_3) and 1832·4, 547·9, 268·9 cm^{-1}, respectively.[87]

The fine structure of the ν_3 (antisymmetric stretching) band of pure gaseous $^{86}KrF_2$ yielded the value of 589·899 cm^{-1} for its centre as well as the B_0 and D_0 rotational constants of $^{86}KrF_2$.[88]

References

1. Tarte, P. (1964). *Spectrochim. Acta* **20**, 238.
2. Tarte, P. (1965). *Spectrochim. Acta* **21**, 313.
3. Tarte, P. (1967). *J. Inorg. Nuclear Chem.* **29**, 915.
4. Rodinova, A. N., Talalaeva, T. V., Shigorin, D. N., Rodinova, G. N. and Kocheshkov, K. A. (1965). *Izvest. Akad. Nauk S.S.S.R. Ser. Khim.* 604; *C.A.* **63**, 3779.
5. Shimizu, K. and Shingu, H. (1966). *Spectrochim. Acta* **22**, 1999.
6. Steinhardt, R. G., Fetsch, G. E. S. and Jordan, M. W. (1965). *J. Chem. Phys.* **43**, 4528.

7. Ginn, S. G. W., Kenney, J. K. and Overend, J. (1968). *J. Chem. Phys.* **48**, 1571.
8. Duncan, J. L. (1967). *J. Mol. Spectroscopy* **22**, 247.
9. Ginn, S. G. W., Brown, C. W., Kenney, J. K. and Overend, J. (1968). *J. Mol. Spectroscopy* **28**, 509.
10. Brown, C. W. and Overend, J. (1968). *Canad. J. Phys.* **46**, 977.
11. Nakane, R. and Ōyama, T. (1966). *J. Phys. Chem.* **70**, 1146.
12. Begun, G. M. and Palko, A. A. (1963). *J. Chem. Phys.* **38**, 2112.
13. Taillandier, M., Liquier, J. and Taillandier, E. (1968). *J. Mol. Struct.* **2**, 437.
14. Taylor, R. C., Gabelnick, H. S., Aida, K. and Amster, R. L. (1969). *Inorg. Chem.* **8**, 605.
15. Begun, G. M. and Palko, A. A. (1967). *J. Chem. Phys.* **47**, 967.
16. Le Calve, J. and Lascombe, J. (1968). *Spectrochim. Acta* **24A**, 737.
17. Lynds, L. and Bass, C. D. (1965). *J. Chem. Phys.* **43**, 4357.
18. Mattes, R. and Becher, H. J. (1968). *Z. phys. Chem. (Frankfurt)* **61**, 177.
19. Bass, C. D., Lynds, L., Wolfram, T. and DeWames, R. E. (1964). *J. Chem. Phys.* **40**, 3611.
20. Lynds, L. (1966). *J. Chem. Phys.* **44**, 1721.
21. Finch, A., Hyams, I. J. and Steele, D. (1965). *Trans. Faraday Soc.* **61**, 398.
22. Lynds, L., Wolfram, T. and Bass, C. D. (1965). *J. Chem. Phys.* **43**, 3775.
23. Hanst, P. L., Early, V. H. and Klemperer, W. (1965). *J. Chem. Phys.* **42**, 1097.
24. (a) Marram, E. P. and Buechler, A. (1962). *Spectrochim. Acta* **18**, 1386.
    (b) Marram, E. P. and Buechler, A. (1963). *J. Chem. Phys.* **39**, 292.
25. Vaško, A. and Srb, I. (1967). *Czech. J. Phys.* **B17**, 1110.
26. Servoss, R. R. and Clark, H. M. (1957). *J. Chem. Phys.* **26**, 1175.
27. Durig, J. R., Green, W. H. and Marston, A. L. (1968). *J. Mol. Struct.* **2**, 19.
28. Grimm, F. A. and Porter, R. F. (1969). *Inorg. Chem.* **8**, 731.
29. Lord, R. C. and Nielsen, E. (1951). *J. Chem. Phys.* **19**, 1.
30. Kaufman, J. J., Koski, W. S. and Anacreon, R. (1963). *J. Mol. Spectroscopy* **11**, 1.
31. Lem Wu, E. C. (1964). *Diss. Abs.* **25**, 1612.
32. Grimm, F. A., Barton, L. and Porter, R. F. (1968). *Inorg. Chem.* **7**, 1309.
33. Latimer, B. and Devlin, J. P. (1965). *Spectrochim. Acta* **21**, 1437.
34. Latimer, B. and Devlin, J. P. (1967). *Spectrochim. Acta* **23A**, 81.
35. Venkateswarlu, K. and Radhakrishnan, M. (1964). *Czech. J. Phys.* **14**, 848.
36. Greenwood, N. N. *J. Chem. Soc.* **1959**, 3811.
37. Kessler, G. and Lehmann, H. A. (1965). *Z. anorg. Chem.* **339**, 87.
38. Kessler, G. (1966). *Z. anorg. Chem.* **348**, 89.
39. Kessler, G., Roensch, E. and Lehmann, H. A. (1969). *Z. anorg. Chem.* **365**, 281.
40. Chopin, F. and Turrell, G. (1969). *J. Mol. Struct.* **3**, 57
41. Lehmann, W. J., Wilson, C. O. Jr. Shapiro, I. (1959). *J. Chem. Phys.* **31**, 1071.
42. Lockhart, J. C. *J. Chem. Soc. (A).* **1966**, 1552.
43. Lehmann, W. J., Weiss, H. G. and Shapiro, I. (1959). *J. Chem. Phys.* **30**, 1222.
44. Greenwood, N. N. and Wright, J. C. *J. Chem. Soc.* **1965**, 448.
45. Lehmann, W. J., Weiss, H. G. and Shapiro, I. (1959). *J. Chem. Phys.* **30**, 1226.
46. Meller, A. and Marecek, H. (1967). *Monatsh.* **98**, 2336.
47. Servoss, R. R. and Clark, H. M. (1957). *J. Chem. Phys.* **26**, 1178.
48. Barcelo, J. R. (1964). *Anales real Soc. españ. Fís. Quim.* **40**, 763.
49. Meller, A. and Schaschel, E. (1967). *Monatsh.* **98**, 390.
50. Funck, E. (1967). *Ber. Bunsenges. phys. Chem.* **71**, 170.
51. Bethke, G. W. and Wilson, M. K. (1957). *J. Chem. Phys.* **26**, 1118.
52. [illegible] J. and Cleveland, F. F. (1960). *J. Chem. Phys.* **32**, 166.
53. Taylor, R. C. (1964). *Advan. Chem. Ser.* No. **42**, 59.

54. Becher, H. J. and Baechle, H. T. (1966). *Z. phys. Chem. (Frankfurt)*. **48**, 359.
55. Whitmer, J. C. (1966). *Diss. Abs.* **27B**, 444.
56. Kartha, V. B., Krishnamachari, S. L. N. G. and Subramanian, C. R. (1957). *J. Mol. Spectroscopy* **23**, 149.
57. Totani, T., Watanabe, H. and Kubo, M. (1969). *Spectrochim. Acta* **25A**, 585.
58. Meller, A. (1967). *Monatsh.* **98**, 1747.
59. Cabana, A., Brault, J. and Lalancette, J. M. (1966). *Spectrochim. Acta* **22**, 377.
60. Heicklen, J. and Knight, V. (1964). *Spectrochim. Acta* **20**, 295.
61. Ruoff, A. (1967). *Spectrochim. Acta* **23A**, 2421.
62. Levin, I. W. and Abramowitz, S. (1968). *J. Res. Nat. Bur. Stand.* **72A**, 247.
63. Pajot, B. and Deltour, J. P. (1967). *Infrared Phys.* **7**, 195.
64. Guenther, A. H. (1959). *J. Chem. Phys.* **31**, 1095.
65. Plyler, E. K. and Humphreys, C. J. (1947). *J. Res. Nat. Bur. Stand.* **39**, 59.
66. Grigg, E. C. M. and Johnston, G. R. (1966). *Austral. J. Chem.* **19**, 1147.
67. van Riet, R. (1964). *Ann. Soc. sci. Bruxelles. Ser. I.* **78**, 237.
68. Brunet, H. and Perez, M. (1969). *J. Mol. Spectroscopy.* **29**, 472.
69. Moret-Bailly, J. (1959). *Cahiers Phys.* **112**, 476.
70. Jones, L. H. (1956). *J. Chem. Phys.* **25**, 1069.
71. Vojta, G. and Koch, D. (1964). *Z. phys. Chem. (Leipzig)*. **225**, 295.
72. Rank, D. H., Rao, B. S. and Wiggins, T. A. (1965). *J. Mol. Spectroscopy* **17**, 122.
73. Herzberg, G. (1950). "Spectra of Diatomic Molecules", p. 142. D. Van Nostrand Company, N.Y.
74. Rank, D. H., Glickman, W. A. and Wiggins, T. A. (1965). *J. Chem. Phys.* **43**, 1304.
75. Rochkind, M. M. and Pimentel, G. C. (1965). *J. Chem. Phys.* **42**, 1361.
76. Robinson, E. A., Lavery, D. S. and Weller, S. (1969). *Spectrochim. Acta* **25A**, 151.
77. Holladay, Th. M. and Nielsen, A. H. (1964). *J. Mol. Spectroscopy* **14**, 371.
78. Jones, E. W., Popplewell, R. J. L. and Thompson, H. W. (1966). *Proc. Roy. Soc.* **A290**, 490.
79. Duncan, J. L. and Allan, A. (1968). *J. Mol. Spectroscopy* **25**, 224.
80. Hadni, A. (1955). *Compt. rend.* **240**, 1983.
81. Antilla, R. and Huhanantti, M. (1968). *Z. Naturforsch.* **23a**, 2098.
82. Hastie, J. W., Hauge, R. and Margrave, J. L. (1968). *J. Phys. Chem.* **72**, 4492.
83. Hill, R. A. and Edwards, T. H. (1965). *J. Chem. Phys.* **42**, 1391.
84. Abramowitz, S. and Levin, I. W. (1967). *Inorg. Chem.* **6**, 538.
85. Rees, C. E. and Thode, H. G. (1966). *Canad. J. Chem.* **44**, 419.
86. Jones, E. W., Popplewell, R. J. L. and Thompson, H. W. (1966). *Spectrochim. Acta* **22**, 639.
87. Laane, J., Jones, L. H., Ryan, R. R. and Asprey, L. B. (1969). *J. Mol. Spectroscopy* **30**, 485.
88. Murchison, C., Reichman, S., Anderson, D., Overend, J. and Schreiner, F. (1968). *J. Amer. Chem. Soc.* **90**, 5690.

CHAPTER 9

Isotope Effects on Absorption Intensity and Width of Bands

9.1 The Theoretically Expected Absorption Intensity of Labelled Compound Bands

It can be shown that the integrated absorption intensity A of the kth fundamental infrared band belonging to a mechanically and electrically harmonic molecular oscillator (i.e. a molecule which its potential energy is a quadratic function of its displacements and its dipole moment is a linear function of them), is given by:[1]

$$A_k = \frac{1}{nl}\int \ln\frac{I_0}{I}\,dv = K\left(\frac{\partial \boldsymbol{\mu}}{\partial Q_k}\right)_0^2 \text{ cm/mole,} \tag{9.1}$$

where n = number of absorbing moles in 1 cm^3, l = optical path (cell thickness) in cm, I_0, I = the intensities of the entering and the emerging light, respectively, v = light frequency in cm^{-1}, K = a universal constant, $\boldsymbol{\mu}$ = molecular dipole moment and Q_k is the kth normal co-ordinate of this molecule.

The A^i/A ratio of the integrated absorption intensities of the only fundamental (stretching) band of two such isotopic (diatomic) R^i—X and R—X molecules can therefore be expressed by:

$$\frac{A^i}{A} = \left[\frac{(\partial \boldsymbol{\mu}^i/\partial Q^i)_0}{(\partial \boldsymbol{\mu}/\partial Q)_0}\right]^2. \tag{9.2}$$

Since it is generally assumed that a molecular dipole moment is practically independent of the nuclear masses which constitute the molecule[2] and since the Q/Q^i ratio for an equal stretching Δr is equal in this case to

$$\rho\ (= [(m_{R^i} + m_X)m_R/m_{R^i}(m_R + m_X)]^{1/2}),$$

it follows that:

$$\frac{A^i}{A} = \rho^2. \tag{9.3}$$

As the isotopic stretching frequencies ratio of these R^iX and RX molecules, ν^i/ν, is also equal to ρ (eqn. 3.6) one gets:

$$\frac{A^i}{A} = \rho^2 = \left(\frac{\nu^i}{\nu}\right)^2 = \frac{\nu^{(i)2}}{\nu^2}. \tag{9.4}$$

Although eqn. 9.4 has been derived only for a fully harmonic oscillator (and when the Born–Oppenheimer approximation[3] can be assumed to be satisfactory) a calculation of the isotopic effect on the integrated absorption intensity of the DCl fundamental band taking into account the anharmonicity of this molecule has yielded practically the same results as those obtained on basis of this equation.[4]

It should be noted here that the isotopic invariant of $A^i/\nu^{(i)2}$ obtained from eqn. 9.4 relates only to the absorption intensity integrated over the *whole* vibration–rotation band in question while the intensity of individual lines of this band is very often affected by an isotopic substitution far beyond eqn. 9.4.

Equation 9.4 was shown to hold theoretically not only for gaseous molecules but also for molecules in dilute solution in a non-polar solvent, as long as the isotopic substitution does not affect the configuration of these molecules.[5]

In the case of fully harmonic gaseous polyatomic molecules which do not possess a permanent dipole moment Crawford has shown[6] that eqn. 9.1 leads to a similar equation for the integrated absorption intensity of the fundamental bands of the isotopic species:

$$\sum^F \equiv \sum_k \frac{I_k}{\omega_k^2} = \sum_k \frac{I_k^i}{\omega_k^{(i)2}}, \tag{9.5}$$

where $\sum_k$ denotes summation over all the fundamentals which belong to any one symmetry type (species) of a normal molecule, $I_k = A_k/K$, $\omega_k =$ the (harmonic) frequency of the kth fundamental of this molecule and the superscript i means that the respective symbol refers to an isotopomer, of the molecule in question, which possesses the same symmetry.

If two isotopic polyatomic molecules do possess a permanent dipole moment and their rotations are therefore also active in absorbing infrared light, eqn. 9.5 holds only for those symmetry types which do not include rotations. For these species which also contain rotations a correction must be applied to the change of the molecular dipole moment with a valence-force co-ordinate of one isotopomer before we can equate it with the parallel change for the second isotopomer. (This need is the result of the fact that vibrations belonging to such species require some compensating rotation in order to conserve angular momentum, this rotation being sensitive to the masses of the vibrating atoms.) This correction introduces a correction term[6] also to eqn. 9.5.

Equation 9.5 is one form of the so called *F sum rule* or *isotope intensity sum rule*.

Decius has shown[7] that for a series of isotopic molecules such as, e.g. acetylene, C_2HD and C_2D_2, the F sum rule yields the following relationship between the respective isotopic I_k sums:

$$\sum_i n_i \left(\sum_k I_{ik} \right) = 0, \tag{9.6}$$

where $\sum_i$ denotes summation over all the members of the isotopic series and n_i is the statistical weighting factor of the ith isotopomer. In case of the Σ^+ species of the $C_{\infty v}$ point group common to all of these acetylenes this rule states:

$$\sum_{k=1}^{3} I_k(C_2H_2) + \sum_{k=1}^{3} I_k(C_2D_2) - 2 \sum_{k=1}^{3} I_k(C_2HD) = 0. \tag{9.7}$$

For bands due to vibration species which also contain rotations of permanent dipole moment molecules (for which the F sum rule must contain a correction term) eqn. 9.6 also has to be properly modified.[7]

Rules similar to eqn. 9.6 and relating to powers of integrated absorption intensity sums of isotopic molecule bands have also been found.[8]

In the case of overtone infrared absorption bands the A/v^2 expression for the isotopic invariant of fundamentals must be substituted by the expression A/nv^{n+1}, where n is 2 for the first overtone of the v fundamental, 3 for the second one, etc.[5]

9.2 The Absorption Intensity of Deuterated Compounds

The integrated absorption intensity of the fundamental D—Cl band, centred at 2091 cm^{-1}, was found to be 1640 darks (1 dark = 1 cm/mmole), as compared with the value of 3200 darks for the integrated intensity of the 2887 cm^{-1} H—Cl fundamental. The A/v^2 quotient thus equals to $0{\cdot}375 \times 10^{-3}$ dark cm^2 for the DCl fundamental and to $0{\cdot}384 \times 10^{-3}$ units for the HCl band, confirming eqn. 9.4 to within 2·4% although the observed, rather than the harmonic, stretching frequencies have been used in this calculation.[9]

The first overtone of the DCl vibration gives rise to a band centred at 4129 cm^{-1} with an integrated absorption intensity of 26 darks, corresponding to the respective HCl 5668 cm^{-1} band which shows an absorption intensity of 70 darks. These results give for the DCl and HCl $A_{2v}/2v^3$ quotients, respectively, the values of $0{\cdot}371 \times 10^{-9}$ and[9] $0{\cdot}379 \times 10^{-9}$ dark cm^3, in agreement with Buckingham's theory.[5]

The strength of the individual isotopic vibrational–rotational lines of the 1–0, 2–0 and 3–0 bands of $D^{35}Cl$ and $D^{37}Cl$ (and the 1–0, 2–0, 3–0, 2–1 and 3–2 bands of $H^{35}Cl$ and $H^{37}Cl$) has also been measured, displaying the theoretically expected relative line strength for the 1–0 band lines.[10]

The strengths of the 1–0 D–^{35}Cl and D–^{37}Cl bands were obtained by summing up separately the strengths of the lines of each isotopomer. The value for $D^{35}Cl$ has thus been found to be *ca.* 1% higher than the respective $D^{37}Cl$ strength while the fundamental band strength of $H^{35}Cl$ was observed to be *lower* by about 4% than the respective $H^{37}Cl$ value. The overtone of the $H^{35}Cl$ band showed a strength-*decrease* of about 2%, relative to the corresponding $H^{37}Cl$ value, but $D^{35}Cl$ exhibited only a decrease of 1% for its 2–0 band strength, as compared to that of the $D^{37}Cl$ 2–0 overtone.[10]

A comparison of the integrated absorption intensity of the D—Cl stretching band of the gaseous hydrogen-bonded complex between DCl and dimethyl ether with that of the H—Cl stretching fundamental of the respective normal complex showed that the former intensity is about half of the latter. Such a behaviour can be expected from an anharmonic, linear triatomic Y····H—X model.[11]

The molar extinction coefficient ε $(=(1/cl)\log_{10}(I_0/I)_{\text{maxim}}$, where C = concentration in mole per litre, l = cell thickness in cm and I_0 and I = light intensity of the entering and the emerging light, respectively) of the ν_3, out-of-phase O—D stretching band of liquid D_2O, centred at *ca.* 2500 cm^{-1}, has been found to be 57 1 $mole^{-1}$ cm^{-1}, as compared with 81 units in the case of the ν_3 band of H_2O.[12]

The integrated absorption intensity of the $\nu_1 + \nu_3$ combination band of liquid D_2O was observed to be $(60 \pm 15) \times 10^6$ cm per mole as compared with the respective calculated value of 49 units.[13]

Acetonitrile solutions of D_2O exhibit an integrated intensity[14] of $(2{\cdot}8 \pm 0{\cdot}3) \times 10^6$ cm/mole for its ν_2 band, at 1200 cm^{-1}, corresponding to a value of $5{\cdot}8 \times 10^6$ units in the case of the ν_2 fundamental of H_2O.

The absolute infrared absorption intensities of D_2O solutions in carbon tetrachloride have also been reported.[15]

In the case of the ν_3 out-of-phase stretching band of $D_2{}^{16}O$ in a nitromethane solution the integrated absorption intensity was observed to be $(8{\cdot}0 \pm 0{\cdot}35) \times 10^6$ cm/mole. The change of the D_2O dipole moment with the change in its S_3 symmetry co-ordinate can thus be calculated to be $-2{\cdot}00$ Debeye/Å, a value which when corrected by 0·063 Debeye/Å, for the difference between the D_2O and H_1O compensating rotation connected with the ν_3 fundamental, agrees well with the $\partial\mu/\partial S_3$ rate of $-1{\cdot}885$ Debeye/Å calculated for H_2O from the $15{\cdot}9 \times 10^6$ cm/mole intensity observed for its ν_3 band in this solvent.[16]

The molar extinction coefficients of the O—H and O—D stretching bands of HDO were observed to be appreciably higher than the corresponding coefficients of H_2O and D_2O, respectively, the $\varepsilon_{OH}^{H_2O}/\varepsilon_{OH}^{HDO}$ and $\varepsilon_{OD}^{D_2O}/\varepsilon_{OD}^{HDO}$ ratios found being equal to 0·62 and 0·50, correspondingly.[12]

The integrated absorption intensity of the HDO fundamentals was found to have the values of $22{\cdot}7 \times 10^6$ (ν_1), $3{\cdot}7 \times 10^6$ (ν_2) and 49×10^6 (ν_3) cm/mole while its $\nu_1 + \nu_3$ band exhibited an intensity of 72 units.[13]

In D_3-acetonitrile solution the integrated absorption of the ν_2 fundamental of HDO is equal to $6{\cdot}7 \times 10^6$ cm/mole, it being much higher than the absorption intensity of the respective H_2O and D_2O bands owing to the high contribution of the compensating molecular rotation of HDO to the integrated absorption intensity of the ν_2 fundamental of the polar HDO molecule.[14]

The integrated intensity of the HDO ν_1 and ν_2 bands in the case of various salt solutions has also been investigated.[17]

The absorption intensity of the ν_3 and ν_4 pressure broadened bands of gaseous SiD_4 has been found to be 11,100 and 30,900 cm^2/mole, correspondingly, as compared with the respective values of 13,900 and 43,700 units in the case of SiH_4.[18] The integrated absorption intensities observed for SiD_4 were useful in the choice of the correct sign for the dipole moment of the Si—H bond and obeyed the F sum rule to within[18] 2%.

The intensities found for $SiHD_3$ were 19,900, 8300 and 4400 cm^2/mole for its bands at 680, 1600 and 2200 cm^{-1}, respectively.[18]

The integrated absorption of other SiH_nD_{4-n} fundamentals has also been reported.[19]

The absorption intensities of the ν_3 and ν_4 fundamentals of GeD_4 have the values of 11,300 and 23,000 cm^2/mole, corresponding to those of 14,800 and 34,400 units for the ν_3 and ν_4 bands of GeH_4, respectively.[20]

In the case of SnD_4 these fundamentals have the integrated intensities of 33,600 and 86,000 cm^2/mole, as compared with the parallel SnH_4 values of 48,500 and 115,000 units, correspondingly.[21]

The integrated absorption intensity of the various deuterated methane bands is given in Table 9.1.

The absorption intensity observed for the deuterated methane fundamentals is always within 40% of the intensity calculated for them on basis of their bond moments, etc.[22] The integrated intensity of the tritiated methane bands has also been computed.[24]

The absorption intensity of the deuterated halogenoform C—D fundamentals under various conditions was investigated extensively.[25–30] Table 9.2 reports some of the results obtained thus.

The strong dependence of the integrated absorption intensity of the C—D stretching band (but not that of the C—D deformation fundamental) of the CDX_3 solutions upon the type of their solvent is due to the fact that the CDX_3 molecules form strongly polar, hydrogen bonded complexes with suitable solvent molecules.[30, 31]

TABLE 9.1 Integrated Absorption Intensity of Deuterated Methane Fundamentals

Isotopomer	Integrated bands	Frequency (cm^{-1})	Intensity	Units	Reference
CD_4	ν_3	2259	1446	cm^2/mole	22
	ν_4	996	1906	cm^2/mole	22
CH_4	ν_3	3019	2310	cm^2/mole	22
	ν_4	1306	2554	cm^2/mole	22
CH_3D	ν_1	2200	84×10^{10}	$sec^{-1}\ cm^{-1}$	23
	ν_2, ν_4	*ca.* 3000	658×10^{10}	$sec^{-1}\ cm^{-1}$	23
	ν_3, ν_5, ν_6	*ca.* 1300	391×10^{10}	$sec^{-1}\ cm^{-1}$	23
CH_2D_2	ν_1, ν_6	*ca.* 3000	466×10^{10}	$sec^{-1}\ cm^{-1}$	23
	ν_2, ν_8	*ca.* 2250	185×10^{10}	$sec^{-1}\ cm^{-1}$	23
	ν_4, ν_7, ν_9	*ca.* 1100	287×10^{10}	$sec^{-1}\ cm^{-1}$	23
CHD_3	ν_1	*ca.* 3000	210×10^{10}	$sec^{-1}\ cm^{-1}$	23
	ν_2, ν_4	*ca.* 2200	280×10^{10}	$sec^{-1}\ cm^{-1}$	23
	ν_3, ν_6	*ca.* 1100	204×10^{10}	$sec^{-1}\ cm^{-1}$	23
	ν_5	*ca.* 1300	$89{\cdot}3 \times 10^{10}$	$sec^{-1}\ cm^{-1}$	23

TABLE 9.2 The Integrated Absorption Intensity of CDX_3 C—D Fundamentals

Compound	Band	Phase	Frequency (cm^{-1})	Intensity	Reference
$CDCl_3$	stretching	liquid	2252	260 l $mole^{-1}$ cm^{-2}	27
$CHCl_3$	stretching	liquid	3018	650	27
$CDCl_3$	stretching	CCl_4 solution	2253	190	27
$CHCl_3$	stretching	CCl_4 solution	3018	460	27
$CDCl_3$	deformation	in CS_2	905	3600[a]	27
$CHCl_3$	deformation	in CS_2	1211	9400[a]	27
$CDCl_3$	stretching	in acetone	2253	1310	28
$CDCl_3$	stretching	in $HCON(CH_3)_2$	2235	2850	28
$CDCl_3$	stretching	in cyclohexane	2253	83	28
$CDBr_3$	stretching	in CCl_4	2258	630	27
$CHBr_3$	stretching	in CCl_4	3031	1390	27
CDI_3	stretching	in CCl_4	2243	1230	27
CHI_3	stretching	in CCl_4	3010	1930	27
CDI_3	deformation	in CS_2	786	9500	27
CHI_3	deformation	in CS_2	1068	9900	27

[a] These values seem to have been interchanged in the original paper.

The integrated intensities of, practically, all the fundamentals of gaseous and liquid CD_2Cl_2, $CHDCl_2$ and CH_2Cl_2 have been measured and compared with the respective values calculated from their possible intensity parameters in order to choose from them the correct values of these parameters.[32]

The F sums for the A_1 species fundamentals of gaseous CD_3Cl, CD_3Br and CD_3I were observed to be equal to $4{\cdot}480 \pm 0{\cdot}200$, 3·377 and 2·033 cm^3/mole, respectively, as compared with the parallel CH_3Cl, CH_3Br and CH_3I F sums of $4{\cdot}918 \pm 0{\cdot}088$, 3·307 and 2·100 cm^3/mole, correspondingly.[33] The discrepancy between the CH_3Cl and CD_3Cl F sums is believed to be due either to the anharmonicity of these molecules or else to some resonance interaction between vibrational levels.[33]

The integrated absorption of the other fundamentals of these gaseous methyl halides,[33] as well as the absorption intensity of the ν_1, ν_2 and ν_3 bands of CD_3I in various solvents,[34] have also been measured while calculated values for the integrated absorption intensity of all the $CDCl_3$, $CDBr_3$ and CDI_3 fundamentals have been reported.[35]

The band areas of the ν_5–ν_9 C_2D_6 fundamentals (centred at 2095, 1077, 2236, 1082 and 594 cm^{-1}) were found to be equal to 1155, 270, 2880, 843 and 558 cm^2/mole, respectively.[36] These absorption intensities[37] as well as those of the D_6-cyclopropane fundamentals[38] have also been evaluated theoretically.

The integrated absorption intensity of the C—D stretching band of D_{12}-cyclohexane was observed to be 21,900 l $mole^{-1}$ cm^{-2} for solutions in CCl_4 and 24,000 units for its acetone solutions.[39]

The investigation of the infrared spectra of a number of gaseous saturated deuterocarbons showed that the integrated absorption intensity of their C—D stretching band is equal to $n\ (230 \pm 15) \times 10^{10}$ cycles/cm sec atmosphere (at STP), where n is the number of their C—D bonds, while the respective normal hydrocarbon band has an intensity of $n \times 429 \times 10^{10}$ units.[40]

The F sums of the out-of-plane, B_{1u} fundamentals and the total in-plane normal frequencies of gaseous D_4-ethylene were found to be about 10 and 13%, respectively, lower than the parallel sums of C_2H_4, these facts being probably due to the anharmonicity of these compounds.[41] The absorption intensity of the *cis*- and *trans*-$C_2H_2D_2$ bands has also been reported,[41] as well as the calculated integrated intensity of all the deuterated ethylene fundamentals.[42,43]

The 2245–2365 cm^{-1} C—D stretching band of a carbon tetrachloride solution of 1,3-D_2-azulene exhibited an integrated absorption of 0·117 cm^3/mole, i.e. 0·058 units per C—D bond, while its 2925–3135 cm^{-1} C—H stretching band displayed an absorption intensity of 0·105 units per C—H bond.[44]

The integrated intensity of the infrared fundamentals of D_4-allene has also been evaluated.[45]

The D_2-acetylene C—D stretching band at 2430 cm^{-1} displays an absorption intensity of 3750 darks, as compared with the value of 6800 darks observed for the respective C—H fundamental, while its C—D bending fundamental appearing at 539 cm^{-1} shows an intensity of 8300 darks corresponding to the value of 16,200 darks found for the respective CH absorption peak.[9] The calculated absorption intensity of these bands[46] as well as the integrated absorption observed for the C_2HD fundamentals[47] have also been reported.

In the case of $CH_3C{\equiv}CD$ the integrated intensity of the C—D stretching band, at 2618 cm^{-1}, was observed to be equal to 3000 darks, while the absorption intensity of its 498 cm^{-1} CD deformation fundamental was determined as 3600 darks. Its C≡C stretching band shows an absorption intensity of 28 darks only, as compared with the value of 390 darks found for the parallel $CH_3C{\equiv}CH$ band.[9]

The "true" integrated intensity of the C—D stretching band of *n*-$C_6H_{13}C{\equiv}CD$ was estimated as 4130 l $mole^{-1}$ cm^{-2} and that of its C≡C stretching peak as 73·5 units.[48]

D_6-benzene which, as C_6H_6, belongs to the $\boldsymbol{D_{6h}}$ point group exhibits only four infrared fundamentals: at 496 (species A_{2u}), 814, 1335 and 2287 cm^{-1} (all three: E_{1u}). Their integrated intensities for gaseous C_6D_6 were observed to have the values of 5000, 800, 290 and 3500 cm/mmoles, in good agreement with the respective calculated intensities.[49] The F sums for these vibration species are thus equal to 20,200 (A_{2u}) and 2040 (E_{1u}) 10^{-6} darks cm^2, as compared with the corresponding C_6H_6 values of 19,400 and 2050 units, respectively.[49]

Investigating the absorption intensity of the C—H and C—D stretching bands of carbon tetrachloride solutions of C_6H_6, C_6D_6, C_6H_5D, *p*-$C_6H_4D_2$ and 1,3,5-D_3-benzene showed that each of their C—H or C—D bonds gives a constant contribution of 0·099 ± 0·004 cm^3/mole to the A/v^2 quotients of these bands.[50] A similar situation was found to exist in the case of 4-D-phenyl-X and 2,4,6-D_3-phenyl-X compounds (X = F, Cl, Br, I, CN, NH_2, NO_2, CH_3, CH_2Cl, $CHCl_2$, CH_2OH, $COOCH_3$, C_6H_5) but here the 0·099 cm^3/mole contribution is given only by a C—D bond in a *para* position to a CH_3 group while this constant contribution is lower for C—D bonds in a *para* position to other X substituents and higher for a C—D bond in an *ortho* or *meta* position to this same CH_3 group. The exact amount of this contribution tends to decrease with a rise in the σ parameter of the substituent (i.e. with a rise in its electron attraction) and for a C—D bond *ortho* to a CN group is equal to 0·014 units.[51]

This dependence of the absorption intensity of a C—D bond upon the σ constant of the substituent bound to it was also found for $CH_3CDXC_6H_{13}$ (X = OH, SH, $N(CH_3)_2$, $N(CH_3)_3I$, SOC_6H_5, $SO_2C_6H_5$, $PO(C_6H_5)_2$, $\frac{1}{2}S$, $\frac{1}{2}(C_6H_4(COO)_2)$; their C—D bands appear at 2118, 2160, 2140, 2160 (in $CHCl_3$) 2160, 2160 (in CCl_4), 2117 (in CCl_4), 2158 and 2190 (in CCl_4) cm^{-1}, respectively) and $CH_3CD(C_6H_5)X$ (X = OH, CN, COOH, OCH_3, C_2H_5,

CH_2COOH; their bands are centred at 2122, 2145, 2140, 2117, 2123, 2150 cm^{-1}) compounds. Thus the root of the integrated intensity of the C—D stretching band of these compounds (as well as that of $CH_3CDBrCH_2COOH$) was observed to decrease linearly with the rise in the σ value of their X substituent.[52]

The absorption intensity of the 2285 cm^{-1} C—D stretching band of a nitrobenzene solution of *m*-deuteronitrobenzene and the respective 2270 cm^{-1} band of an aniline solution of *m*-deuteroaniline were both reported.[53]

The molar extinction coefficient of the 2669 cm^{-1} O—D stretching band of a C_6H_5OD solution in carbon tetrachloride was determined as 160 l $mole^{-1}$ cm^{-1} as compared with the value of 230 l $mole^{-1}$ cm^{-1} observed for the O—H stretching band of phenol.[27]

The band intensity of the fundamentals of various deuterated furans has been evaluated recently.[54]

The integrated absorption of the parallel ν_1 (1921 cm^{-1}) and ν_3 (2629 cm^{-1}) (stretching) vibrations of gaseous DCN was observed to have the values of 0·64 and 241·0 cm^{-1}/cm atmos. as compared with the respective HCN values of 12·2 and 136·0 units for its corresponding 2089 and 3312 cm^{-1} bands.[6] The F sum rule is thus obeyed in this case to within 5%. The absorption intensity of liquid DCN[55] and its solutions in $CHCl_3$, $CHBr_3$, CH_2Cl_2, etc.[56] has also been reported.

The integrated absorption of several bending vibrations of gaseous CD_3CN has been measured in connection with the evaluation of the methyl cyanide bond moments.[57]

The absorption intensity of N-deuterated pyrrole[27] and $(C_2H_5)_3SnC\equiv CD$[58] was also studied.

9.3 THE ABSORPTION INTENSITY OF OTHER LABELLED COMPOUNDS

The integrated absorption intensity of the nearly-symmetrical ν_1 (stretching) and ν_4 (deformation) fundamentals of $N^{12}C—^{13}CN$, appearing at 2298 and 495 cm^{-1}, respectively, was used in the calculation of the bond moments of cyanogen.[59]

The absorption intensity of the C=O stretching band of a dilute chloroform solution of *ca.* 50 atom % (monomeric) $C_6H_5{}^{13}COOH$ was found to be within 2% of the integrated intensity calculated for it on basis of eqn. 9.4, assuming that the C=O group can be taken as an isolated diatomic oscillator[60] (see Section 7.1).

The integrated intensity of all the C=O stretching bands of a dimethylsulphoxide solution of 99 atom % ^{15}N-barbituric acid was found to be equal to 115,000 l $mole^{-1}$ cm^{-2}, as compared with the respective 113,500 ± 8000 units value for normal barbituric acid.[61]

The ν_1 in-phase stretching fundamental of CO_2 is in Fermi resonance with its $2\nu_2$ overtone frequency, giving rise, in the case of $C^{16}O^{18}O$, to two bands located at 1259 and 1367 cm^{-1}. Their integrated absorption intensity was found to be equal to $0{\cdot}752 \times 10^{10}$ and $0{\cdot}919 \times 10^{10}$ cycles per second per cm at STP, correspondingly.[62]

The absorption intensity of various X—^{18}O(X = H,C,P,S,As) stretching bands of groups which are either doubly bonded or else contain a very light atom (and hence can according to Section 3.2 be represented quite well by a diatomic XO oscillator for which, theoretically, eqn. 9.4 should hold), has been compared to the absorption intensity of the respective X—^{16}O bands. The results obtained thus are given in Table 9.3.

It can be seen from Table 9.3 that as a rule ^{18}O-labelled groups do not obey eqn. 9.4, since the $\nu^{(i)2}/\nu^2$ ratio for these groups is never below $840^2/880^2 = 0{\cdot}91$ (as in the case of the As=O group[74]) while usually it is equal to about $1637^2/1667^2 = 0{\cdot}96$ (as for the C=O groups of benzophenone[66]) and for the OH groups[64] even as high as $3610^2/3620^2 = 0{\cdot}994$. These ratios mean that the ^{18}O-induced intensity differences expected here according to eqn. 9.4 vary between $-0{\cdot}6\%$ and -9%, as compared with the -9 to -41% and $+22$ to $+88\%$ differences observed experimentally.

The F sum (see eqn. 9.5) for all the A_1 species fundamentals of 97 atom % $P^{18}OCl_3$ solutions in cyclohexane has also been shown to be lower by about 5·3% from the respective $P^{16}OCl_3$ sum, in disagreement with eqn. 9.5. The difference between this F sum for $P^{18}OCl_3$ in CCl_4 and the respective $P^{16}OCl_3$ sum seems to be even appreciably larger, as can be judged from the fact that the isotopic integrated intensities for the main A_1 fundamental (their PO stretching band) are 4345 ± 60 and 4885 ± 80 l $mole^{-1}$ cm^{-2} for their C_6H_{12} solutions but 4060 ± 60 and 5360 ± 80 l $mole^{-1}$ cm^{-2} for their CCl_4 solutions, respectively.[75]

In the case of the ν_3 band of gaseous 97 atom % $C^{18}O_2$, which is the only member of its Σ_u^+ vibration species, the A/ω^2 quotient was found to have the value of $(1{\cdot}56 \pm 0{\cdot}03) \times 10^7$ cm sec^{-1} atm^{-1}, as compared with the corresponding $C^{16}O_2$ value of $1{\cdot}52 \times 10^7$ units, respectively.[16, 76]

$H_2{}^{18}O$ solutions in nitromethane also showed the same integrated absorption intensity for their ν_3 fundamental as $H_2{}^{16}O$ (6970 ± 160 and 6900 ± 180 l $mole^{-1}$ cm^{-2}, respectively), the small increase in the $H_2{}^{18}O$ band intensity being due to the contribution of the compensating rotation connected with this vibration.[16, 63]

The fact that the absorption intensity of the ν_3 band of $H_2{}^{18}O$ and $C^{18}O_2$ is *not* materially different from the corresponding band intensity of $H_2{}^{16}O$ and $C^{16}O_2$ is not necessarily a contradiction of the results reported in Table 9.3 since, during their ν_3 stretching vibration, one X—O group is stretched while another X—O group is compressed, so that some compensation of the difference between the stretching ^{18}O atom and the ^{16}O atom can be anticipated in this case. The observation that the integrated intensity of the sym-

TABLE 9.3 The Relative Integrated Intensity (l mole^{-1} cm^{-2}) of Various X—^{18}O Stretching Bands

Group	% ^{18}O	Compound	Solvent	X^{18}O Intensity	X^{16}O Intensity	Difference
OH	29	^{18}O—$CH_3CHOHCH_2OH$[63]	$CHCl_3$	1370 ± 50	1920 ± 100	−29%
OH	74	$C_6H_5{}^{18}OH$[64]	CCl_4	4050 ± 200	5300 ± 100	−24%
OH	78	$(C_6H_5)_3C^{18}OH$[65]	CCl_4	1510 ± 200	2560 ± 120	−41%
C=O	89	$C_6H_5C^{18}OC_6H_5$[66]	CCl_4	6400 ± 300	8700 ± 400	−26%
C=O	90	$C_6H_5C^{18}O_2H$[67]	$CHCl_3$	17,100 ± 100[a]	18,800 ± 500[a]	− 9%
C=O	84	$C_6H_5C^{18}OOCH_3$[67]	CCl_4	6250 ± 100[a]	7000 ± 200[a]	−11%
C=O	50	$HC^{18}ON(CH_3)_2$[53]	CCl_4	13,000 ± 1400	25,000 ± 400	−48%
C=O	79	$C_6H_5C^{18}ONH_2$[68]	CCl_4	12,800[a]	10,400[a]	+23%
C=O	89	$C_6H_5C^{18}ONHCH_3$[69]	CCl_4	10,900 ± 300[a]	8900 ± 300[a]	+22%
C=O	69	$C_6H_5C^{18}ON(CH_3)_2$[70]	CCl_4	11,300 ± 700[a]	9100 ± 400[a]	+24%
C=O	90	$(C_6H_{11}NH)_2C^{18}O$[71]	CCl_4 + $(CH_3)_2SO$	17,200 ± 1000[a]	22,000 ± 500[a]	−22%
C=O	79	$(p\text{-}CH_3C_6H_4NH)_2C^{18}O$[71]	$(CH_3)_2SO$	19,000 ± 3500[a]	10,100 ± 500[a]	+88%
P=O	86	$(C_6H_5)_3P^{18}O$[72]	CCl_4	5900[a]	7000[a]	−16%
S=O	81	$(C_6H_5)_2S^{18}O$[73]	CCl_4	4250[a]	2700[a]	[b]
As=O	90	$(C_6H_5)_3As^{18}O$[74]	$CHCl_3$	8400 ± 300	10,300 ± 300	−18%

[a] Obtained by multiplying the maximum molecular extinction coefficient by the half-width.
[b] The exact difference cannot be evaluated owing to resonance effects.

metrical 1598 cm^{-1}, ν_2 (bending) fundamental of $H_2{}^{18}O$ solutions in CH_2Cl_2 displays an A/ν^2 quotient of 2·12 cm^3/mole corresponding to the respective 2·18 cm^3/mole value of the $H_2{}^{16}O$ 1605 cm^{-1} band, seems to indicate in this direction although when the F sum is taken over all their ν_1 (3600 cm^{-1}) ν_2 and ν_3 (3679 cm^{-1}) bands this effect is smeared out to a small difference between 2·96 cm^3/mole for $H_2{}^{18}O$ and 2·99 cm^3/mole for $H_2{}^{16}O$ only.[16,63]

The reason for the unexpected integrated absorption intensities of $X^{18}O$ stretching bands is not yet clear although many similar phenomena have been observed which all appear to indicate that the electronic structure of ^{18}O is somewhat different from that of ^{16}O.[77–80]

9.4 The Width of Labelled Compound Absorption Bands

Buckingham has shown that the decrease $\Delta\omega$ brought about in the frequency of a diatomic oscillator upon going over from the gas phase to the solution phase should, theoretically, obey the equation:

$$\Delta\omega^i/\omega^i = \Delta\omega/\omega, \tag{9.8}$$

where ω^i is the frequency of an isotopic molecule.[5] Since the width of a solution band is a function of the interaction between the solvent and the solute, an interaction which is also responsible for $\Delta\omega$, eqn. 9.8 should, theoretically, also hold for its *half-width*, i.e. for the distance (in cm^{-1}) between two frequencies in it for which $\log (I_0/I)$ is 50% of its maximum value. In practice, however, eqn. 9.8 is often not satisfactory due to the fact that this interaction is sometimes strongly dependent upon the exact isotopic structure of the interacting solute groups.

The half-width of the 2500 cm^{-1} O—D stretching band of dilute solutions of HDO in liquid H_2O was observed to be equal to 150 cm^{-1} as compared with a value of 31 cm^{-1} for the ice phase.[81]

The half-width of the N—D stretching bands of ND_4F, ND_4Cl, ND_4Br and ND_4I (this salt at -15°C; Phase II) was found to be 88, 80, 78 and 113 cm^{-1}, respectively, corresponding to the 110 cm^{-1} half-width of the N—H stretching fundamental of NH_4F.[82]

The C—D stretching band of $CDCl_3$ shows a half-width of 9·3 and 11·8 cm^{-1} for its neat and CCl_4 solution phases, respectively, as compared with the corresponding $CHCl_3$ values of 15·6 and 18·8 cm^{-1} and the parallel $CDBr_3$ half-widths of 8·3 and 10·8 cm^{-1}, respectively. The half-width of the C—D valence band of a CCl_4 solution of CDI_3 was observed to be 11·8 cm^{-1} while the corresponding C—H fundamental displayed a half-width of 20·3 cm^{-1}.

The half-width of the C—D stretching band of $CDCl_3$ in various other solvents was also reported[28,83] while in the case of its 905 cm^{-1} C—D deformation band a CS_2 solution was found[27] to exhibit practically the same half-width as for the respective C—H deformation fundamental at 1211 cm^{-1} (5·2 versus 5·3 cm^{-1}). A CS_2 solution of CDI_3, however, shows a much smaller half-width for its C—D deformation band at 786 cm^{-1} (3·1 cm^{-1}) than that observed for the analogous 1068 cm^{-1} C—H deformation band of CHI_3[27] (7·6 cm^{-1}).

The half-width of the CS_2 solution C—D stretching fundamental of C_2DCl_5 is *ca.* 12 cm^{-1}, as compared with the respective value of 22 cm^{-1} observed for C_2HCl_5.[83]

The widths of the six fundamentals of CD_3Cl solutions in CS_2 or CCl_4 have the values of 8·6 (ν_1), 11·5 (ν_2), 16 (ν_3), 30 (ν_4), 87 (ν_5) and 33 (ν_6) cm^{-1}, corresponding to the CH_3Cl values of 10·5, 10, 13, 45, 48 and 33 cm^{-1}, respectively.

TABLE 9.4 The Half-Width (cm^{-1}) of X—^{18}O Stretching Bands

Group	Compound	Solvent	Frequency	Half-Width	Reference
^{18}O—H	$(C_6H_5)_3C^{18}OH$	CCl_4	3609	27	65
^{16}O—H	$(C_6H_5)_3COH$	CCl_4	3621	23	65
C=^{18}O	$C_6H_5C^{18}OC_6H_5$	CCl_4	1637	10·5	66
C=^{16}O	$C_6H_5COC_6H_5$	CCl_4	1667	10·5	66
C=^{18}O	$C_6H_5C^{18}O_2H$	$CHCl_3$	1688	49	67
C=^{16}O	$C_6H_5CO_2H$	$CHCl_3$	1715	49	67
C=^{18}O	$C_6H_5C^{18}OOCH_3$	CCl_4	1696	20	67
C=^{16}O	$C_6H_5COOCH_3$	CCl_4	1727	18	67
C=^{18}O	$C_6H_5C^{18}ONH_2$	CCl_4	1666	22	68
C=^{16}O	$C_6H_5CONH_2$	CCl_4	1690	20	68
C=^{18}O	$C_6H_5C^{18}ONH_2$	$CHCl_3$	1655	21	68
C=^{16}O	$C_6H_5CONH_2$	$CHCl_3$	1675	25	68
C=^{18}O	$C_6H_5C^{18}ONHCH_3$	CCl_4	1651	35	69
C=^{16}O	$C_6H_5CONHCH_3$	CCl_4	1675	33	69
C=^{18}O	$C_6H_5C^{18}ONHCH_3$	$CHCl_3$	1633	35	69
C—^{16}O	$C_6H_5CONHCH_3$	$CHCl_3$	1657	33	69
C=^{18}O	$C_6H_5C^{18}ON(CH_3)_2$	CCl_4	1616	21·5	70
C=^{16}O	$C_6H_5CON(CH_3)_2$	CCl_4	1640	22·5	70
C=^{18}O	$C_6H_5C^{18}ON(CH_3)_2$	$CHCl_3$	1602	24·5	70
C=^{16}O	$C_6H_5CON(CH_3)_2$	$CHCl_3$	1622	28	70
P=^{18}O	$P^{18}OCl_3$	CCl_4	1261	9·9	75
P=^{16}O	$POCl_3$	CCl_4	1306	10·1	75
P=^{18}O	$P^{18}OCl_3$	C_6H_{12}	1265	8·4	75
P=^{16}O	$POCl_3$	C_6H_{12}	1310	6·9	75
P=^{18}O	$(C_6H_5)_3P^{18}O$	CCl_4	1172	19	72
P=^{16}O	$(C_6H_5)_3PO$	CCl_4	1202	17	72
S=^{18}O	$(C_6H_5)_2S^{18}O$	CCl_4	1021	14	73
S=^{16}O	$(C_6H_5)_2SO$	CCl_4	1051	10·5	73
As=^{18}O	$(C_6H_5)_3As^{18}O$	$CHCl_3$	840	22·5	64
As=^{16}O	$(C_6H_5)_3AsO$	$CHCl_3$	880	24·5	64

The widths of the observed bands of CD_3Br and CD_3I are generally fairly similar to those of the respective CD_3Cl bands.[84]

The apparent half-width of the C≡N stretching band of DCN solutions in $CHCl_3$ and $CHBr_3$ has the value of 7·0 cm^{-1}, as compared with those values of 8·0 and 9·0 cm^{-1} in the case of HCN. The width of its CD stretching band was found[56] to be 12·5, 13, 25 and 31 cm^{-1} for its solutions in $CHCl_3$, $CHBr_3$, *m*-xylene and $C_2H_5NO_2$, respectively, corresponding to the parallel HCN half-widths of 22, 27, 50 and 69 cm^{-1}.

While CCl_4 solutions of N-deuterated imidazole exhibit the same (15 cm^{-1}) half-width for their N—D stretching band as the respective NH band, solvents forming strong hydrogen bonds with the imidazole NH(D) groups give rise to an N—D band which is, relatively, much narrower than the corresponding N—H band. Dimethyl sulphoxide solutions, e.g. of the ND isotopomer thus display a half-width of 77 cm^{-1} only, as compared with the 170 cm^{-1} width observed for the N—H band of the normal imidazole solutions in this solvent.[85]

The half-width found for various X—^{18}O stretching bands is reported in Table 9.4.

It can be seen from Table 9.4 that most of the X—^{18}O stretching bands do not obey eqn. 9.8.

References

1. Wilson, E. B. Jr., Decius, J. C. and Cross, P. C. (1955). "Molecular Vibrations", p. 191. McGraw-Hill Book Company, N.Y.
2. Brown, T. H. (1965). *J. Chem. Phys.* **43**, 2780.
3. Born, M. and Oppenheimer, J. R. (1927). *Ann. Physik* **84**, 457.
4. Cashion, K. (1963). *J. Mol. Spectroscopy* **10**, 182.
5. Buckingham, A. D. (1958). *Proc. Roy. Soc.* **248A**, 169; (1960). **255A**, 32.
6. Crawford, B. Jr. (1952). *J. Chem. Phys.* **20**, 977.
7. Decius, J. C. (1952). *J. Chem. Phys.* **20**, 1039.
8. Sverdlov, L. M. (1960). *Optics and Spectroscopy* **8**, 129; (1964). **17**, 515.
9. Jouve, P. (1966). *Compt. rend.* **262C**, 441.
10. Benedict, W. S., Herman, R., Moore, G. E. and Silverman, S. (1957). *J. Chem. Phys.* **26**, 1671.
11. Bertie, J. E. and Millen, D. J. *J. Chem. Soc.* **1965**, 514.
12. Thompson, W. K. (1965). *Trans. Faraday Soc.* **61**, 2635.
13. Swenson, C. A. (1965). *Spectrochim. Acta* **21**, 987.
14. Swenson, C. A. (1968). *Spectrochim. Acta* **24**, 721.
15. Scrocco, M. and de Luca, B. (1967). *Ricerca sci.* **37**, 24.
16. Laulicht, I. (1966). Ph.D. Thesis, The Weizmann Institute of Science, Rehovoth.
17. Kęcki, Z., Dryjański, P. and Kozłowska, E. (1968). *Roczniki Chem.* **42**, 1749.
18. Ball, D. F. and McKean, D. C. (1962). *Spectrochim. Acta* **18**, 1019; see also Levin, I. W. and King, W. T. (1962). *J. Chem. Phys.* **37**, 1375.
19. Kovalev, I. F. (1964). *Izvest. vȳssh. ucheb. Zaved.* (*Tomsk*) *Ser. Fiz.* **No. 5**, 102.
20. Levin, I. W. (1965). *J. Chem. Phys.* **42**, 1244.
21. Levin, I. W. (1967). *J. Chem. Phys.* **46**, 1176.

22. Sverdlov, L. M. (1962). *Optics and Spectroscopy* **12**, 66.
23. Hiller, R. E. Jr. and Straley, J. W. (1960). *J. Mol. Spectroscopy* **5**, 24.
24. Sverdlov, L. M. (1961). *Optics and Spectroscopy* **10**, 33.
25. Huggins, C. M. and Pimentel, G. C. (1955). *J. Chem. Phys.* **23**, 896.
26. Lord, R. C., Nolin, B. and Stidham, H. D. (1955). *J. Amer. Chem. Soc.* **77**, 1365.
27. Forel, M. T., Leicknam, J. P. and Josien, M. L. (1960). *J. Chim. phys.* **57**, 1103.
28. Heess, R. and Kriegsmann, H. (1968). *Spectrochim. Acta* **24A**, 2121.
29. Morcillo, J., Biarge, J. F., Heredia, J. M. V. and Medina, A. (1969). *J. Mol. Struct.* **3**, 77.
30. Boobyer, G. J. and Orville-Thomas, W. J. (1966). *Spectrochim. Acta* **22**, 147.
31. Kagarise, R. E. (1963). *Spectrochim. Acta* **19**, 629.
32. Saeki, S. and Tanabe, K. (1969). *Spectrochim. Acta* **25A**, 1325.
33. Dickson, A. D., Mills, I. M. and Crawford, B. Jr. (1957). *J. Chem. Phys.* **27**, 445.
34. Higuchi, S., Tanaka, S. and Kamada, H. (1968). *Spectrochim. Acta* **24A**, 1929.
35. Sverdlov, L. M. (1959). *Optics and Spectroscopy* **6**, 477.
36. Nyquist, I. M., Mills, I. M., Person, W. B. and Crawford, B. Jr. (1957). *J. Chem. Phys.* **26**, 552.
37. Sverdlov, L. M. (1959). *Optics and Spectroscopy* **7**, 21.
38. Prokofeva, N. I. and Sverdlov, L. M. (1965). *Optics and Spectroscopy* **19**, 363.
39. Broun, E. V. and Iogansen, A. V. (1968). *Optika i Spektroskopiya* **24**, 399.
40. Amick, J. A. (1955). *Diss. Abs.* **15**, 339.
41. Golike, R. C., Mills, I. M., Person, W. B. and Crawford, B. Jr. (1956). *J. Chem. Phys.* **25**, 1266.
42. Sverdlov, L. M. (1958). *Optics and Spectroscopy* **4**, 697.
43. Sverdlov, L. M. (1960). *Optics and Spectroscopy* **8**, 594.
44. Hoffmann, V. and Mecke, R. (1968). *Z. Naturforsch.* **A23**, 1043.
45. Borisov, M. G. and Sverdlov, L. M. (1963). *Optics and Spectroscopy* **15**, 14.
46. Sverdlov, L. M. (1961). *Optics and Spectroscopy* **11**, 18.
47. Eggers, D. F. Jr., Hisatsune, I. C. and van Alten, L. (1955). *J. Phys. Chem.* **59**, 1124.
48. Romanet, R. and Wojtkowiak, B. (1960). *Compt. rend.* **251**, 364.
49. Dows, D. A. and Pratt, A. L. (1962). *Spectrochim. Acta* **18**, 433.
50. Langenbucher, F. and Mecke, R. (1965). *Spectrochim. Acta* **21**, 1287.
51. Schmid, E. D. and Langenbucher, F. (1966). *Spectrochim. Acta* **22**, 1621.
52. Pine, S. H. and Steele, D. R. (1967). *Spectrochim. Acta* **23A**, 1509.
53. Dauben, W. G., Pimentel, G. C. and Vaughan, C. W. Jr. (1955). *J. Amer. Chem. Soc.* **77**, 2886.
54. Evseeva, L. A., Finkel, A. G., Sverdlov, L. M. and Pronina, L. V. (1969). *Zhur. priklad. Spektroskopii* **10**, 614.
55. Girin, O. P., Akopyan, S. Ch. and Bachshiev, N. G. (1968). *Optika i Spektroskopiya* **25**, 207.
56. Caldow, G. L., Cunliffe-Jones, D. and Thompson, H. W. (1960). *Proc. Roy. Soc.* **254A**, 17.
57. Hendricks, R. W. (1956). *Diss. Abs.* **16**, 2323.
58. Gastilovic, E. A., Shigorin, D. N., Zukova, K. V. and Akcurina, I. S. (1969). *Zh. fiz. Khim.* **43**, 5.
59. Schultz, J. W. and Eggers, D. F. Jr. (1958). *J. Mol. Spectroscopy* **2**, 113.
60. Pinchas, S. (1965). *J. Phys. Chem.* **69**, 2256.
61. Pinchas, S. (1966). *Spectrochim. Acta* **22**, 1889.
62. Eggers, D. F. Jr. and Arends, C. B. (1957). *J. Chem. Phys.* **27**, 1405.

63. Pinchas, S., Laulicht, I. and Cohen, J. *Abstracts of 9th European Congress on Molecular Spectroscopy*, Madrid, 1967. p. 226.
64. Pinchas, S., Sadeh, D. and Samuel, D. (1965). *J. Phys. Chem.* **69**, 2259.
65. Pinchas, S., Samuel, D. and Wasserman, I. (1967). *Spectrochim. Acta* **23A**, 1307.
66. Laulicht, I. and Pinchas, S. (1963). *Israel J. Chem.* **1**, 404.
67. Pinchas, S., Samuel, D. and Weiss-Broday, M. *J. Chem. Soc.* **1961**, 2382.
68. Pinchas, S., Samuel, D. and Weiss-Broday, M. *J. Chem. Soc.* **1961**, 1688.
69. Pinchas, S., Samuel, D. and Weiss-Broday, M. *J. Chem. Soc.* **1961**, 2666.
70. Pinchas, S., Samuel, D. and Weiss-Broday, M. *J. Chem. Soc.* **1961**, 3063.
71. Lapidot, A., Pinchas, S. and Samuel, D. *J. Chem. Soc.* **1963**, 1128.
72. Halmann, M. and Pinchas, S. *J. Chem. Soc.* **1958**, 3264.
73. Pinchas, S., Samuel, D. and Weiss-Broday, M. *J. Chem. Soc.* **1962**, 3968.
74. Bernstein, J., Halmann, M., Pinchas, S. and Samuel, D. *J. Chem. Soc.* **1964**, 821.
75. Laulicht, I., Pinchas, S., Sadeh, D. and Samuel, D. (1964). *J. Chem. Phys.* **41**, 789.
76. Pinchas, S. and Laulicht, I. "XIII Colloquium Spectroscopicum Internationale", p. 211. Adam Hilger Ltd, London, 1968.
77. Pinchas, S. and Sadeh, D. (1968). *J. Inorg. Nucl. Chem.* **30**, 1785.
78. Pinchas, S. (1969). *J. Chem. Phys.* **51**, 2284.
79. Pikman, H. and Pinchas, A. (1970). *J. Inorg. Nucl. Chem.* **32**, 2441; Pinchas, S. and Meshulam, E. (1970). *Chem. Commun.* 1147.
80. Pinchas, S. and Shamir, J. (1969). *Israel J. Chem.* **7**, 805.
81. Yellin, W. and Courchene, W. L. (1968). *Nature* **219**, 852.
82. Plumb, R. C. and Hornig, D. F. (1955). *J. Chem. Phys.* **23**, 947.
83. Bernstein, R. B. and Tamres, M. (1955). *J. Chem. Phys.* **23**, 2201.
84. Bulanin, M. O. and Tonkov, M. V. (1964). *Optics and Spectroscopy* **16**, 234.
85. Perchard, C. and Josien, M. L. (1965). *J. Chim. phys.* **62**, 423.

CHAPTER 10

Labelled Molecule Spectra in the Condensed Phases

The infrared absorption spectrum of a molecule in the liquid, solid or adsorbed (on a solid surface) phase is usually quite different from its absorption spectrum in the gas phase. This difference is mainly due to the following reasons:

(a) The potential field acting upon the condensed molecule as a whole and upon its various constituent atoms is (especially in the adsorbed phase) quite different from that acting upon it in the gaseous state. This fact usually brings about a shift of its intramolecular vibrations upon condensation and interferes strongly with its rotations and translations.

(b) Many infrared-inactive molecular vibrations become infrared-active in the condensed phases, especially in the crystalline one, as a result of a decrease in the molecular site symmetry brought about by the crystalline lattice or fluctuations in the constitution of the liquid around the vibrating molecule, etc. Such a decrease in the site symmetry of a crystalline molecule also often removes the degeneracy of molecular vibrations which are degenerate in the gaseous phase, thus causing the splitting of the corresponding crystalline absorption bands.

(c) The frequency of many normal vibrations of crystalline or adsorbed molecules depends appreciably upon the concomitant vibration of their (coupled) adjacent neighbours, whether it is in-phase or out-of-phase in regards to them. This oscillator coupling thus splits the respective gas phase absorption bands in many crystalline or adsorbed phase spectra.

(d) The near proximity between the molecules in the condensed phases very often results in strong interactions between them which are also reflected in their absorption spectrum. The most important example of such interactions is the case of hydrogen-bonds formation (e.g. in solid amides). These interactions can even result in the formation of a completely different chemical species in the case of the solid phase as compared to the situation in the gaseous phase (e.g. dimerization of NO_2, etc.).

The intermolecular potential energy of isotopic molecules in the condensed phase is slightly different from that of the corresponding normal molecules since the intermolecular force is in each case averaged over slightly different internal vibrations. This isotopic change in the intermolecular potential energy was found to be mainly due to the coupling of the internal molecular vibrations with the (hindered) rotations and translations of the condensed phase molecules,[1] a coupling which was indeed reported to have an isotopic effect upon their internal frequencies.[2]

A crystal-induced splitting of a degenerate XH_n band of a normal compound can be expected to be reproduced by the respective XD_n fundamental of a dilute solid solution of its deuterated isotopomer in this compound if the reason for this splitting is a low crystalline site symmetry (as described above in (b)) for these groups. If, however, such a splitting is *not* found for the labelled compound the splitting observed in the case of the normal compound must be due to a coupling of *similar* adjacent oscillators (as described in (c)) which is not possible for the labelled compound since each labelled molecule in the dilute solid solution is surrounded only by normal isotopomers.[3]

The fact that crystalline 15 atom % deuterated *n*-decanol shows a split OH stretching and a single OD stretching band while the reverse is true for a 75 atom % deuterated sample of this alcohol is explained similarly.[4]

An intermolecular vibration coupling was also observed for the ν_2 (out-of-plane B-motion) and ν_3 (antisymmetric B—F stretching) fundamentals of crystalline BF_3 of various isotopic constitutions. Both the frequency and the width of the absorption peaks due to these vibrations were found to be affected by this coupling as can be judged by the different values obtained for them in samples which contained varying amounts of isotopic $^nBF_3$ neighbours around the oscillating $^mBF_3$ molecule.[5]

Such interactions between the nearest neighbouring CO_3^{2-} ions of a Nujol mull of 60 atom % $Ba^{13}CO_3$ give rise to a fine structure for their out-of-plane bending fundamental which is split in this case into branches at 832·7, 835·0, 839–840 (*sh*; all three due to $^{13}CO_3^{2-}$ ions), 860·8, 862·6, 866·7 and 874·0 cm^{-1} ($^{12}CO_3^{2-}$), belonging to isotopically isolated ions, ion pairs, ion triplets etc.[6]

A mull of 50 atom % $K^{15}NO_3$ shows its respective absorption peaks[6] at 807·3 (strongest), 810·4 ($^{15}NO_3^-$), 828, 830·6 and 836·5 cm^{-1}.

Isotopic mixtures of 36% ^{12}CO and 64% ^{13}CO, chemisorbed on platinum, give rise to a normal chemisorbed ^{12}CO band, at *ca.* 2070 cm^{-1}, and another absorption band near 2000 cm^{-1}, the relative intensities of which depending strongly upon the surface coverage. The experimental results regarding these intensities could be explained on basis of an assumed dipole–dipole interaction between the chemisorbed molecules.[7]

A study of the 500–960 cm^{-1} absorption spectra of crystals containing: a) 1% C_6H_6 in C_6D_6, b) 0·1% C_6D_6 in C_6H_6, or c) solid solutions of C_6H_5D, *m*-, *p*-$C_6H_4D_2$, *sym.* $C_6H_3D_3$, *m*-$C_6H_2D_4$ or C_6HD_5 in C_6H_6, has shown that the crystal-induced splittings of absorption bands of these compounds as a result of a low site symmetry or their positioning in different (as a result of an incomplete labelling) orientations in crystal sites, does not usually exceed[8] 10 cm^{-1}. Splittings due to intermolecular interactions between adjacent molecules of a different isotopic constitution have also been observed for these compounds.[8]

The far infrared spectra of various crystalline labelled compounds, being mainly due to their lattice modes of vibration, have also been investigated extensively,[9] in order to shed light on their crystalline structure.

The main research carried out on the infrared absorption of individual, condensed, labelled molecules is reviewed in Table 10.1.

TABLE 10.1 Reported Infrared Absorption Spectra of Individual Labelled Molecules in the Condensed Phases

Compound	Labelling isotope	Phase	Remarks	Reference
Hydrogen	D	Liquid	Induced absorption spectrum	10
Hydrogen	D	Solid	Induced absorption spectrum	11, 12
Hydrogen	D	Adsorbate	Chemisorbed	13
Hydrogen fluoride	D	Solid	40–3800 cm^{-1} region	14
Hydrogen chloride	D	Solid	Pure and mixed with HCl	9c, 15, 16, 17
Hydrogen chloride $(CH_3)_2O$	D	Solid	Hydrogen-bond study	18
Hydrogen bromide	D	Solid	Pure and mixed with HBr	16, 19
CO	^{13}C, ^{17}O, ^{18}O	Solid	Fundamental and overtones	20, 21, 22
CO	^{13}C, ^{18}O	Solid	Calculations	23
$C^{16}O^{18}O$	^{18}O	Solid	Intensity measurements	24
$C^{18}O_2$	^{18}O	Adsorbate	On alumina	25
N_2O	^{15}N, ^{18}O	Solid	400–5000 cm^{-1} region	26
HDO and D_2O	D	Solid	Intensity measurements	27 (28, 29)
Water	D	Ice II, V, IX	30–360 cm^{-1} region	30
Water	D	Ice VI	50–4000 cm^{-1} region	31
Water	D	Adsorbate	On NH_4-montmorillonite	32
Cl_2O	^{18}O	Solid	200–4000 cm^{-1}	33
Hydrogen sulphide	D	Solid	From 80 cm^{-1} and upwards	34, 35
BF_3	^{10}B	Liquid	400–3000 cm^{-1}	36
Ammonia	D	Solid	From 80 cm^{-1} and upwards	35, 37, 38
Ammonia	D	Adsorbed	On silica or alumina	39, 40
Hydrazoic acid	D	Solid (I, II)	Strongly hydrogen-bonded	41
Hydrogen peroxide	D	Solid; glass	From 50 cm^{-1} and upwards	42, 43
Nitric acid	D	Solid	400–4000 cm^{-1} region	44
Silane	D	Solid	Fully deuterated species	45
Germane	D	Solid	Fundamentals only	46

TABLE 10.1—*continued*

Compound	Labelling isotope	Phase	Remarks	Reference
Perchloric acid	D	Solid (two states)	300–5000 cm^{-1} region	47
Hydrazine	D	Solid	33–4000 cm^{-1}	48, 49
Biphosphine	D	Solid	At 93° K	50
Methane	D	Liquid; solid	ν_3 fundamental	51
CF_4	^{13}C	Liquid; solid	Natural abundance	52
Chloroform	D	Solid	700–3100 cm^{-1} region	53, 54
Bromoform	D	Solid	400–3100 cm^{-1}	55
Methylene chloride	D	Liquid	Band intensity measured	56
Methylene bromide	D	Solid	200–3075 cm^{-1}	57
Methyl chloride	D, ^{35}Cl	Solid	At 77° K	58
Methyl bromide	D	Solid	Relative intensities	58
Methyl iodide	D	Solid	High resolution	58
CD_2Br—CD_2Br	D	Solid	Study of rotational isomerism	59
$CD_3(CH_2)_6CD_3$	D	Solid	600–1500 cm^{-1}	60
Cyclohexane	D	Solid		61
Ethylene	D	Solid	Also: intensities; 60–120 cm^{-1} region	62, 63, 64
Acetylene	D	Solid (2 states)	Also: far infrared region	65, 66, 67
Acetylene	D	Adsorbed	On zeolites	68
Benzene	D	Solid	Also: mixed crystals	69, 70
Benzene	D	Solid solutions	Intensities	71
Naphthalene	D	Solid	Spectrum interpretation	72
Anthracene	D	Mixed crystals	D_{10}-anthracene	73
C_2D_5OD	D	Rigid glass	O—D stretching band	74
C_3D_7OD	D	Rigid glass	Concentration of ~0·1 molar	74
C_3H_7OD	D	Rigid glass	At 88° K	74
C_3D_7OH	D	Rigid glass	O—H stretching band	74
C_6D_5OD	D	Rigid glass	O—D band at 2384 cm^{-1}	74
Tetrahydrofuran	D	Solid	350–3200 cm^{-1} region of D_8-compound	75
Ketene	D	Solid	300–3600 cm^{-1} of D- and D_2-isotopomers	76
O=C=C=C=O	^{13}C, ^{18}O	Liquid	20–4000 cm^{-1} region	77
Formic acid	D	Solid	80–3000 cm^{-1}	78
Acetic acid	D	Solid	CD_3COOD, CH_3COOD, CD_3COOH	79
Methylamine	D	Solid	CH_3ND_2 and CH_3NHD	80
Trimethylamine	D	Solid	200–3000 cm^{-1} region	81
HCN	^{13}C	Solid		82
CH_3CN	D	Solid	300–3500 cm^{-1}	83

$(CN)_2$	^{15}N	Solid	600–4000 cm^{-1}	84
Dimethyl-silane	D	Solid solution	Assignments	85
COS	^{13}C, ^{18}O, ^{33}S, ^{34}S	Solid solutions	At 80°K	86
Isothiocyanic acid	D	Solid	250–4000 cm^{-1} region	87

References

1. Gordon, R. G. (1966). *J. Chem. Phys.* **44**, 576.
2. Stern, M. J., Van Hook, W. A. and Wolfsberg, M. (1963). *J. Chem. Phys.* **39**, 3179.
3. Hrostowski, H. J. and Pimentel, G. C. (1951). *J. Chem. Phys.* **19**, 661.
4. Jakobsen, R. J., Mikawa, Y. and Brasch, J. W. (1967). *Nature* **215**, 1071.
5. Dows, D. A. (1959). *J. Chem. Phys.* **31**, 1637.
6. Decius, J. C. (1955). *J. Chem. Phys.* **23**, 1290.
7. Hammacker, R. M., Francis, S. A. and Eischens, R. P. (1965). *Spectrochim. Acta* **21**, 1295.
8. Bernstein, E. R. (1969). *J. Chem. Phys.* **50**, 4842.
9. (a) Kirby, R. D., Nolt, I. G., Alexander, R. W. Jr. and Sievers, A. J. (1968). *Phys. Rev.* **168**, 1057.
   (b) Klein, M. V. and Macdonald, H. F. (1968). *Phys. Rev. Lett.* **20**, 1031.
   (c) Brunel, L. C. and Peyron, M. (1969). *Compt. rend.* **268B**, 1746.
10. Bulanin, M. O., Melnik, M. G. and Tonkov, M. V. (1967). *Optika i Spektroskopiya* **23**, 714.
11. Crane, A. and Gush, H. P. (1966). *Canad. J. Phys.* **44**, 373.
12. Jean-Louis, M., Bahreini, M. and Vu, H. (1969). *Compt. rend.* **268B**, 390; 479.
13. Smith, W. H., Eckstrom, H. C. and Baer, F. (1968). *J. Phys. Chem.* **72**, 369.
14. Kittelberger, J. S. and Hornig, D. F. (1967). *J. Chem. Phys.* **46**, 3099.
15. Hiebert, G. L. and Hornig, D. F. (1957). *J. Chem. Phys.* **27**, 1216.
16. Savoie, R. and Anderson, A. (1966). *J. Chem. Phys.* **44**, 548.
17. Rank, D. H., Glickman, W. A. and Wiggins, T. A. (1965). *J. Chem. Phys.* **43**, 1304.
18. Seel, R. M. and Sheppard, N. (1969). *Spectrochim. Acta* **25A**, 1287.
19. Hornig, D. F. and Hiebert, G. L. (1957). *J. Chem. Phys.* **27**, 752.
20. Ewing, G. E. and Pimentel, G. C. (1961). *J. Chem. Phys.* **35**, 925.
21. Maki, A. G. (1961). *J. Chem. Phys.* **35**, 931.
22. Vu, H., Atwood, M. R. and Vodar, B. (1963). *J. Chem. Phys.* **38**, 2671.
23. Johansen, H. (1966). *Z. Naturforsch.* **A21**, 836.
24. Eggers, Jr, D. F. and Arends, C. B. (1957). *J. Chem. Phys.* **27**, 1405.
25. Parkyns, N. D. *J. Chem. Soc.* (*A*). **1969**, 410.
26. Jouve, P. and Le Roy, A. (1967). *Compt. rend.* **265B**, 1017.
27. Ikawa, Sh. and Maeda, Sh. (1968). *Spectrochim. Acta* **24A**, 655.
28. Leadbetter, A. J. (1965). *Proc. Roy. Soc.* **A287**, 403.
29. Hornig, D. F., White, H. F. and Reding, F. P. (1958). *Spectrochim. Acta* **12**, 338.
30. Bertie, J. E., Labbe, H. J. and Whalley, E. (1968). *J. Chem. Phys.* **49**, 775.
31. Bertie, J. E., Labbe, H. J. and Whalley, E. (1968). *J. Chem. Phys.* **49**, 2141.
32. Tarasevich, Yu, I. (1969). *Ukrain. khim. Zhur.* **35**, 431.
33. Rochkind, M. M. (1966). *Diss. Abs.* **27**, 785B.
34. Reding, F. P. and Hornig, D. F. (1957). *J. Chem. Phys.* **27**, 1024.
35. Anderson, A. and Walmsley, S. H. (1965). *Mol. Phys.* **9**, 1.
36. Steinhardt, R. G., Fetsch, G. E. S. and Jordan, M. W. (1965). *J. Chem. Phys.* **43**, 4528.

37. Reding, F. P. and Hornig, D. F. (1951). *J. Chem. Phys.* **19**, 594.
38. Reding, F. P. and Hornig, D. F. (1955). *J. Chem. Phys.* **23**, 1053.
39. Pichat, P., Mathieu, M. V. and Imelik, B. (1969). *J. Chim. Phys.* **66**, 845.
40. Davydov, A. A. and Shchekocichin, Yu, M. (1969). *Kinet. Katal.* **10**, 163.
41. Dows, D. A. and Pimentel, G. C. (1955). *J. Chem. Phys.* **23**, 1258.
42. Bain, O. and Giguère, P. A. (1955). *Canad. J. Chem.* **33**, 527.
43. Giguère, P. A. and Chapados, C. (1966). *Spectrochim. Acta* **22**, 1131.
44. Kozlowska, E. (1968). *Compt. rend.* **267B**, 696.
45. Ball, D. F. and McKean, D. C. (1962). *Spectrochim. Acta* **18**, 1029.
46. McKean, D. C. and Chalmers, A. A. (1967). *Spectrochim. Acta* **23A**, 777.
47. Giguère, P. A. and Savoie, R. (1962). *Canad. J. Chem.* **40**, 495.
48. Durig, J. R., Bush, S. F. and Mercer, E. E. (1966). *J. Chem. Phys.* **44**, 4238.
49. Baglin, F. G., Bush, S. F. and Durig, J. R. (1967). *J. Chem. Phys.* **47**, 2104.
50. Nixon, R. (1956). *J. Phys. Chem.* **60**, 1055.
51. Ewing, G. E. (1964). *J. Chem. Phys.* **40**, 179.
52. Fournier, R. P., Savoie, R., Bessette, F. and Cabana, A. (1968). *J. Chem. Phys.* **49**, 1159.
53. Denariez, M. M. (1965). *J. Chim. Phys.* **62**, 323.
54. (a) Maklakov, L. I., Nikitin, V. N. and Purkina, A. V. (1963). *Optika i Spektroskopiya* **15**, 332.
 (b) Kimoto, A. and Yamada, H. (1968). *Bull. Chem. Soc. Japan* **41**, 1096.
55. Maklakov, L. I. and Nikitin, V. N. (1963). *Optics and Spectroscopy* **15**, 445.
56. Saëki, S. and Tanabe, K. (1969). *Spectrochim. Acta* **25A**, 1325.
57. Marzocchi, M. P., Manzelli, P., Schettino, V. and Califano, S. (1968). *J. Chem. Phys.* **49**, 5438.
58. Jacox, M. E. and Hexter, R. M. (1961). *J. Chem. Phys.* **35**, 183.
59. Kukina, V. S. and Sverdlov, L. M. (1966). *Zhur. fiz. Khim.* **40**, 2837.
60. Pimentel, G. C. and Klemperer, W. A. (1955). *J. Chem. Phys.* **23**, 377.
61. Le Roy, A. (1965). *Compt. rend.* **261**, 4022.
62. (a) Jacox, M. E. (1962). *J. Chem. Phys.* **36**, 140.
 (b) Dows, D. A. (1962). *J. Chem. Phys.* **36**, 2833.
63. Brith, M. and Ron, A. (1969). *J. Chem. Phys.* **50**, 3053.
64. Wieder, G. M. and Dows, D. A. (1962). *J. Chem. Phys.* **37**, 2990.
65. Bottger, G. L. and Eggers, D. F. Jr. (1964). *J. Chem. Phys.* **40**, 2010; (1966). **44**, 4366.
66. Anderson, A. and Smith, W. H. (1966). *J. Chem. Phys.* **44**, 4216.
67. Smith, W. H. (1969). *Chem. Phys. Lett.* **3**, 464.
68. Cicishvili, G. V., Bagratishvili, G. D. and Oniashvili, N. I. (1969). *Zhur. fiz. Khim.* **43**, 950.
69. Pasquier, B., Sourisseau, C. and Josien, M. L. (1969). *Compt. rend.* **268B**, 1366.
70. Bernstein, E. R. and Robinson, G. W. (1968). *J. Chem. Phys.* **49**, 4962.
71. Hollenberg, J. L. and Dows, D. A. (1963). *J. Chem. Phys.* **39**, 495.
72. Pimentel, G. C., McClellan, A. L., Person, W. B. and Schnepp, O. (1953). *J. Chem. Phys.* **23**, 234.
73. Califano, S. (1962). *Mol. Phys.* **5**, 601.
74. Asselin, M., Bélanger, G. and Sandorfy, C. (1969). *J. Mol. Spectroscopy* **30**, 96.
75. Palm, A. and Bissell, E. R. (1960). *Spectrochim. Acta* **16**, 459.
76. Moore, C. B. and Pimentel, G. C. (1963). *J. Chem. Phys.* **38**, 2816.
77. Smith, W. H. and Leroi, G. E. (1966). *J. Chem. Phys.* **45**, 1767.
78. Mikawa, Y., Brasch, J. W. and Jakobsen, R. J. (1967). *J. Mol. Spectroscopy* **24**, 314.

79. Haurie, M. and Novak, A. (1965). *Spectrochim. Acta* **21**, 1217.
80. (a) Wolff, H. and Ludwig, H. (1966). *Ber. Bunsen Gesellschaft Phys. Chem.* **70**, 474.
(b) Wolff, H. (1967). *Ber. Bunsen Gesellschaft Phys. Chem.* **71**, 1107.
81. Goldfarb, T. D. and Khare, B. N. (1967). *J. Chem. Phys.* **46**, 3379.
82. Hoffman, R. E. and Hornig, D. F. (1949). *J. Chem. Phys.* **17**, 1163.
83. (a) Milligan, D. E. and Jacox, M. E. (1962). *J. Mol. Spectroscopy* **8**, 126.
(b) Noe, L. J. (1968). *Diss. Abs.* **28B**, 3673.
84. Nixon, E. R. and Verderame, F. D. (1964). *J. Chem. Phys.* **41**, 1682.
85. Ball, D. F. and McKean, D. C. (1965). *Spectrochim. Acta* **21**, 204.
86. Verderame, F. D. and Nixon, E. R. (1966). *J. Chem. Phys.* **44**, 43.
87. Durig, J. R. and Wertz, D. W. (1967). *J. Chem. Phys.* **46**, 3069.

CHAPTER 11

Absorption in Matrices

The infrared absorption spectrum of numerous labelled molecules and, especially, isotopic free radicals has recently been obtained for their dilute suspensions inside a matrix made up of a frozen inert gas, such as nitrogen or one of the noble gases which are all transparent to the infrared light. Although a molecule in this matrix isolated phase is, in principle, in the solid state, in reality it is very similar to a gaseous molecule since it is almost free from intermolecular interactions and can usually even carry out nearly free rotations inside the cavity in which it is trapped. As a result of this fact as well as owing to the inevitably low temperature of such matrices matrix-isolated molecules exhibit very sharp absorption bands. This makes it possible to observe fundamentals of isotopomers present in the investigated suspended sample only to an extent of 1–2%, even if they are shifted by less than 2 cm^{-1} from the respective bands of its main isotopic species. Typical results for such cases are collected in Table 11.1 which gives the isotopic splitting pattern of a number of absorption bands observed for various chlorinated tetrahedral molecules in a frozen argon matrix[1] (argon/compound ratio = 200/1000).

TABLE 11.1 The Isotopic Splitting of Various Matrix-Isolated $XYCl_3$ Bands[1] (cm^{-1})

Compound	Band	Approximate description	$^{35}Cl_4$	$^{35}Cl_3$	$^{35}Cl_2$	^{35}Cl	$^{37}Cl_3$
$CHCl_3$	ν_2	Symmetric C—Cl stretching		673·3	671·7	670·1	668·6
	ν_5	Asymmetric C—Cl stretching		767·6	765·3	763·9	
					767·6	766·3	
	abundance:			42·8%	42%	13·7%	1·5%
CCl_4	$\nu_1 + \nu_4$	C—Cl stretching–bending combination	767·4	762·1	758·0	750·8	
				764·5	759·5	754·2	
					761·1		
$SiCl_4$	ν_3	Asymmetric C—Cl stretching	616·5	612·5	610·8	614·8	
					613·6		

As well as for obtaining highly resolved spectra, the matrix isolation technique is also very valuable in investigating the characteristic absorption of refractory compounds that have a very low vapour pressure and hence cannot be studied at temperatures which do not yet produce too broad spectra (as a result of highly excited vibrational and rotational levels). This method is even more important in this case since these substances are usually also insoluble in solvents suitable for infrared absorption measurements and do not yield satisfactory spectra in the crystalline phase. The infrared absorption of an appreciable number of labelled compounds belonging to this class has therefore been measured under these conditions since isotopic modifications of such substances are often also included in their spectroscopic study. This facilitates the assignment of their observed absorption bands and the calculation of their force constants. The infrared spectrum of $^{26}MgF_2$ suspended in solid argon,[2] which exhibits absorption bands at 240, 450, 477, 483, 740 and 840 cm^{-1}, can be taken as a proper representative of this type of spectra.

The matrix technique has also been used successfully to isolate radicals, produced *in situ* by photolysis or a chemical reaction, from reacting further to form stable molecules. The absorption spectrum of many isotopic radicals has thus been investigated in parallel to that of the corresponding normal radicals, with the aim of aiding in the solution of the sometimes very difficult problem of assigning the absorption bands observed for the latter. In the case of the ND radical, e.g. produced by photolyzing DN_3 suspended in frozen argon at 4°K, the fundamental band frequency was found[3] to be 2323 ± 2 cm^{-1}, as compared with the respective matrix isolated N—H frequency of 3125·6 cm^{-1}.

Photolyzing DI suspended in a CO matrix at 14°K produced the DCO radical which exhibited its fundamentals at 852 (bending), 1800 (C=O stretching) and 1937 cm^{-1} (C—D stretching), corresponding to the HCO radical bands at 1090, 1861 and 2488 cm^{-1}, respectively.[4] $D^{13}CO$ showed these absorption peaks at 845, 1780 and 1910 cm^{-1} while the C=O stretching band of $H^{13}CO$ appeared at[4] 1821 cm^{-1}.

The new molecular species of methyl alkali halides (CH_3····MX), which exists in alkali halide matrices in which methyl radicals were produced, was found to display its CD_3 bending fundamental at 567 cm^{-1} in the case of CD_3····LiI and at 533 cm^{-1} in the case of CD_3····NaI, as compared with the parallel CH_3····MI frequencies of 730 and 696 cm^{-1}, respectively.[5] CH_3····^{6}LiI was observed to exhibit this band at 730·1 cm^{-1} while CH_3····^{7}LiI showed it at 729·6 cm^{-1}. CD_3 radicals suspended in solid N_2 displayed, however, this fundamental at 463 cm^{-1} since these radicals are no longer bonded in this matrix.[5]

Table 11.2 collects the references to most of the literature papers dealing with the infrared absorption spectra of labelled molecules or isotopic radicals isolated in matrices. A number of papers on the absorption spectrum of various labelled ions isolated in single salt crystals are also included in this table.

TABLE 11.2 Reported Infrared Absorption Spectra of Labelled Compounds, Ions or Radicals in Host Matrices

Absorbing species	Labelling isotope(s)	Remarks	Reference
Hydrogen	D	In solid argon or neon	6, 7
NH radical	D, ^{15}N	In solid argon at 4°K	3
OH radical	D, ^{18}O	In solid neon (evidence for rotation)	8, 9
DF	D	17–70 cm^{-1} region	10
Hydrogen chloride	D, ^{35}Cl	In Ar, Kr, Xe; SF_6; 33–300 cm^{-1} region	11–16
Dimeric (and trimeric) HCl	D	Far infrared	17, 18
Hydrogen bromide	D	At 4–20°K	11, 13, 15
LiO	^7Li	Also $^7Li_2O$ and $^7Li_2O_2$	19
LiF	^nLi	Also $^nLi_2F_2$ and $^nLi_3F_3$; 150–1000 cm^{-1} region	20, 21
LiCl	^nLi	In solid argon	22
LiBr	^nLi	Also $^nLi_2Br_2$	22
CN radical	^{13}C, ^{15}N	Also halogen cyanides	23
CO	^{13}C, ^{18}O	In N_2 and Ar	24, 25
CF radical	^{13}C	Also HCF and H_2CF radicals	26
NX radical (X = F, Cl, Br)	^{15}N, ^{35}Cl, ^{37}Cl	In solid argon	27
OF radical	^{18}O	In solid N_2 and Ar	28
ScO radical	^{18}O	In solid neon at 4°K	29
TiO	^{18}O	In solid neon or argon	30
ZrO radical	^{18}O		30
HfO radical	^{18}O		30
TaO radical	^{18}O	In solid argon	31
$W_xO_y (x = 1,2,3,4; y = 1,2,3,6,8,9,12)$	^{18}O	In solid neon or argon	32
MgF radical	^{26}Mg	In solid argon or krypton	2
Li_2O	^6Li, ^7Li	50–500 cm^{-1} region	33
LiO_2	^6Li, ^7Li	200–4000 cm^{-1} region	33, 34
HCCl radical	^{13}C, ^nCl	400–3400 cm^{-1} region	35
C_3 radical	^{13}C	In solid argon at 20°K	36
CCO radical	^{13}C, ^{18}O		37
NCN radical	^{13}C, ^{15}N	In solid argon at 14°K	38
NCO radical	^{13}C, ^{15}N	400–3500 cm^{-1} region	39
FCO radical	^{13}C, ^{18}O		40
COS	^{13}C	In solid argon or nitrogen	41
COCl radical	^{13}C		42
CF_2 radical	^{13}C	450–1750 cm^{-1} region	43
CCL_2 radical	^{13}C	In solid argon	44
NH_2 radical	D, ^{15}N	At 14°K in solid N_2 (NHD and ND_2) or krypton	45, 46

HNO radical	D	At 20 and 53°K	47
HNF radical	D, ^{15}N	At 14°K in solid argon	48
CNN radical	^{13}C, ^{15}N	In solid argon or N_2	49
HDO, D_2O	D	In solid argon at 4–27°K	50
D_2O	D	In solid nitrogen	51
HO_2 radical	D, ^{18}O	In solid argon	52
HOF	D, ^{18}O	In solid argon at 15°K; also $(DOF)_2$ and $(H^{18}OF)_2$	53
HOCl	D, ^{18}O, ^{35}Cl, ^{37}Cl	In solid argon	54
HOBr	D, ^{18}O	At 4°K	54
CsOD	D	200–350 cm^{-1} region	55
LiON	^{n}Li, ^{15}N, ^{18}O	In solid argon	56
Ozone	^{18}O	In solid O_2; 700–3040 cm^{-1} region	57
OF_2	^{18}O	Evaluation of force constants	58
OCl_2	^{18}O	350–410, 700–1050 cm^{-1} regions	59
FO_2 radical	^{18}O	In solid N_2 or O_2; also $(O_2F)_n$ absorption	60, 61
Al_2O radical	^{18}O	In solid argon or krypton	62
SiF_2	^{n}Si	300–2000 cm^{-1} region	63, 64
$SiCl_2$ radical	^{n}Si	400–2300 cm^{-1} region	65
SO_2	^{n}S, ^{18}O	In solid krypton	66, 67
SeO_2	^{n}Se, ^{18}O	500–1020 cm^{-1} region	66
ClOO radical	^{18}O	In solid argon at 4°K; 300–1500 cm^{-1}	68
Cl_3 radical	^{n}Cl	365–375 cm^{-1} zone	69
CaF_2	^{44}Ca	In neon, argon or krypton; 200–4000 cm^{-1}	70
NiF_2	^{n}Ni	At 14°K in solid argon	71
$NiCl_2$	^{n}Cl, ^{n}Ni	400–4000 cm^{-1} region	71
$HBCl_2$	D, ^{n}B	Comparison with the calculated fundamentals	72
Alkali metal metaborates	^{n}B	60–2100 cm^{-1} region	73
BF_3, $(BF_3)_2$	^{n}B	In solid argon or krypton at 20–50°K	74
BCl_3	^{n}B, ^{n}Cl	ν_3 fundamental	75
Methyl radical	D, ^{13}C	In solid argon or N_2; 300–3400 cm^{-1}	76, 77
$CHCl_2$ radical	D, ^{13}C, ^{n}Cl	At 15°K in solid argon	78
$CHBr_2$ radical	D	300–1250 cm^{-1} region	79
CHN_2 radical	D	650–1200, 1700–1800 cm^{-1}	80
DOCN	D	In solid argon or N_2	81
DNCS	D	1:1000 in solid argon	82
FNCN radical	^{13}C, ^{15}N	400–2850 cm^{-1} region	83
CO_3 radical	^{13}C, ^{18}O	At 50–77°K	84
CF_3 radical	^{13}C	450–1750 cm^{-1} region	43
CCl_3 radical	^{13}C, ^{n}Cl	In solid argon at 15°K	85
Ammonia	D, ^{15}N	In solid argon or N_2	86, 87
trans HN=NH	D	HN=ND and DN=ND; in N_2 at 20°K	87
Hydrazoic acid	D, ^{15}N	In solid N_2 (6 isotopomers)	88, 89
$(DN_3)_2$	D	At 20°K	89
N_3F	^{15}N	At 4–10°K	27

TABLE 11.2—*Continued*

Absorbing species	Labelling isotope(s)	Remarks	Reference
N_3Cl	^{15}N, ^{n}Cl	In solid argon	27
N_3Br	^{15}N		27
$(NO)_2$	^{15}N, ^{18}O	1000–1100, 1700–2300 cm^{-1}	90
Nitrous acid	D	Both *cis* and *trans* isomers; at 20°K in solid N_2	91
Li_2O_2	^{n}Li		33
Cl_2O_2	^{18}O	350–410, 700–1050 cm^{-1} regions	59
Si_2O_2	^{n}Si	730–810 cm^{-1}; also: Ge_2O_2 and Sn_2O_2	92
$SiCl_3$ radical	^{n}Si, ^{n}Cl	400–2300 cm^{-1} region	93
B_2O_3	^{n}B	350–4000 cm^{-1}; also B_2O_2 and BO_2	94
Methane	D	At 5–40°K; in solid argon, krypton or xenon	95, 96
Methyl lithium	^{n}Li	In solid argon; 380–620 cm^{-1} region	97
CH_3Cl	^{n}Cl	In solid argon	1
CH_2Cl_2	^{n}Cl	At 20°K	1
$FCCl_3$	^{n}Cl	500–1300 cm^{-1} region	1
$BrCCl_3$	^{n}Cl	Ar/compound ratio = 200–1000	1
Ketene	D	Both CD_2CO and CHDCO	98
Methylene imine	D, ^{15}N	In solid N_2	99
Diazomethane	D, ^{15}N	In solid N_2 or Ar; both CD_2N_2 and $CHDN_2$	100
$LiCCl_3$	^{n}Li, ^{13}C, ^{n}Cl	In solid argon at 15°K	101
$LiCBr_3$	^{n}Li	370–590 cm^{-1} region	101
CCl_4	^{13}C, ^{n}Cl	(See also ref. 103)	101
Carbon suboxide	^{13}C, ^{18}O	Both $O{=}^{13}C{=}^{12}C{=}^{13}C{=}O$ and $O{=}^{13}C{=}^{12}C{=}^{12}C{=}O$	102
NF_2CN	^{13}C, ^{15}N	400–2850 cm^{-1} region	83
CF_4	^{13}C	In solid argon, krypton or N_2	103
$(DCN)_2$	D	70–3320 cm^{-1} region	104
Methyl azide	D	In solid argon	99
Silyl acetylene	D	In solid argon at 4°K	105
1,1,1-Trichloroethane	^{n}Cl	In solid argon	1
1-Chloropropane	^{n}Cl	500–1300 cm^{-1} region	1
Trimethylamine	D	200–3000 cm^{-1} region	106
Cyanate ion	^{13}C, ^{15}N, ^{18}O	In KBr; 1150–1350 cm^{-1} region	107
Nitrite ion	^{15}N, ^{18}O	In KBr crystals	108
Nitrate ion	^{15}N, ^{18}O	In KBr crystals	108
Borohydride ion	D	In alkali-halide single crystals	109
Chlorate ion	^{n}Cl	In rubidium halide lattices	110
Perchlorate ion	^{n}Cl	In alkali-halide crystals	110

References

1. King, S. T. (1968). *J. Chem. Phys.* **49**, 1321.
2. Mann, D. E., Calder, G. V., Seshadri, K. S., White, D. and Linevsky, M. J. (1967). *J. Chem. Phys.* **46**, 1138.
3. Milligan, D. E. and Jacox, M. E. (1964). *J. Chem. Phys.* **41**, 2838.
4. Milligan, D. E. and Jacox, M. E. (1964). *J. Chem. Phys.* **41**, 3032.
5. Tan, L. Y. and Pimentel, G. C. (1968). *J. Chem. Phys.* **48**, 5202.
6. Jean-Louis, M., Bahreini, M. and Vu, H. (1969). *Compt. rend.* **B268**, 41.
7. Chen, H. (1968). *Diss. Abs.* **28B**, 4454.
8. Tinti, D. S. (1968). *J. Chem. Phys.* **48**, 1459.
9. Acquista, N., Schoen, L. J. and Lide, D. R. Jr. (1968). *J. Chem. Phys.* **48**, 1534.
10. Robinson, D. W. and Von Holle, W. G. (1966). *J. Chem. Phys.* **44**, 410.
11. Bowers, M. T. and Flygare, W. H. (1966). *J. Chem. Phys.* **44**, 1389.
12. Keyser, L. F. and Robinson, G. W. (1966). *J. Chem. Phys.* **44**, 3225.
13. Mann, D. E., Acquista, N. and White, D. (1966). *J. Chem. Phys.* **44**, 3453.
14. Ranganath, R., Whyte, Jr., T. E., Theophanides, T. and Turrell, G. C. (1967). *Spectrochim. Acta* **23A**, 807.
15. Katz, B., Ron, A. and Schnepp, O. (1967). *J. Chem. Phys.* **46**, 1926.
16. Friedmann, H. and Kimel, S. (1967). *J. Chem. Phys.* **47**, 3589.
17. Keyser, L. F. and Robinson, G. W. (1966). *J. Chem. Phys.* **45**, 1694.
18. Katz, B., Ron, A. and Schnepp, O. (1967). *J. Chem. Phys.* **47**, 5303.
19. White, D., Seshadri, K. S., Dever, D. F., Mann, D. E. and Linevsky, M. J. (1963). *J. Chem. Phys.* **39**, 2463.
20. Snelson, A. (1967). *J. Chem. Phys.* **46**, 3652.
21. Abramowitz, S., Acquista, N. and Levin, I. W. (1968). *J. Res. Nat. Bur. Stand.* **A72**, 487.
22. Schlick, S. and Schnepp, O. (1964). *J. Chem. Phys.* **41**, 463.
23. Milligan, D. E. and Jacox, M. E. (1967). *J. Chem. Phys.* **47**, 278.
24. Vu, H., Atwood, M. R. and Vodar, B. (1963). *J. Chem. Phys.* **38**, 2671.
25. Leroi, G. E. and Pimentel, G. C. (1964). *J. Chem. Phys.* **40**, 2298.
26. Jacox, M. E. and Milligan, D. E. (1969). *J. Chem. Phys.* **50**, 3252.
27. Milligan, D. E. and Jacox, M. E. (1964). *J. Chem. Phys.* **40**, 2461.
28. Arkell, A., Reinhard, R. R. and Larson, L. P. (1965). *J. Amer. Chem. Soc.* **87**, 1016.
29. Weltner, W. Jr., McLeod, D. Jr. and Kasai, P. H. (1967). *J. Chem. Phys.* **46**, 3172.
30. Weltner, W. Jr. and McLeod, D. Jr. (1965). *J. Phys. Chem.* **69**, 3488.
31. Weltner, W. Jr. and McLeod, D. Jr. (1965). *J. Chem. Phys.* **42**, 882.
32. Weltner, W. Jr. and McLeod, D. Jr. (1965). *J. Mol. Spectroscopy* **17**, 276.
33. Seshadri, K. S., White, D. and Mann, D. E. (1966). *J. Chem. Phys.* **45**, 4697.
34. Andrews, L. (1968). *J. Amer. Chem. Soc.* **90**, 7368.
35. Jacox, M. E. and Milligan, D. E. (1967). *J. Chem. Phys.* **47**, 1626.
36. Weltner, W. Jr. and McLeod, D. Jr. (1964). *J. Chem. Phys.* **40**, 1305.
37. Milligan, D. E. and Jacox, M. E. (1965). *J. Chem. Phys.* **43**, 3734.
38. Milligan, D. E. and Jacox, M. E. (1966). *J. Chem. Phys.* **45**, 1387.
39. Milligan, D. E. and Jacox, M. E. (1967). *J. Chem. Phys.* **47**, 5157.
40. Milligan, D. E. and Jacox, M. E. (1965). *J. Chem. Phys.* **42**, 3187.
41. Verderame, F. D. and Nixon, E. R. (1966). *J. Chem. Phys.* **44**, 43.
42. Milligan, D. E. and Jacox, M. E. (1965). *J. Chem. Phys.* **43**, 866.

43. Milligan, D. E. and Jacox, M. E. (1968). *J. Chem. Phys.* **48**, 2265.
44. Andrews, L. (1968). *Tetrahedron Letters* 1423; (1968) *J. Chem. Phys.* **48**, 979.
45. Milligan, D. E. and Jacox, M. E. (1965). *J. Chem. Phys.* **43**, 4487.
46. Keyser, L. F. (1967). *Diss. Abs.* **28B**, 630.
47. Milligan, D. E., Jacox, M. E., Charles, S. W. and Pimentel, G. C. (1962). *J. Chem. Phys.* **37**, 2302.
48. Jacox, M. E. and Milligan, D. E. (1967). *J. Chem. Phys.* **46**, 184.
49. Milligan, D. E. and Jacox, M. E. (1966). *J. Chem. Phys.* **44**, 2850.
50. Redington, R. L. and Milligan, D. E. (1962). *J. Chem. Phys.* **37**, 2162.
51. Harvey, K. B. and Shurvell, H. F. (1968). *J. Mol. Spectroscopy* **25**, 120.
52. Milligan, D. E. and Jacox, M. E. (1963). *J. Chem. Phys.* **38**, 2627.
53. Noble, P. N. and Pimentel, G. C. (1968). *Spectrochim. Acta* **24A**, 797.
54. Schwager, I. and Arkell, A. (1967). *J. Amer. Chem. Soc.* **89**, 6006.
55. Acquista, N., Abramowitz, S. and Lide, D. R. (1968). *J. Chem. Phys.* **49**, 780.
56. Andrews, W. L. S. and Pimentel, G. C. (1966). *J. Chem. Phys.* **44**, 2361.
57. le Brumant, J., Barbe, A. and Jouve, P. (1969). *Compt. rend.* **B268**, 549.
58. Ogden, J. S. and Turner, J. J. (1967). *J. Chem. Soc.* (*A*) 1483.
59. Rochkind, M. M. and Pimentel, G. C. (1967). *J. Chem. Phys.* **46**, 4481.
60. Spratley, R. D., Turner, J. J. and Pimentel, G. C. (1966). *J. Chem. Phys.* **44**, 2063.
61. Noble, P. N. and Pimentel, G. C. (1966). *J. Chem. Phys.* **44**, 3641.
62. Linevsky, M. J., White, D. and Mann, D. E. (1964). *J. Chem. Phys.* **41**, 542.
63. Milligan, D. E. and Jacox, M. E. (1968). *J. Chem. Phys.* **49**, 4269.
64. Hastie, J. W., Hauge, R. H. and Margrave, J. L. (1969). *J. Amer. Chem. Soc.* **91**, 2536.
65. Milligan, D. E. and Jacox, M. E. (1968). *J. Chem. Phys.* **49**, 1938.
66. Hastie, J. W., Hauge, R. H. and Margrave, J. L. (1969). *J. Inorg. Nuclear Chem.* **31**, 281.
67. Allavena, M., Rysnik, R., White, D., Calder, G. V. and Mann, D. E. (1969). *J. Chem. Phys.* **50**, 3399.
68. Arkell, A. and Schwager, I. (1967). *J. Amer. Chem. Soc.* **89**, 5999.
69. Nelson, L. Y. and Pimentel, G. C. (1967). *J. Chem. Phys.* **47**, 3671.
70. Snelson, A. (1966). *J. Phys. Chem.* **70**, 3208.
71. Milligan, D. E., Jacox, M. E. and McKinley, J. D. (1965). *J. Chem. Phys.* **42**, 902.
72. Bass, C. D., Lynds, L., Wolfram, T. and DeWames, R. E. (1964). *Inorg. Chem.* **3**, 1063.
73. Seshadri, K. S., Nimon, L. A. and White, D. (1969). *J. Mol. Spectroscopy* **30**, 128.
74. Bassler, J. M., Timms, P. L. and Margrave, J. L. (1966). *J. Chem. Phys.* **45**, 2704.
75. Comeford, J. J., Abramowitz, S. and Levin, I. W. (1965). *J. Chem. Phys.* **43**, 4536.
76. Milligan, D. E. and Jacox, M. E. (1967). *J. Chem. Phys.* **47**, 5146.
77. Andrews, L. and Pimentel, G. C. (1967). *J. Chem. Phys.* **47**, 3637.
78. Granville, T., Carver, T. G. and Andrews, L. (1969). *J. Chem. Phys.* **50**, 4235.
79. Carver, T. G. and Andrews, L. (1969). *J. Chem. Phys.* **50**, 4223.
80. Ogilvie, J. F. (1968). *Canad. J. Chem.* **46**, 2472.
81. Jacox, M. E. and Milligan, D. E. (1964). *J. Chem. Phys.* **40**, 2457.
82. Durig, J. R. and Wertz, D. W. (1967). *J. Chem. Phys.* **46**, 3069.
83. Milligan, D. E. and Jacox, M. E. (1968). *J. Chem. Phys.* **48**, 4811.
84. Moll, N. G., Clutter, D. R. and Thompson, W. E. (1966). *J. Chem. Phys.* **45**, 4469.
85. Andrews, L. (1968). *J. Chem. Phys.* **48**, 972.

86. Milligan, D. E., Hexter, R. M. and Dressler, K. (1961). *J. Chem. Phys.* **34**, 1009.
87. Rosengren, K. and Pimentel, G. C. (1965). *J. Chem. Phys.* **43**, 507.
88. Moore, C. B. and Rosengren, K. (1966). *J. Chem. Phys.* **44**, 4108.
89. Pimentel, G. C., Charles, S. W. and Rosengren, K. (1966). *J. Chem. Phys.* **44**, 3029.
90. Guillory, W. A. and Hunter, C. E. (1969). *J. Chem. Phys.* **50**, 3516.
91. Hall, R. T. and Pimentel, G. C. (1963). *J. Chem. Phys.* **38**, 1889.
92. Anderson, J. S., Ogden, J. S. and Ricks, M. J. *Chem. Commun.* **1968**, 1585.
93. Jacox, M. E. and Milligan, D. E. (1968). *J. Chem. Phys.* **49**, 3130.
94. Sommer, A., White, D., Linevsky, M. J. and Mann, D. E. (1963). *J. Chem. Phys.* **38**, 87.
95. Cabana, A., Savitsky, G. B. and Hornig, D. F. (1963). *J. Chem. Phys.* **39**, 2942.
96. Hopkins, H. P. Jr., Curl, R. F. Jr. and Pitzer, K. S. (1968). *J. Chem. Phys.* **48**, 2959.
97. Andrews, L. (1967). *J. Chem. Phys.* **47**, 4834.
98. Moore, C. B. and Pimentel, G. C. (1963). *J. Chem. Phys.* **38**, 2816.
99. Moore, C. B., Pimentel, G. C. and Goldfarb, T. D. (1965). *J. Chem. Phys.* **43**, 63.
100. Moore, C. B. and Pimentel, G. C. (1964). *J. Chem. Phys.* **40**, 342; **41**, 3504.
101. Andrews, L. and Carver, T. G. (1968). *J. Phys. Chem.* **72**, 1743.
102. Jacox, M. E., Milligan, D. E., Moll, N. G. and Thompson, W. E. (1965). *J. Chem. Phys.* **43**, 3734.
103. Abramowitz, S. and Comeford, J. J. (1965). *Spectrochim. Acta* **21**, 1479.
104. King, C. M. and Nixon, E. R. (1968). *J. Chem. Phys.* **48**, 1685.
105. Reeves, R. B., Wilde, R. E. and Robinson, D. W. (1964). *J. Chem. Phys.* **40**, 125.
106. Goldfarb, T. D. and Khare, B. N. (1967). *J. Chem. Phys.* **46**, 3379.
107. Decius, J. C. and Gordon, D. J. (1967). *J. Chem. Phys.* **47**, 1286.
108. Kato, R. and Rolfe, J. (1967). *J. Chem. Phys.* **47**, 1901.
109. Coker, E. H. and Hofer, D. E. (1968). *J. Chem. Phys.* **48**, 2713.
110. Krynauw, G. N. and Schutte, C. J. H. (1967). *Z. phys. Chem. (Frankfurt)* **55**, 113.

CHAPTER 12

Analytical Applications

12.1 Introduction

Since the infrared absorption spectrum of isotopic molecules is generally quite distinct from that of the corresponding normal species, this spectrum has very frequently been used to merit both in their qualitative and in their quantitative analysis, in the presence of the respective normal modification as well as in its absence. This method of analysis has the important advantages that it is non-destructive, very specific, rapid, easy to follow and needs only very minute samples, often as little as a few micrograms of material. Its high specificity enabled, e.g. its use in the detection of small amounts of deuteroethylene in the presence of deuteroethanes[1] or the determination of the carbonylic (as differing from the etheric) oxygen-18 content of ^{18}O-labelled cinnamyl *p*-nitrobenzoate.[2]

The presence of deuterated water on Mars has also been proved by infrared spectroscopy.[3]

Good results (a relative accuracy of 0·4%) have similarly been obtained in the quantitative determination of isotopic gases such as $C^{18}O$, $^{13}CO_2$, $^{15}N_2O$, etc. by means of analysers.[4a, 4b]

12.2 Analysis of Deuterated Compounds

Most of the research done so far on the application of infrared spectroscopy to the analysis of labelled compounds has been connected with the problem of the quantitative determination of the deuterium (or protium) content of various heavy water samples. It has thus been found that the infrared methods can be considered as the most suitable ones for the analysis of 0–3 and 97–100% deuterated water samples.[5] In this case the determination is made on basis of their characteristic absorption in the near infrared region.[5]

The analysis of samples having an intermediate deuterium content (3–97 atom %) has been carried out by mixing 1 ml of the water sample with 7 ml of methyl formate, acidifying the mixture and making it homogeneous (formation of methanol and CH_3OD), dissolving it in carbon tetrachloride

(1:20) and measuring the absorbance of this solution between 2000 and 4000 cm^{-1}. The ratio between the area of the O—H stretching band at 3333 cm^{-1} and that of the OD band, at *ca.* 2480 cm^{-1}, yields the H:D molar ratio of the original sample.[6]

Samples containing 3–25 atom % deuterium have also been analysed by dissolving them in dimethyl sulphoxide and observing the absorbance of the OH and the OD deformation bands of the H_2O (1595 cm^{-1}) and D_2O (1179 cm^{-1}) species present in such a solution (see Fig. 12.1). From the ratio of

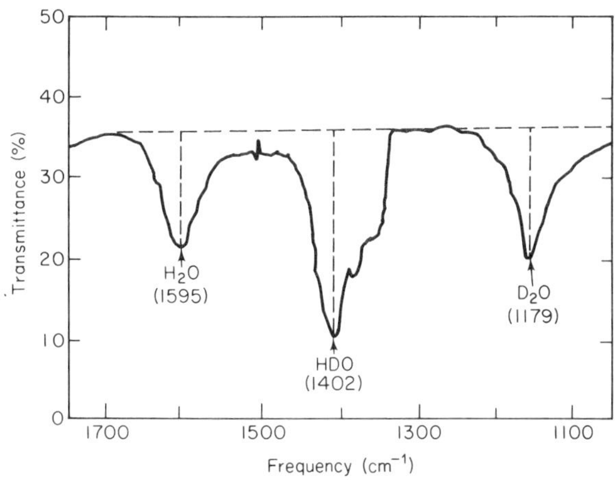

Fig. 12.1 Infrared spectrum of a normal water–deuterium oxide mixture.

these absorbances the D_2O content of the original samples has been determined by reading it from a calibration curve.[7]

The deuterium concentration of enriched water samples containing 0·09–0·24 % D has been estimated[8] with a precision of ±3% from their HO—D stretching band intensity at *ca.* 2500 cm^{-1}.

This HOD absorption peak has also been used to detect differences as low as 0·0002 atom % from the natural concentration level (≈0·0148 atom %) of deuterium in water.[9]

The determination of the deuterium content of D_2O—H_2O mixtures has also been carried out in acetone solutions (using the ratio between the absorbance of their O—H stretching band at 3534 cm^{-1} and that of the O—D stretching band at 2564 cm^{-1}), with an accuracy of[10] ±2%. Another method of analysing for D_2O converts it into deuterated acetylene the infrared absorption of which is then measured.[11] Further work on this determination has also been reported.[12–15]

The spectrophotometric determination of the D-enrichment of body water to which a small (known) amount of D_2O has been added in the body has been used to find out its total content.[16]

The deuterium enrichment of liquid ammonia (under pressure) containing 80–100 atom % deuterium can be determined from the absorbance of its NHD_2 deformation band at 1462 cm^{-1} which is lower the higher is this enrichment. In the case of a 30–80 atom % deuterium content the absorption intensity of the corresponding NH_2D band at 1599 cm^{-1} must also be taken into account in this determination. For still lower deuterium enrichments the deuterium analysis[17] is carried out on basis of the H_2N—D stretching band appearing at *ca.* 2452 cm^{-1}.

The terminal deuterium atoms content of deuterated decaborane samples was determined satisfactorily ($\pm 2\%$) from their B—D stretching band absorption intensity at *ca.* 1934 cm^{-1}, while the amount of their bridging deuterium atoms can be estimated from their absorption at *ca.* 1150 cm^{-1}, due to their B—D—B stretching mode.[18]

The quantitative analysis of deuterated methanes and the three isomeric D_2-ethylenes was carried out by condensing their diluted gaseous samples onto a cold plate, transparent to the infrared light, and recording the absorbance of the pseudo-matrix isolated isotopomer (or isotopomers) at its (theirs) key-wavelength(s).[19]

The characteristic absorption of γ-D_6-benzene hexachloride was made use of in the determination of γ-benzene hexachloride by adding a known amount of the deuterated isotopomer to the weighed sample to be analysed and measuring spectrophotometrically the D-enrichment of the γ-benzene hexachloride isolated from this mixture.[20]

A satisfactory quantitative analysis of a mixture of toluene and its *o*-, *m*- and *p*-deuterated isotopomers was effected on basis of their characteristic 728, 631, 645 and 608 cm^{-1} bands, respectively.[21]

The relative amount of RCDO aldehydes in a mixture with the corresponding normal aldehydes can be determined from their absorbance at *ca.* 2070 and 2100 cm^{-1}, where the CDO group shows its specific absorption bands.[22]

The 2000–2400 cm^{-1} integrated absorption intensity of a number of D_1-D_6-methyl stearates, all deuterated at the 6–16 carbon atoms, exhibits a quantitative correlation with their deuterium content so that this absorption intensity provides a good basis for an infrared spectroscopic analysis of the deuterium enrichment of such stearates.[23]

Acrylonitrile deuterated at the central carbon atom displayed a linear relationship between its absorbance at 1165 cm^{-1} and its deuterium content.[24]

The rate of the deuterium-protium exchange (i.e. the formation of the N—D bond) in the case of 2-pyrrolidone was followed by observing the increase in its absorption at *ca.* 7090 cm^{-1} while the similar formation of $CH_3CONDCH_3$ and $CH_3CONDC_2H_5$ was determined[25] on basis of their band at 7000 cm^{-1}.

The nicotinic acid in complex mixtures was assayed by diluting it with (a known amount of) ring-deuterated nicotinic acid, esterifying the acid, crystallizing its (methyl) ester, measuring its absorbance at *ca.* 741 cm^{-1} and reading the D-enrichment corresponding to this absorbance from a calibration curve.[26]

The average ratio of the extinction coefficient of the 3360 cm^{-1} O—H stretching band of the amorphous portion of regenerated cellulose to that of the corresponding OD band at 2530 cm^{-1} was found[27] to be 1·11.

12.3 Analysis of Other Labelled Compounds

A grating infrared spectrometer was found to resolve completely the ν_{14} and ν_{18} fundamentals of the $^{10}B_2$, $^{10}B^{11}B$ and $^{11}B_2$ isotopomers of diborane containing 1·1, 18·8, 56·4, 73·5 or 96 atom % ^{10}B, this resolution being good enough for the estimation of each of these isotopic species separately. Even by calibrating an ordinary infrared spectrometer with samples of known ^{10}B-content it is possible to determine the boron isotopic composition of diborane to within 1%, from the exact position of its (now appearing merged together) ν_{14} band[28] (shown by normal B_2H_6 at 973 cm^{-1}).

The $^{13}C/^{12}C$ ratio of carbon dioxide samples between 0·011 and 0·10 was determined[29] by an analyser (a non-dispersive infrared absorption instrument) with an accuracy of $\pm$0·002.

The ^{15}N-enrichment of nitrate ions dispersed in crystalline KBr was determined from the absorption intensity of the 817 cm^{-1} ν_2 band of the $^{15}NO_3^-$ ions, corresponding to the 839 cm^{-1} band of the $^{14}NO_3^-$ ions under these conditions.[30]

The ^{15}N-content of labelled ammonium chloride samples was determined by converting it into 2,4-dinitroaniline, measuring its N—H stretching band intensity, in a CH_2Cl_2 solution, against that of $^{14}N_3$-2,4-dinitroaniline and the use of a calibration curve.[31]

The ^{18}O-enrichment of water and a number of oxy-acids (*p*-$CH_3C_6H_4SO_3H$ and $C_6H_5CH_2NHPO(OH)_2$) was estimated by reacting them with carbodiimides to form ^{18}O-labelled NN′-disubstituted ureas, dissolving these ureas in a mixture of 75% CCl_4 and 25% dimethyl sulphoxide and measuring the optical density of the fully resolved C=^{16}O stretching band of this solution. By comparing this absorption intensity with that of a similar solution of the corresponding normal urea the C=^{16}O content of the analysed solution is obtained and from it, by difference, that of its C=^{18}O groups, which is equal to the ^{18}O-enrichment of the parent water or oxy-acid sample.[32]

The concentration and distribution of ^{18}O in films of labelled silica could be determined by observing its ^{18}O-characteristic band[33] at *ca.* 1040 cm^{-1}, corresponding to the normal silica band appearing at *ca.* 1085 cm^{-1}

The quantitative analysis of $BaS^{16}O_3{}^{18}O$ in the presence of an excess of normal barium sulphate was carried out on basis of its 961 cm^{-1} ν_1, S—O stretching, band which is shifted by 20 cm^{-1} in respect to the analogous $BaS^{16}O_4$ band. This determination was made on a Nujol mull of the labelled salt using the base-line technique.[34]

The formation of ^{18}O-labelled acetone, acetophenone and benzophenone (by an exchange reaction occurring between $H_2{}^{18}O$ and the respective normal ketones, in dilute CH_3CN) was followed by measuring the optical density of their $C{=}^{18}O$ stretching bands at *ca.* 1695, 1668 and 1646 cm^{-1}, respectively.[35]

An infrared spectroscopic submicrodetermination method for ^{18}O-benzophenone, capable of giving results accurate to $\pm 0{\cdot}5\%$ (absolute) for 0·1–0·7 mg samples containing 1–61 atom % of oxygen-18, has also been described. This method is again based on the optical density of (CCl_4) solutions of the labelled ketone at the $C{=}^{18}O$ and $C{=}^{16}O$ stretching frequencies.[36]

The $C{=}^{18}O$ content of samples of ^{18}O-cinnamyl *p*-nitrobenzoate (dissolved in CCl_4) was determined on basis of their optical density at 1700 cm^{-1} due to their $C{=}^{18}O$ stretching band.[2]

References

1. Van Riet, R. (1955). *Ann. Soc. sci. Bruxelles, Ser. I.* **69**, 117.
2. Braude, E. A. and Turner, M. (1958). *J. Chem. Soc.* 2404.
3. Shirk, J. S., Haseltine, W. A. and Pimentel, G. C. (1965). *Science* **147**, 48.
4. (a) Meier, G. and Hermann, H. D. (1964). *Abh. dt. Akad. Wiss. Berl. Kl. Chem. Geol. Biol. No. 7*, 407.
   (b) Bosselaar, C. A. (1963). *Appl. sci. Res., Hague* **10B**, 137.
5. Keder, W. E. and Kalkwarf, D. R. (1966). *Analyt. Chem.* **38**, 1288.
6. Kimbrough, Jr., R. D. and Askins, R. W. (1969). *Analyt. Chem.* **41**, 1147.
7. Mahadevan, E. G. (1967). *Analyst* **92**, 717.
8. Fydelor, P. J. and Rawson, D. S. (1965). *Spectrochim. Acta* **21**, 1957.
9. Bayly, J. G., Stevens, W. H. and Thurston, W. M. "An Infrared Method for Determining Variations In The Deuterium Content of Natural Water", Atomic Energy of Canada Limited CRDC–1141 Report, Chalk River, Ontario (1964).
10. Fochler, H. S. (1963). *Appl. Spectroscopy* **17**, 105.
11. Balachovskii, I. S. and Pavelkovskii, S. A. (1969). *Lab. Delo* **5**, 274.
12. Demidenkova, I. V. (1969). *Zhur. priklad. Spektrosk.* **11**, 824.
13. Knezhevits, Z. V. (1965). *Isotopenpraxis* **1**, 69.
14. Hallaba, E. and Hamza, M. S. (1966). *Isotopenpraxis* **2**, 394.
15. Roth, E. and Ravoire, J. (1964). *Atom. Energy Rev.* **2**, 55.
16. Stansell, M. J. and Mojica, Jr., L. (1968). *Clin. Chem.* **14**, 1112.
17. Corset, J., Guillermet, J. and Lascombe, J. *Bull. Soc. Chim. France* **1966**, 1231.
18. Kaufman, J. J. and Koski, W. S. (1956). *J. Amer. Chem. Soc.* **78**, 5774.
19. Rochkind, M. M. (1968). *Analyt. Chem.* **40**, 762.
20. Trenner, N. R., Walker, R. W., Arison, B. H. and Buhs, R. P. (1949). *Analyt. Chem.* **21**, 285.
21. Van Dyke Tiers, G. (1951). *J. Chem. Phys.* **19**, 1072.
22. Wiberg, K. B. (1954). *J. Amer. Chem. Soc.* **76**, 5372.

23. Rohwedder, W. K., Scholfield, C. R., Rakoff, H., Nowakowska, J. and Dutton, H. J. (1967). *Analyt. Chem.* **39**, 820.
24. Cockburn, W. F. and Hubley, C. E. (1957). *Appl. Spectroscopy* **11**, 188.
25. Klotz, I. M. and Feidelseit, Ph. L. (1966). *J. Amer. Chem. Soc.* **88**, 5103.
26. Trenner, N. R., Walker, R. W., Arison, B. H. and Trumbauer, C. (1951). *Analyt. Chem.* **23**, 489.
27. Mann, J. and Marrinan, H. J. (1956). *Trans. Faraday Soc.* **52**, 492.
28. Kaufman, J. J., Koski, W. S. and Anacreon, R. (1963). *J. Mol. Spectroscopy* **11**, 1.
29. Milatz, J. M. W., Kluyver, J. C. and Hardebol, J. (1951). *J. Chem. Phys.* **19**, 887.
30. Knezevic, Z. V. (1967). *Isotopenpraxis* **3**, 452.
31. Schmidt, H. L., Schmelz, E. and Rohns, G. (1968). *Z. Analyt. Chem.* **243**, 59, (1968). *Isotopenpraxis* **4**, 133.
32. Lapidoth, A., Pinchas, S. and Samuel, D. (1962). *Proc. Chem. Soc.*, 109.
33. Spitzer, W. G. and Ligenza, J. R. (1961). *J. Phys. and Chem. Solids* **17**, 196.
34. Spencer, B. (1959). *Biochem. J.* **73**, 442.
35. Aksnes, G., Aksnes, D. and Albriktsen, P. (1966). *Acta Chem. Scand.* **20**, 1325.
36. Laulicht, I. and Pinchas, S. (1964). *Analyt. Chem.* **36**, 1980.

Author Index

Numbers in parentheses are reference numbers and are included to assist in locating references in which the authors' names are not mentioned in the text. Numbers in italics indicate the page on which the reference is listed.

B

C

D

E

F

G

H

I

J

K

L

M

N

O

P

Q

R

S

X

Y

Z

Subject Index

Compounds are indexed according to chemical name or class. Where a particular isotopic species is emphasized it is indicated in brackets, e.g. Methanes, (D), (T), would indicate that on the pages mentioned information will be found relating to the infrared spectroscopy of various deuterated and tritiated methanes.

B

C

K

L

O

T